Early Modern Fire

Intersections

INTERDISCIPLINARY STUDIES IN EARLY MODERN CULTURE

General Editor

Karl A.E. Enenkel (*Chair of Medieval and Neo-Latin Literature
Universität Münster*
e-mail: kenen_01@uni_muenster.de)

Editorial Board

W. de Boer (*Miami University*)
S. Bussels (*University of Leiden*)
A. Dlabačová (*University of Leiden*)
Chr. Göttler (*University of Bern*)
J.L. de Jong (*University of Groningen*)
W.S. Melion (*Emory University*)
A. Montoya (*Radboud University Nijmegen*)
R. Seidel (*Goethe University Frankfurt am Main*)
P.J. Smith (*University of Leiden*)
J. Thompson (*Queen's University Belfast*)
A. Traninger (*Freie Universität Berlin*)
C. Zittel (*Ca' Foscari University of Venice / University of Stuttgart*)
C. Zwierlein (*Bonn*)

VOLUME 95 – 2025

The titles published in this series are listed at *brill.com/inte*

Early Modern Fire

Science, Technology, and the Urban Space

Edited by

Gianenrico Bernasconi
Marco Storni

BRILL

LEIDEN | BOSTON

Cover illustrations: (central image) A furnace used in the processing of iron. Engraving, 18th century. Courtesy Wellcome Collection (30421i). Image in the public domain; (background image) Inferno, canto 37–39, A rain of fire hits the blasphemists.

Library of Congress Cataloging-in-Publication Data

Names: Bernasconi, Gianenrico, 1973– editor. | Storni, Marco, 1990– editor.
Title: Early modern fire : science, technology, and the urban space /
    edited by Gianenrico Bernasconi, Marco Storni.
Description: Leiden ; Boston : Brill, 2025. | Series: Intersections,
    1568–1181 ; volume 95 | Includes bibliographical references and index. |
Identifiers: LCCN 2024035725 | ISBN 9789004521759 (hardback) |
    ISBN 9789004521766 (ebook)
Subjects: LCSH: Fire—Europe—History—17th century. | Fire—Europe—History—
    18th century. | Firemaking—Europe—History—17th century. | Firemaking—
    Europe—History—18th century. | Firefighting—Europe—History—17th century. |
    Firefighting—Europe—History—18th century. | Urbanization—Europe—
    History—17th century. | Urbanization—Europe—History—18th century.
Classification: LCC GN416 .E37 2025 |
    DDC 541/.36109409032–dc23/eng/20240906
LC record available at https://lccn.loc.gov/2024035725

Typeface for the Latin, Greek, and Cyrillic scripts: "Brill". See and download: brill.com/brill-typeface.

ISSN 1568-1181
ISBN 978-90-04-52175-9 (hardback)
ISBN 978-90-04-52176-6 (e-book)
DOI 10.1163/9789004521766

Copyright 2025 by Koninklijke Brill BV, Leiden, The Netherlands.
Koninklijke Brill BV incorporates the imprints Brill, Brill Nijhoff, Brill Schöningh, Brill Fink, Brill mentis, Brill Wageningen Academic, Vandenhoeck & Ruprecht, Böhlau and V&R unipress.
All rights reserved. No part of this publication may be reproduced, translated, stored in a retrieval system, or transmitted in any form or by any means, electronic, mechanical, photocopying, recording or otherwise, without prior written permission from the publisher. Requests for re-use and/or translations must be addressed to Koninklijke Brill BV via brill.com or copyright.com.

This book is printed on acid-free paper and produced in a sustainable manner.

Contents

Acknowledgments   VII
List of Figures   VIII
Notes on the Editors   XI
Notes on the Contributors   XII

Introduction: the Early Modern Fire   1
    Gianenrico Bernasconi and Marco Storni

PART 1
Fire in Early Modern Science

1   Purity, Purification, and Fire in the Seventeenth-Century Chymical
    Texts of Nicaise Le Febvre and Michael Sendivogius   23
    Hannah Elmer

2   'Atoms of Fire': Galileo's Unachieved Theory of Heat and the
    Beginnings of Thermometry (*c.*1603–1638)   49
    Stefano Salvia

3   Without a Thermometer: the Technical Knowledge of Heat in the
    Early Modern Age   90
    Marco Storni

4   Working with Fire: Remedy Making from the Shop to the Garden
    in the Apothecaries' Guild (Eighteenth Century, Paris)   117
    Bérengère Pinaud

PART 2
Early Modern Fire Technologies

5   The Kitchen Fire (Eighteenth and Early Nineteenth Century)   147
    Gianenrico Bernasconi

6   Fire Mechanics: Inventors and Promoters of Heating Systems
    (Eighteenth and Early Nineteenth Century)   172
    Olivier Jandot

VI   CONTENTS

7   'The Manner of Conducting Fire': Firing Architectural Terracotta in the Modern Era, between Know-How, Wood Shortage, and Innovations   191
     Cyril Lacheze

8   John Smeaton's Fire Engine Trials   225
     Andrew M.A. Morris

PART 3
Fire in the Urban Space

9   The Outbreak of Fire: Inventions, Materials, and Combustion Knowledge in the Eighteenth Century   257
     Marie Thébaud-Sorger

10   What Firefighting Tells Us about Eighteenth-Century Urban Police   282
     Catherine Denys

11   Organising the Chaos: Firefighting in Upper Lusatia in the Early Modern Period   302
     Cornelia Müller

12   Spectacle, Enthusiasm, Objectivity – Managing Fire as an Emoterial   329
     Simon Werrett

     Conclusion: the Technicity of Fire in Modern Europe – a Historiographical Crossroads   345
     Liliane Hilaire-Pérez

     Index Nominum   361

Acknowledgments

The present volume is the result of the international workshop *Fire Management in the Early Modern Age: Knowledge, Technology, Economy*, held at the University of Neuchâtel on 16–17 September 2021. The event was supported by the Swiss National Science Foundation (SNSF) as part of the research project *Mesure du temps, chimie et cuisine: formalisation des pratiques au XVIIe et au XVIIIe siècle* (grant number 100011_184856).

This book contains a selection of the papers presented at the conference as well as new contributions (chapters 2, 8 and 10). Chapter 10 is the English version of an article originally published in French under the title "Ce que la lutte contre l'incendie nous apprend de la police urbaine au XVIIIe siècle", *Orages, Littérature et culture (1760–1830)* 10 (2011) 17–36. We are grateful to Catherine Denys and the editors of *Orages* for agreeing to publish an updated version of this text in our volume.

We would like to thank the University of Neuchâtel for its practical help in organising the workshop, the editorial staff of Brill's *Intersections* series for their support, and all the libraries and institutions that allowed us to reproduce images from their collections.

We are grateful to Mark F. Rogers for revising the English texts of the non-native authors. Special thanks to Nadège Barbezat and Catherine Herr-Laporte for their help in organising the workshop, and to Aude Monié for her assistance in proofreading the volume.

Gianenrico Bernasconi and Marco Storni
January 2024

Figures

0.1   *An alchemist using bellows at a furnace in his laboratory* [*The Chymist*]. Etching by T. Major, after D. Teniers the Younger (1750). Wellcome Collection, London. Public domain   8

0.2   Franz Kessler's stove, in Kessler, Franz, *Holzsparkunst: Das ist, Ein solche new, zuvorn niemahln gemein, noch am Tag gewesen Invention etlicher unterschiedlichere Kunstofen* (Frankfurt a. M.: 1618)   11

0.3   [Unknown artist] *The Great Fire of London, with Ludgate and Old St. Paul's* (ca. 1670). Yale Centre for British Art, Paul Mellon Collection. Public domain   15

2.1   The thermoscope, as illustrated by Benedetto Castelli in his letter to Ferdinando Cesarini in Rome (September 20, 1638). From: Favaro, Del Lungo, Marchesini (eds.), *Opere di Galileo Galilei* XVII (1906) 378. Copyright: Library of the Museo Galileo, Florence   55

2.2   Nineteenth century replica of 'Galileo's' thermoscope. Museo Galileo (Florence), Room VII. Unknown maker. Inventory: 2444. Materials: glass, cork. Height: 440 mm. Copyright: Museo Galileo (Florence)   56

2.3   Sanctorius's *pulsilogium* and closed-air thermoscope, from his *Commentaria in primam Fen primi libri Canonis Avicennae* (Venice, Giacomo Sarcina: 1625) 22. Copyright: Wellcome Library, London   60

2.4   Sanctorius's *pulsilogium* and closed-air thermoscope (detail), from his *Commentaria in primam Fen primi libri Canonis Avicennae* (Venice, Giacomo Sarcina: 1625) 22. Copyright: Wellcome Library, London   61

2.5   Sanctorius's *marginalia* to col. 406C–D of his own copy of the *Commentaria in primam Fen primi libri Canonis Avicennae* (Venice, Giacomo Sarcina: 1625) 406. British Library (Sloane Collection), 542.h.11. Image from Bigotti F., 'A Previously Unknown Path to Corpuscularianism in the Seventeenth Century: Santorio's Marginalia to the *Commentaria in Primam Fen Primi Libri Canonis Avicennae* (1625)', *Ambix* 64.1 (2017) 33. Copyright: British Library, London   63

2.6   Examples of Florentine thermometers and a hygrometer. Image from Magalotti L., *Saggi di naturali esperienze fatte nell'Accademia del Cimento* VIII. Copyright: Library of the Museo Galileo, Florence   83

2.7   Mid-seventeenth-century spiral thermometers of the Accademia del Cimento. Museo Galileo (Florence), Room VIII. Unknown maker. Copyright: Museo Galileo, Florence   84

4.1   Nicolas Lémery, *Cours de chimie* (Paris, Théodore-Hyacinthe Baron: 1757), 790–1. Copyright: Bibliothèque Nationale de France   130

FIGURES IX

4.2   Nicolas Lémery, *Cours de chimie* (Paris, Théodore-Hyacinthe Baron: 1757), 790–1. Copyright: Bibliothèque Nationale de France   131

4.3   Michel-Étienne Turgot (patron) and Louis Brétez (designer), "Jardin des apothicaires", *Plan de Paris*, (1734–1739), plate 7. Copyright: Wikimedia Commons   133

8.1   Illustration to the French edition of William Watson's *Expériences et observations pour server à l'explication de la nature et des propriétés de l'électricité* (Paris, Sébastien Jorry: 1748) plate 3. The friction machine on the right (of the bottom image) charges the horizontal rod (the prime conductor) held by the man on the left, who is holding a sword in his other hand, that he uses to ignite the liquid in the spoon. The man is insulated from the ground so that the charge is transmitted from the prime conductor, across his body and into the sword   229

8.2   The table of results for Smeaton's experiments on mortars from John Smeaton, *A Narrative of the Building and a Description of the Construction of the Edystone Lighthouse with Stone* (London, H. Hughs: 1791) 122   232

8.3   A technical drawing of the workings of Smeaton's waterwheel apparatus from John Smeaton, "An experimental enquiry concerning the natural powers of water and wind to turn mills, and other machines, depending on a circular motion", *Philosophical Transactions* (1759) 102, tab V   234

8.4   The table of results for Smeaton's experiments on windmills from John Smeaton, "An experimental enquiry concerning the natural powers of water and wind to turn mills, and other machines, depending on a circular motion", *Philosophical Transactions* (1759) 144, tab. V   236

8.5   Schematic sketch of a Newcomen engine, in Max de Nansouty, *Chaudières et machines à vapeur* (Paris, Boivin et Cie: 1911) 61   240

8.6   John Smeaton, *Chacewater engine*, 1775, drawing, 63 × 44.5 cm, Smeaton Collection, vol. 3, fol. 111, Royal Society Archive   246

8.7   Table of fire engine dimensions from John Farey Jr., *A Treatise on the Steam Engine*, vol. 1 (London, Longman: 1827) 183   250

11.1   Hand drawn view of Löbau in Samuel Grosser, Lausitzische Merkwürdigkeiten Darinnen von beyden Marggrafthümern in fünf verschiedenen Theilen […] (Leipzig – Budissin, David Richter – Immanuel Tietze: 1714) after page 83 in part III. Görlitz, Oberlausitzische Bibliothek der Wissenschaften. Copyright: Oberlausitzische Bibliothek der Wissenschaften Görlitz   306

11.2   Hand drawing of the 1634 fire in Bautzen in Samuel Grosser, Lausitzische Merkwürdigkeiten Darinnen von beyden Marggrafthümern in fünf verschiedenen Theilen […] (Leipzig – Budissin, David Richter – Immanuel Tietze: 1714) between page 250 and 251. Görlitz, Oberlausitzische Bibliothek der

Wissenschaften. Copyright: Oberlausitzische Bibliothek der Wissenschaften, Görlitz   307

11.3   Fire ordinances of Bautzen, Görlitz and Löbau in the seventeenth and eighteenth century, own illustration   309

11.4   Scope of the fire ordinances with a share on the topic of firefighting, own illustration   310

11.5   Section of the plan of the town of Görlitz from 1790. The draughtsman Liebsch has marked the various quarters within the town walls here: Dark the Frauenviertel, above it the Reichenbach quarter, to the right of the Frauenviertel the Neisse quarter and finally slightly above the Reichenbach and Neisse quarters the Nicolai quarter. See Liebsch "Plan der Churfürstlich Sächsischen Sechs Stadt Görliz", copper engraving from 1790, Görlitz, Oberlausitzische Bibliothek der Wissenschaften. Copyright: Oberlausitzische Bibliothek der Wissenschaften Görlitz   313

11.6   Map of Löbau, 1843, Löbau municipal Archive; Own processing. Legend: *Blue*: Zittau Quarter, *Yellow*: Görlitz Quarter, *Green*: Budissin (aka Bautzen) quarter   314

11.7   Section of the ground plan of Bautzen drawn by Johann Gottlob Krause, Grund-Riss der im Marggrafthum Oberlausitz gelegene Haupt Sechs-Stadt Budissin nebst der umher liegente Gegend, Handdrawing from 1781. Own adaptation of the map: Sächsische Landesbibliothek – Staats- und Universitätsbibliothek Dresden (SLUB) / Deutsche Fotothek. Copyright: SLUB / Deutsche Fotothek   315

11.8   Willow fire bucket from the 19th century, Bautzen municipal Museum, Inventory no. L Opp. 333/1. Copyright: Municipal Museum of Bautzen. These buckets had been made watertight on the inside with pitch   318

11.9   Public buildings such as schools or churches should also keep extinguishers on site. This is an example from Angermünde with a leather fire extinguisher bucket from the 1830s, Angermünde municipal Museum, Inventory No. 271. Copyright: Municipal Museum of Angermünde   319

11.10   2 wooden hand sprayers from the first half of the 19th century, left in front the tip of a fire hook, municipal museum of Bautzen, Inventory no. 11196 a & b. Copyright: Municipal Museum of Bautzen   320

12.1   Anon, *The Grand Whim For Posterity To Laugh At*, 1749, etching and letterpress printing, 469 × 351 mm, British Museum Department of Prints and Drawings. Copyright: The Trustees of the British Museum   330

Notes on the Editors

Gianenrico Bernasconi

is associate professor and "directeur de recherche" in History of Technology at University of Neuchâtel. He has been principal investigator of the SNSF-project *Mesure du temps, chimie et cuisine: formalisation des pratiques au XVIIe et au XVIIIe siècle* (2018–2023). His current work explores the history of technology, of timekeeping, of chronometrical observatories, and of cooking. He is the author of *Objets portatifs au siècle des Lumières* (Paris: 2015), the co-editor, with Guillaume Carnino, Liliane Hilaire-Pérez and Olivier Raveux, of *Les Réparations dans l'Histoire. Cultures techniques et savoir-faire dans la longue durée* (Paris: 2022), and the co-editor, with Susanne Thürigen, of *Material Histories of Time: Objects and Practices, 14th–19th Centuries* (Berlin – Boston: 2020).

Marco Storni

is a postdoctoral fellow at the Université libre de Bruxelles. After completing his PhD at the École Normale Supérieure of Paris and the University of Bologna, he has been a postdoctoral fellow at Ca' Foscari University of Venice and at the University of Neuchâtel. His research focuses on the history of science and of philosophy of science in the early modern period, particularly in the eighteenth century. He is the author of *Maupertuis. Le philosophe, l'académicien, le polémiste* (Paris: 2022; winner of the Prize of the Fondation Del Duca-Institut de France 2023).

Notes on the Contributors

Catherine Denys

is Professor Emeritus of Modern History at the University of Lille. She has been director of the Institut de Recherche Historique du Septentrion, vice-president of the scientific board and director of Presses Universitaires du Septentrion. A specialist of the history of eighteenth-century police, she has studied security in the cities of northern France and Belgium and is currently working on the colonial police of the Old Regime.

Hannah Elmer

is currently a post-doctoral researcher in early modern history at the Historisches Seminar at the Leibniz Universität Hannover, Germany. She received her PhD in 2019 from Columbia University. Her research interests include later medieval and early modern cultural, religious, and intellectual history, and she focuses especially on Central Europe. She is currently working on a monograph that investigates interactions of supernatural, natural, and human agency in problems of miraculous reanimation in the late fifteenth century.

Liliane Hilaire-Pérez

is professor of modern history at the University of Paris (laboratoire ICT-Les Europes dans le Monde), director of studies at EHESS (Centre Alexandre-Koyré) and a senior member of the Institut Universitaire de France. Her books include *L'invention technique au siècle des Lumières* (Paris: 2000) and *La pièce et le geste. Artisans, marchands et savoir technique à Londres au XVIIIᵉ siècle* (Paris: 2013). She co-edited with Catherine Lanoë *Les sciences et les techniques, laboratoire de l'histoire. Hommage à Patrice Bret* (Paris: 2022), and is preparing the publication, with Guillaume Carnino and Jérôme Lamy, of *Global History of Techniques (19th–21st c.)* (Turnhout: 2024). She is an editor of the journal *Artefact. Techniques, histoire et sciences humaines*, and the co-director of the Groupement de recherche 2092 TPH-"Techniques et production dans l'histoire".

Olivier Jandot

holds the Agrégation and a PhD in History from the University of Lyon. He teaches in literary preparatory classes for Grandes Écoles (Lycée Notre-Dame de la paix, Lille), at Sciences Po Lille and at the Université d'Artois, where he is associate fellow at the history research centre (UR 4027 CREHS, Arras, France).

His research focuses on the intersection between the history of material culture, the history of the body and the history of sensibilities, especially in the domain of thermal comfort. He is the author of *Les délices du feu: l'homme, le chaud et le froid à l'époque moderne* (Ceyzérieu: 2017).

Cyril Lacheze

is a post-doctoral fellow at the University of Technology of Belfort-Montbéliard (FEMTO-ST/RECITS). He defended his PhD thesis in 2020, within the history of techniques team of the Institute for Modern and Contemporary History at the Paris 1 Panthéon-Sorbonne University, where he was Temporary Teaching and Research Assistant (ATER). He devotes his research to technical systems of the early modern era, starting with architectural terracotta production, using both historical and archaeological sources. He seeks in particular to reconstitute via a systemic approach dynamic of everyday technical productions, intrinsically linked to each other within a given socio-geographical space. He is the author of *L'art du briquetier, XVIᵉ–XIXᵉ siècle. Du régime de la pratique aux régimes de la technique* (Paris: 2023).

Andrew M.A. Morris

recently completed his PhD in the history and philosophy of science at the Centre for Logic and Philosophy of Science of the Vrije Universiteit Brussel. His research is funded by the FWO – Research Foundation Flanders, grant number 148531. His dissertation explores eighteenth-century English engineer John Smeaton's methods as an instrument maker, experimental philosopher and engineer, covering Smeaton's early electrical experiments, trials on model waterwheels and windmills, and his design for an air pump and innovative pressure gauge. The dissertation focuses, in particular, on the interactions between these different practices in Smeaton's work, and how these interactions might throw new light on the science-technology distinction.

Cornelia Müller

is a research associate and lecturer at the University of Applied Sciences of Zittau/Görlitz, Faculty of Social Science. Her dissertation is supervised by Susanne Rau (University of Erfurt) and Raj Kollmorgen (University of Applied Sciences Zittau/Görlitz). After an MA degree in Early Modern and Modern History, Political Science and Saxonian regional History at the University of Dresden in 2002, she worked for several educational institutions in Saxony and Saxony-Anhalt. She was president of the German Korczak Association (2012–2015) and member of the Board of the International Korczak Association

Bérengère Pinaud

is a PhD candidate at the École des hautes études en sciences sociales, Paris, and affiliated to the Centre Alexandre-Koyré. By undertaking a material and social history of science, her thesis dissertation focuses on material culture in the apothecaries' daily practice in eighteenth-century Paris. She examines the logistics and supply of the trade, the spaces of remedy making and knowledge transmission, and the shop as a social space through its arrangement, in order to shed new light on apothecaries as medical artisans, whose knowledge was at the crossroads of natural history, chemistry, artisanship, and commerce.

Stefano Salvia

PhD, is a research assistant in History of Science at the Department of Civilizations and Forms of Knowledge, University of Pisa. Currently, he is also high-school teacher of Philosophy and History in Turin and a scientific collaborator at the Museo Galileo – Institute and Museum of History of Science, Florence. Former member of the International Research Network *History of Scientific Objects*, coordinated by the Max Planck Institute for the History of Science in Berlin, he was granted a Lerici Scholarship and a visiting fellowship at the Center for History of Science of the Royal Swedish Academy of Sciences. His research interests range from early modern natural philosophy to history of modern physics, from historiography of science to historical epistemology and STS-oriented museology. He is co-editor and contributor of the volume *Scienza e filosofia della complessità* (Rome: 2020).

Marie Thébaud-Sorger

is a tenured research fellow at the French National Centre for Scientific Research (CNRS) since 2014, based at the Centre Alexandre-Koyré (CNRS/EHESS/MNHN) in Paris, and an associate of the *Maison française* in Oxford where she ran the History of Science programme (2017–2020). She has been Marie Curie Fellow (2008–2010) at Warwick University, visiting fellow at the MPIWG (2016) and associate member at the History faculty of Oxford university (2019–2021). Her research seeks to explore the entanglement between materialities and mediations, the culture of the "arts", technical improvements and inventive practices in Eighteenth century Europe especially focusing on fire and stale air management. She is also currently Co-PI of the ANR FabLight *The making of light in the visual arts during the Enlightenment* (2022–2026). She published extensively on the reception of the lighter-than-air machines in French and European societies, *L'aérostation au temps des Lumières* (Rennes: 2009), and co-edited with Liliane Hilaire-Perez and Fabien Simon, *L'Europe des sciences et des techniques. Un dialogue des savoirs XVᵉ–XVIIIᵉ siècle* (Rennes: 2016).

Simon Werrett
is Professor of the History of Science in the Department of Science & Technology Studies at University College London. He joined UCL after ten years in the Department of History at the University of Washington. His work explores the history of science and various arts, spectacular, domestic, and navigational. He is the author of *Fireworks: Pyrotechnic Arts and Sciences in European History* (Chicago: 2010) and *Thrifty Science: Making the Most of Materials in the History of Experiment* (Chicago: 2019), and edited, with Lissa Roberts, *Compound Histories: Materials, Governance, and Production, 1760–1840* (Leiden: 2017).

Introduction: the Early Modern Fire

Gianenrico Bernasconi and Marco Storni

In one of the most famous Greek myths, recounted by Hesiod in the *Theogony* (507–616) and in the *Works and Days* (42–105), the titan Prometheus steals fire from the Olympian Gods to donate it to humankind. Zeus punishes Prometheus for his crime by binding him in chains, and making an eagle eat his liver every day, for all eternity: 'And he [Zeus] set upon him [Prometheus] a long-winged eagle which ate his immortal liver, but this grew again on all sides at night just as much as the long-winged bird would eat during the whole day'.[1] The importance of the gift of fire for the birth of civilisation is well illustrated in a philosophical version of the myth, namely that of Socrates in Plato's dialogue *Protagoras* (320d–322a). Socrates explains to an unnamed friend that Prometheus's brother Epimetheus had distributed the gifts of nature to all the animals on Earth, but had forgotten humans; to remedy this, 'Prometheus [...] stole from Hephaestus and Athena wisdom in the arts, along with fire – for without fire there was no means for anyone to possess or use art itself – and so gave them to man. In this way man got wisdom for the needs of life [...]'.[2] Fire is presented here as the driving force of the arts, and therefore central to the development of all the techniques and utensils necessary for human life.

Modern readings of the myth offer different interpretations of Prometheus's gift, and thus indirectly of the fundamental relationship of humankind with fire. On the one hand, some authors have emphasised that fire has endowed humans with a power that makes them almost divine: not only did Prometheus commit the sin of *hubris* by stealing fire from the Gods, but fire also encouraged humans to transgress their limits and exercise – in Nietzschean terms – "a will to power" over the natural world.[3] On the other hand, fire is also seen as an ordinary tool, essential for sustaining life, which is at the centre of many daily

---

1 Hesiod, *Theogony*, 523–525, in *Theogony, Works and Days, Testimonia*, ed. by Glenn W. Most (Cambridge, MA – London: 2006) 45.

2 Plato, *Protagoras*, 321d, in *The Dialogues of Plato*, vol. 3: *Ion, Hippias Minor, Laches, Protagoras*, ed. by R.E. Allen (New Haven – London: 1996) 181.

3 An example of a 'heroic reading' of Prometheus's myth is provided by Johann Wolfgang Goethe's poem *Prometheus*, which inspired other works, such as Massimo Cacciari's *libretto* for Luigi Nono's non-orthodox opera (in fact a series of cantatas) *Prometeo. Tragedia dell'ascolto* (1981–84). The myth of Prometheus inspired many other composers, especially from the Romantic age, amongst whom Ludwig van Beethoven (the ballet *The Creatures of Prometheus*, 1801) and Franz Liszt (the symphonic poem *Prometheus*, 1850).

© KONINKLIJKE BRILL BV, LEIDEN, 2025 | DOI:10.1163/9789004521766_002

activities and has to be cared for. In the dialogue between Prometheus and the eagle imagined by Piero Bevilacqua, while Prometheus claims that 'fire is good for every human being', the eagle reminds him that 'only some can enjoy it [the fire]', while 'others have to feed it day and night, so that it does not go out'.[4] Fire is presented here as a constraint, at least for those who are not fortunate enough as to enjoy its pleasures, and who have to work to keep the flame alive. The burden represented by fire, and the inequalities created by its use (or misuse), especially in the industrial age, have recently led Peter Sloterdijk – inspired by the philosopher Günther Anders – to speak of the modern world as 'the era of "Promethean shame"'.[5]

The myth of Prometheus, in its ancient and modern versions, offers insights into the many facets of fire and its multiple meanings in human life. Fire is a symbol of human growth and perfection, the basis of any society. Beyond the symbolic dimension, fire is at the origin of arts and crafts, a powerful force that gives the power to manipulate nature, but is also potentially devastating.[6] However, fire is also a daily obligation that humans must meet in order to survive. In the domestic space, it is essential for heating and cooking, so much so that the hearth is often used as a metonym for the home itself.

The polysemy of fire, and its omnipresence, explain the difficulty of conceiving of it as a separate subject, and thus of making it an object of historical study.[7] Paradoxically, fire is so present in so many areas of human activity that it ends up being trivialised. Traditional historiography has focused on industrialisation, presenting fire as the natural agent that feeds machinery, especially the steam engine, as well as the effect of the combustion of fossil fuels, which are at the basis of Western economic development while representing a major source of pollution.[8] Even in recent times, fire has rarely been front and centre of historiographical accounts that mention it. The concept of "energy" is more

---

4   Bevilacqua P., *Prometeo e l'aquila. Dialogo sul dono del fuoco e i suoi dilemmi* (Roma 2005) 24.

5   Sloterdijk P., *Die Reue des Prometheus. Von der Gabe des Feuers zur globalen Brandstiftung* (Berlin 2023) 33–36. Sloterdijk refers to Günther Anders's book *Die Antiquiertheit des Menschen*, vol. 1: *Über die Seele im Zeitalter der zweiten industriellen Revolution* (München 1956).

6   In environmental history, fire is often taken as an emblem of the anthropic impact on Earth, materialized in combustion of fossil fuels, and more generally in the devastating effects of fire on ecosystems. See Pyne S., *The Pyrocene: How We Created an Age of Fire, and What Happens Next* (Oakland, CA: 2021).

7   The multifaceted nature of fire is showcased, for instance, by the variety of contributions on fire in the Middle Ages gathered in *Il fuoco nell'Alto Medioevo: Spoleto, 12–17 aprile 2012* (Spoleto 2013).

8   See e.g. Landes D.S., *The Unbound Prometheus: Technological Change and Industrial Development in Western Europe from 1750 to the Present* (Cambridge: 1969).

often seen (explicitly or implicitly) as the 'only universal currency',[9] and fire, while closely associated with energy production, is conceived as a natural force or basic resource that studying it is not relevant per se. Some scholars have gone so far as to see energy as a key to reading the whole of human history: as Vaclav Smil puts it, 'both prehistoric human evolution and the course of history may be seen fundamentally as the quest for controlling greater energy stores and flows'.[10] Even more local and modest perspectives, such as that epitomised by a recent work on energy in the early modern home, do not question the centrality and universality of the very notion of energy.[11] However, it was not until after 1800 that this concept emerged as the result of a complex process of scientific reflection and technological research.[12] While it is true that all human beings throughout history have been confronted with the problem of managing energy sources, it was not until the nineteenth century that the increase of energy became an 'obsession'[13] as this period witnessed the rise of 'energy system[s]', namely broad-ranging plans for energy development underpinned by 'political, economic, social, and cultural structures and decisions'.[14]

The aim of this present volume is to renew the history of fire by treating it as an object worth studying in its own right. Rather than lurking in the background of particular histories (of energy, of the environment, of domesticity, etc.), fire is here brought to the fore of the narrative. The focus of all the chapters in this book is on the period that preceded the advent of industrial societies, that is, before fire was widely seen as a force to enhance the power of machines and feed the quest for energy, as well as a threat to civil life and to ecosystems. Far from endorsing a view of the genesis of fire's "modernity" – that is, its history after 1800 – that insists solely on progress and success, or on grand dramatic choices such as the wide-scale adoption of fossil fuels,[15] this volume offers a multifaceted perspective on this historical process, emphasising

---

9   Smil V., *Energy in World History* (Boulder, CO: 1994) 1.

10 Ibidem.

11 Saelens W. – Blondé B. – Ryckbosch W. (eds.), *Energy in the Early Modern Home: Material Cultures of Domestic Energy Consumption in Europe, 1450–1850* (London – New York: 2023).

12 Guedj M. – Mayrargue A., "Éclairages historiques sur l'émergence du concept d'énergie", *Recherches en didactique des sciences et des technologies* 10 (2014) 35–61.

13 Jarrige F. – Vrignon A., *Face à la puissance. Une histoire des énergies alternatives à l'âge industriel* (Paris: 2020) 50.

14 Mathis C.F., *La civilisation du charbon. En Angleterre, du règne de Victoria à la Seconde Guerre mondiale* (Paris: 2021) 18.

15 In an influential study, Alain Gras has defined the modern preference for fossil fuels as the 'choice of fire (*choix du feu*)', giving to it a political connotation: Gras A., *Le choix du feu. Aux origines de la crise climatique* (Paris: 2007).

the importance of continuities, the persistence of traditional know-how, small-scale innovations, and the coexistence of different paradigms.

We propose that fire be considered first and foremost as a phenomenological object of which every human being has an experience: fire is a process that produces flames, heat, and light, and that transforms matter as it burns. The first question raised by the encounter with fire is how to handle its power. As the anthropologist Johan Goudsblom has noted, 'the ability to handle fire is a universal human achievement, found in every known society'.[16] The notion of "handling (fire)" (or "(fire) management" – we consider the two to be equivalent) is not a simple one: in fact, handling fire has several epistemic and practical levels, and is tethered to an idea of economy, intended both as a wise management of resources and as an optimisation of production processes.[17] First, handling fire requires some knowledge: both the practical knowledge needed to physically handle the process of combustion, and a more theoretical knowledge, namely an understanding of the nature of fire and its effects on natural objects.[18] Second, handling has a technical dimension: governing fire is an operation that is often mediated by artefacts of various kinds, that can be used to contain its power, amplify it, protect natural bodies or artefacts from its effects, or better expose them to its action. Third, fire management is a matter of social management, because fire is a natural force, potentially devastating, that is at the centre of many social activities and that, more generally, can arouse strong emotions in the population, ranging from amazement to fright. The management of fire in society requires appropriate political deliberations, but also the collective efforts of different social groups.

The seventeenth and eighteenth centuries witnessed transformations and continuities in the modes of handling fire, which point to the non-linearity of scientific and technological development, and to the complex genesis of the modern economy, politics, and society. While new physical theories of heat were elaborated, new measuring instruments (notably the thermometer) were invented, and the foundations of modern chemistry – to which the interpretation of combustion is central – were laid, alchemical practice and vernacular

---

16   Goudsblom J., *Fire and Civilization* (London: 1992) 1.

17   To refer to the idea of economy as the wise management of resources, historians often use the lemmas 'oeconomy' or 'thrift'. See Werrett S., *Thrifty Science: Making the Most of Materials in the History of Experiment* (Chicago – London: 2019).

18   The definition of 'practical knowledge' that we adopt is inspired by Matteo Valleriani's one: 'Practical knowledge is the knowledge needed to obtain a certain product [...] that follows a defined workflow' (Valleriani M., "The Epistemology of Practical Knowledge", in Valleriani M. (ed.), *The Structures of Practical Knowledge* (Cham: 2017) 1).

INTRODUCTION: THE EARLY MODERN FIRE

knowledge persisted.[19] The knowledge of craftsmen and other practitioners working outside academic circles was still seen as a viable alternative to expert knowledge, and instances of 'hybrid expertise' were far from rare.[20]

Similarly, the development of machines, such as the first attempts to channel and control fire, to increase its power – from the first pressure cookers to the steam engine – went hand-in-hand with work on everyday technologies. This work on ordinary objects and practices was aimed at making the most of available resources, especially fuels, an aspect of central importance in an age constantly threatened with a dearth of energy supplies.[21] In parallel with artisanal workshops, the eighteenth century also saw the emergence of manufactures, where the technical mastery of fire was put at the service of the enhancement of production with a view to profit.

At the level of social organisation, the early modern period saw the emergence of specific policies of fire management in the sense of prevention and firefighting, which were massively implemented throughout Europe, including the creation of professional fire brigades. However, this did not mean that traditional configurations were *ipso facto* eliminated. On the contrary, several European cities – especially in peripheral contexts – testify to the long survival of traditional firefighting policies.[22] However, firefighting was by no means the only instance of fire management in the urban context: the example of fireworks shows that fire management could also be management of public passions, in which the symbolic dimension played a crucial role.[23]

The volume is divided into three sections corresponding to the main facets of the notion of 'fire handling' that have just been mentioned, namely the scientific, the technological and the social. The specific contributions that

---

19 See Smith P.H., *From Lived Experience to the Written Word: Reconstructing Practical Knowledge in the Early Modern World* (Chicago – London: 2022).

20 On hybrid expertise, see Klein U., "Hybrid Experts", in Valleriani M. (ed.), *The Structures of Practical Knowledge* (Cham: 2017) 287–306.

21 On the scarcity of energy supplies in early modern Europe, and the role of dearth in the elaboration of knowledge, see Mukherjee A., *Penury into Plenty: Dearth and the Making of Knowledge in Early Modern England* (Abingdon – New York: 2015). On the French case, see Boissière J., "La grande disette de bois à Paris des années 1783–1785", in Higounet C. (ed.), *L'approvisionnement des villes. De l'Europe occidentale au Moyen Âge et aux Temps Modernes* (Toulouse: 1985) 237–242.

22 Zwierlein C., *Der gezähmte Prometheur: Feuer und Sicherheit zwischen früher Neuzeit und Moderne* (Göttingen: 2011); Garrioch D., "Towards a Fire History of European Cities (Late Middle Ages to Late Nineteenth Century)", *Urban History* 46.2 (2019) 202–224.

23 Werrett S., *Fireworks: Pyrotechnic Arts and Sciences in European History* (Chicago – London: 2010).

the authors of the chapters make to each of these fields are presented in the remainder of this introduction.

1 Fire in Early Modern Science

In Aristotle's *Physics*, fire is one of the simple bodies or elements, along with earth, water, and air, which are combined to form any natural body. These elements are defined by a different combination of two fundamental properties from the alternatives hot/cold and wet/dry (fire is hot and dry). Each element naturally tends to occupy its natural place: earth and water tend downwards, towards the centre of the world, while air and fire always tend upwards.[24] As the Aristotelian paradigm remained dominant up until the seventeenth century, these categories had a long-lasting influence on Western natural-philosophical thought. In the early modern period, alternative views of the nature of matter began to emerge in the context of a profound change in European scientific culture, usually referred to as the 'Scientific Revolution'.[25] One might consider the rise of corpuscular philosophy and mechanical philosophy (often combined): the former held that physical events are the result of the interaction of homeomeric particles; the latter proposed to explain natural phenomena exclusively in terms of matter and motion. With regard to fire, the application of these new philosophical perspectives had two major consequences. The first is that fire was no longer seen as an element, but as a process. This opened up to a modern scientific consideration not only of the phenomenon of combustion – first resolved with the introduction of 'phlogiston', then radically reinterpreted by Antoine-Laurent Lavoisier, marking a key step in the so-called 'Chemical Revolution' – but also of the effects of fire on other natural bodies. The second consequence is that fire was no longer characterised by a set of essential properties of a qualitative nature, but could be subjected to quantification. In fact, this was the first step towards the elaboration of a modern science of heat, which culminated in the formulation of thermodynamics in the early nineteenth century.[26] However, the development

---

24   See e.g. *Physics* I, 6, 189a36–189b16. See also *De Caelo* I, 3, 269b18–270a12.

25   On the "Scientific Revolution" see Floris Cohen H., *How Modern Science Came into the World: Four Civilizations, One 17th-Century Breakthrough* (Amsterdam: 2012).

26   On fire in seventeenth- and eighteenth-century science (physics, chemistry) see Locqueneux R., *Sur la nature du feu aux siècles classiques. Réflexions des physiciens et des chimistes* (Paris: 2014).

INTRODUCTION: THE EARLY MODERN FIRE

of the scientific study of fire in the early modern period should not be seen as linear. The knowledge of fire developed by alchemists, for example, continued to influence the debate on heat and combustion for decades, and was deeply intertwined with the studies of fire that would be crucial to the rise of modern chemistry.[27]

In the first chapter, Hannah Elmer focuses on purity and purification in the early seventeenth-century Paracelsian tradition, topics that are almost always discussed in connection with fire in the historical sources. Indeed, fire was a crucial agent of purification, a technical tool of a processual nature that could be activated in different situations to induce changes in matter towards higher degrees of purity. However, fire did not cease to be an element as well, a substance that itself possessed the highest degree of purity in the sublunary world. Elmer shows that these two aspects of fire cannot be separated if one is to understand the status of fire in the Paracelsian tradition: the purity embodied by fire as an element was crucial in conceptualising fire's action on matter and its ability to remove impurity from natural bodies.

In the second chapter, Stefano Salvia examines a further step in the distinction that gradually emerged in early modern scientific circles between fire as an element characterised by qualitative properties – luminosity, subtlety, mobility, etc. – and fire as a physical-chemical reaction. The new scientific concept of fire made it possible to distinguish between the subjective sensations of hot and cold and the objective measurements of heat intensity, which were carried out with specific technical tools (thermoscopes and thermometers). The case study chosen by Salvia is that of Galileo Galilei, who attempted to quantify, albeit indirectly – in terms of its effects on different substances – the amount of fire acquired or released by a given body. Interestingly, Galileo's achievements depended not only on his adoption of a new natural-philosophical approach, but also on his technical expertise in managing fire in the laboratory.

The history of fire in early modern science has several other facets that are explored in the remaining chapters of the first section. Consider the series of paintings depicting an alchemist or chemist at work that David Teniers the Younger (and followers) produced in the mid-seventeenth century. One of these, entitled *The Chymist* [Fig. 0.1] – actually an etching after Teniers made by Thomas Major in 1750 – represents an old man feeding fire with a bellows in order to carry out an experiment. The natural-philosophical nature of his

---

27   Newman W.R. – Principe L.M., *Alchemy Tried in the Fire: Starkey, Boyle, and the Fate of Helmontain Chemistry* (Chicago: 2002); Rampling J.M., *The Experimental Fire: Inventing English Alchemy, 1300–1700* (Chicago: 2020).

FIGURE 0.1 *An alchemist using bellows at a furnace in his laboratory* [*The Chymist*]. Etching by T. Major, after D. Teniers the Younger (1750)
WELLCOME COLLECTION, LONDON. PUBLIC DOMAIN

INTRODUCTION: THE EARLY MODERN FIRE

practice is symbolised by the presence of an open book in the bottom left corner of the picture; however, the practitioner does not pay attention to the book, concentrating solely on the fire in front of him. Other practitioners and various tools complete the scene, which is full of references to the dimension of production rather than of pure natural-philosophical research. Indeed, in the early modern period, the use of fire in experimental settings often combined natural-philosophical pursuits with goal-oriented practice, embodying the promise of useful results.[28] In chapter 4, Bérengère Pinaud examines the knowledge and use of fire amongst the Parisian apothecaries in the eighteenth century. Fire was central to the daily work of all members of the apothecaries' guild. Pinaud suggests that apothecaries' fire can be called 'artisanal fire' in that it was both a tool for preparing medicines and a force that had to be controlled to avoid damaging materials, and therefore had to be used in specific settings. However, the apothecary fire was also an epistemic object, as practitioners gained knowledge from observing its effects on substances and instruments, which was then codified and disseminated, for example, in pharmacy courses.

In the early modern period, the 'artisanal fire' was increasingly distinguished from the 'scientific fire'. In the eighteenth century, the emergence of an academic community of expert natural philosophers (physicists, chemists), and the advent of new measuring instruments, such as the thermometer, which allowed access to precise quantification, turned fire into an epistemic object of great complexity, far removed from the know-how of craftsmen. In chapter 3, Marco Storni shows that this received narrative should be partly revised, since the sources demonstrate the persistence of a 'technical knowledge of fire' well into the eighteenth century. This knowledge consisted, on the one hand, of a qualitative vocabulary that distinguished the types of fire used in laboratories and workshops, identifying each one by the technical method used to generate heat or by the effects of heat on materials. Technical knowledge also included informal strategies for measuring the intensity of heat. Reference to thermometric scales was avoided in favour of other measurement techniques – which Storni characterises as 'quasi-quantification' – closer to the everyday experience of practitioners, such as the number of coals used to feed the fire or the effect of fire on materials.

---

28   For examples of the intertwined nature of natural-philosophical research and the sphere of production and commerce in the early modern period see, e.g., Spary E.C., *Eating the Enlightenment: Food and the Sciences in Paris, 1670–1760* (Chicago – London: 2012).

2   Early Modern Fire Technologies

In 1618, the German painter and inventor Franz Kessler published the *Holzsparkunst* (*The Art of Saving Wood*), in which he described a new type of stove, made of several tiers, which was designed to optimise the use of fire, particularly for heating purposes [Fig. 0.2].

Kessler described fire as a force that was extremely difficult to control: 'There is no element more subtle, or more rapid, or which disappears more quickly than fire'.[29] Heat dissipation was a constant problem with household appliances (stoves, ovens) that, in Kessler's view, needed to be solved through technological improvements. The main objective of his technological work was economic: at a time when energy resources were scarce, it was crucial not to waste fuel, especially wood. As Kessler pointed out, 'it has happened that wood, because of its high cost in various places, has often been more difficult for the common people to obtain than even bread'.[30] The dearth of energy supplies remained a constant problem throughout the early modern age, as evidenced by the widespread concern in the historical sources about the management of forests, and by the reports of the dangers of deforestation, which affected European countries to varying degrees.[31] The problem of fuel scarcity became more pressing with the expansion of fire-based productive sectors – glassmaking, pottery, metalworking (iron and steel in particular) – and the concomitant growing concern for domestic comfort. The development of new technologies of fire was thus a response to this permanent *disette*, trying either to reduce the consumption of fuels, or to increase their effects. Several attempts were also made to use alternative energy sources to wood, such as peat and coal. Alongside innovations, however, there was a persistence of traditional methods to govern fire thriftily, which were readapted over time, and long considered viable alternatives to new technologies. These forms of persistence point to the non-linearity of innovation in fire technology, which was in fact a discontinuous and irregular process.

In chapter 5, Gianenrico Bernasconi discusses the issues raised by the technological governance of fire in terms of an economy of resources, focusing on the early modern development of cooking. Bernasconi presents the

---

29   Kessler Franz, *Holzsparkunst: Das ist, Ein solche new, zuvorn niemahln gemein, noch am Tag gewesen Invention etlicher unterschiedlichere Kunstofen* (Frankfurt a. M.: 1618) 5.

30   Ibidem, 'Dedication'.

31   Warde P., *Energy Consumption in England and Wales, 1560–2000* (Rome: 2007); Abad R., "L'Ancien Régime à la recherche d'une transition énergétique?", in Bouvier Y. – Laborie L. (eds.), *L'Europe en transitions. Energie, mobilité, communication, XVIIIe–XXIe siècles* (Paris: 2016) 23–84.

INTRODUCTION: THE EARLY MODERN FIRE 11

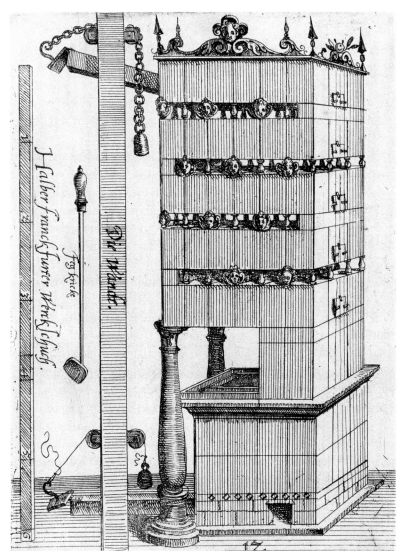

FIGURE 0.2   Franz Kessler's stove, in Kessler, Franz, *Holzsparkunst: Das ist, Ein solche new, zuvorn niemahln gemein, noch am Tag gewesen Invention etlicher unterschiedlichere Kunstofen* (Frankfurt a. M.: 1618)

technological shift from the "open fire" of the hearth to the fire enclosed in the metal pipes of ovens and stoves. This was a crucial transition, not only from the perspective of saving resources, but also for the evolution of the art of cooking. Enclosing the fire in a metal frame gave autonomy to the cooking processes and made it possible to experiment with new forms of food preparation in which the personal skills of the cooks played a marginal role. However, the acceptance of these new technologies in the kitchen was slow: fireplaces were still present in homes well into the nineteenth century, and projects for ovens and stoves took time to develop into a real industry. This shows that, while the history of fire technologies in the seventeenth and eighteenth centuries is one of innovation, it should not be overlooked that there was a persistence of technologies already in use, which were familiar to many practitioners, and deeply rooted in craft practices and everyday habits.

Similar patterns can be seen at work in other aspects of domestic life in which fire played a central role, particularly heating. In chapter 6, Olivier Jandot focuses on the history of the fireplace, which was the main heating system in Western Europe until the nineteenth century. The central figure in this study is the obscure figure of Gauger, the inventor of a discipline called "caminology," or the art of constructing fireplaces in such a way as to save fuel and reduce smoke. In his book *La mécanique du feu* (1713), fire was no longer seen as a subtle element – as in Kessler's work – but as a natural force (heat) that acts on materials: 'By *rays of heat* we mean both the parts of the wood that separate from it when it burns, and the parts of the material that surround the fire and that the fire causes to rotate'.[32] Gauger applied his reflections on the nature of heat and the modes of its propagation to the design of more efficient and economical fireplaces. Following Gauger's pioneering work, other inventors – whom Jandot calls 'fire mechanists' – worked intensively to improve fireplaces. This intense technological activity contributed to the emergence of a new industrial sector at the end of the eighteenth century: that of manufacturers of heating appliances.

In the realm of domesticity, fire management thus had a dual nature: on the one hand, it corresponded to the technological endeavour of creating devices to control the action of fire; on the other hand, it was the ability to save fuel, time, and effort in dealing with fire. The same dual nature can be found in the management of fire at the level of manufacturing activities. The seventh chapter, authored by Cyril Lacheze, deals with the production of architectural terracotta, i.e. bricks and tiles. The "operational chain" for the production of architectural terracotta required a skilful management of fire, especially in the use of the kiln. But fire management also involved the ability to avoid wasting too much wood in the firing process. Lacheze's analysis goes beyond the confines of the early modern period and considers several regions (of France

---

32 Gauger Nicolas, *La mécanique du feu, ou l'art d'en augmenter les effets, & d'en diminuer la dépense* (Paris, J. Estienne & J. Jombert: 1713) 4.

INTRODUCTION: THE EARLY MODERN FIRE

and neighbouring countries), since the slow development of techniques can often only be understood by adopting a comparative perspective, with broad chronological contours. At the same time, Lacheze shows that understanding fire management strategies requires taking into account various factors that go beyond the technology itself: for example, the use of alternative fuels (peat or coal) was always a possibility in the early modern period, but its actual implementation could be hindered by logistical, economic, or cultural factors.

Any consideration of early modern fire technologies must include an important area of research that attracted the interest of craftsmen, scientists, and entrepreneurs: the work on fire pumps and on the steam engine. In particular, traditional historiography has insisted on the steam engine as an invention resulting from cutting-edge research in the physical sciences; this machine has also often been seen as a metaphor for industrial growth.[33] Some recent revisions to this narrative have suggested reconsidering the role of artisans, and of figures with hybrid expertise – scientific and artisanal – in the research on fire engines, and more generally in the process of technological innovation that triggered the first industrialisation. In chapter 8, Andrew M.A. Morris contributes to this line of inquiry by focusing on the work of the British engineer John Smeaton, particularly his efforts to improve the fire engine invented by Thomas Newcomen. Smeaton's approach lies at the intersection of experimental science and technological innovation. Smeaton experimented with the fire engine and was able to significantly improve its efficiency. He took the practices of the craftsmen and turned them into hypotheses to be tested; his method was to optimise the process by experimenting with variations of several parameters. But Smeaton also wanted to draw general conclusions from his trials and provide rules that could be shared by communities of practitioners. Far from the supposed secrecy of artisanal knowledge, Morris shows that Smeaton sought to create an epistemic community that prefigured the principles of "open science."[34]

---

33   Landes, *The Unbound Prometheus*.

34   Some historians, notably Liliane Hilaire-Pérez, have spoken of 'open technique' to refer to dynamics of knowledge-sharing amongst craftsmen. See Hilaire-Pérez L., "Technology as a Public Culture in the Eighteenth Century: the Artisans' Legacy", *History of Science* 45.2 (2007) 135–153; Hilaire-Pérez L. – Garçon A.-F., "'Open Technique' between Community and Individuality in Eighteenth-Century France", in de Goey F. – Veluwenkamp J.W. (eds), *Entrepreneurs and Institutions in Europe and Asia, 1500–2000* (Amsterdam: 2002), 237–256.

3   Fire in the Urban Space

In the early modern age, fire was not only an epistemic object, or a force to be domesticated and enhanced through technology, but also a constant threat that populations had to deal with on a daily basis. On 2 September 1666, a fire broke out in the centre of London and raged for four days. The diarist Samuel Pepys witnessed it and spoke of a fire that 'rage[d] every way, and nobody, to my sight, endeavouring to quench it, but to remove their goods, and leave all to the fire'.[35] There seemed to be no escape from the fury of the fire, which was so strong that 'every thing [...] [was] proving combustible, even the very stones of churches'.[36] Amongst the many visual representations of the Great Fire of London, one provided by an unknown artist a few years after the tragedy (ca. 1670) is particularly revealing [Fig. 0.3]: it shows the devastating effects of the fire on Ludgate, with flames higher than any building, and people challenging the fire to save their own belongings. On the right, however, far from the panic of the first group of people, one can see other people queuing in front of the River Fleet – now an underground river – to fill buckets with water, presumably to help fight the fire. Despite the efforts to save lives and property, the fire caused incalculable damage, especially in material terms: more than 13,200 houses were destroyed.[37] The Great Fire of London and other major fires of this kind, which were not infrequent, explain the urge of early modern people to find different solutions to the constant danger of fire, not only in terms of more efficient organisation of firefighting, but also in terms of preventive measures.

In chapter 9, Marie Thébaud-Sorger discusses two competing inventions developed in the late eighteenth century by the Englishmen David Hartley and Charles Mahon. Both inventions – Hartley's fire plates and Mahon's mortar – aimed to develop incombustible materials that could protect buildings from fire. Thébaud-Sorger shows that these inventions were based on a better knowledge of the combustion process and on the observation that fires occurred despite the observance of preventive behaviour: the solution was therefore to develop technologies to improve the very architectural structures of cities. Hartley and Mahon gave public demonstrations of their methods, and news of their inventions spread throughout Europe. The inventions aroused the interest of governments, learned societies, architects, and others, and fuelled debates that touched on both economic and moral issues,

---

35   Pepys S., *Diary*, vol. 7, ed. by Robert Latham – William Matthews (London: 1972) 269.
36   Ibidem.
37   Field J.F., *London, Londoners and the Great Fire of 1666: Disaster and Recovery* (London – New York: 2018) 69.

INTRODUCTION: THE EARLY MODERN FIRE 15

FIGURE 0.3 [Unknown artist] *The Great Fire of London, with Ludgate and Old St. Paul's* (ca. 1670)
YALE CENTRE FOR BRITISH ART, PAUL MELLON COLLECTION.
PUBLIC DOMAIN

demonstrating their interconnectedness. For example, some authors, such as the Frenchman Augustin Piroux, argued that the cost of reconstruction was demonstrably higher than that of prevention, but that the fragility of buildings also endangered human life, considered a precious social resource. According to Thébaud-Sorger, these discussions can be interpreted as signs of an evolving perception of risk: once it was understood that the threat of fire could not be suppressed, people worked to develop means of controlling fire and limiting the costs to society, both in material and human terms.

In addition to the technological work on materials that could make buildings fireproof, the early modern period saw significant changes in the political organisation of cities to prevent fire and, above all, to fight its outbreak. These changes consisted not only in the development of national or local firefighting policies, but also in the emergence of new agents of social order – including a modern police force – and in the organisation of the social body that was to be prepared to face the emergency. In chapter 10, Catherine Denys focuses on the case of Lille, in French Flanders, as emblematic of the evolution of police practices, organisation, and transformations in the Old Regime. Denys dwells on the regulations, which at first sight appear to be repetitive and monotonous texts: on the contrary, she shows that regulations are part of a 'patient pedagogy' aimed at making the population aware of what could and could not be done. The pedagogy of the police was also directed at officers, who were reminded of what behaviour was and was not allowed, which helped to make them a specialist professional category. The organisation of firefighting is not only a good example of the professionalisation of police tasks, but it also highlights the enduring importance of the contribution of other groups of citizens that the police had to supervise and coordinate. Denys shows that in the event of a fire, the Lille police – but other European cities show similar patterns – had to coordinate the functions of the various figures and groups involved in firefighting, thus contributing to the rationalisation of the rescue effort, while keeping the curious away from the scene, as they were considered an element of disturbance.

In chapter 11, Cornelia Müller provides an analysis of early modern firefighting that complements Denys's by examining a different case than Lille. Müller discusses the regulations and practices of firefighting in three cities – Bautzen, Görlitz, and Löbau – in Upper Lusatia (eastern Germany, near the Polish border). The perspective she adopts is inspired by the French philosopher Henri Lefebvre, suggesting that fire regulations should be read as 'mental spaces' in which different facets of a city's economic, political, and social life are embedded, as well as suggestions for improvement: these texts thus provide an insight into civic organisation, both ideal and real. Müller shows that the urban space was not conceived as a unit, but was divided into different zones, which were assigned different functions and times of intervention in case of fire. She also discusses the visibility and distribution of firefighting equipment (buckets, ladders, syringes, etc.) as a mirror of social organisation. According to Müller,

INTRODUCTION: THE EARLY MODERN FIRE

the organisation of firefighting reveals the nature of power relations within a city, and the existing social hierarchies.

Fire in the urban space, however, was not only a danger to human life and property that had to be prevented and fought. Fire could also be a spectacle, with symbolic and political meanings. One might think of the development of pyrotechnics between the seventeenth and the eighteenth centuries, and its use to celebrate important public events. In 1749, George Frideric Handel (born Georg Friedrich Händel) composed a suite entitled *Music for the Royal Fireworks*, commissioned by King George II to accompany the celebrations for the signing of the Treaty of Aachen (1748), at the end of the War of the Austrian Succession.[38] The spectacle of pyrotechnics depended on the technical skills of expert craftsmen, and was also closely linked to contemporary scientific research: one need only think of the work carried out by eighteenth-century chemists such as Lavoisier on saltpetre (potassium nitrate), one of the basic components of gunpowder.[39] In chapter 12, Simon Werrett discusses pyrotechnics as a crucial example for understanding the connections between materials and emotions in the early modern period. Fireworks created an emotional connection between human bodies (spectators, technicians), materials, and the environment – a configuration that Werrett depicts through the category of 'emoterial'. 'Emoterial' is more than 'material' insofar as it refers to an emotional process that unfolds in the interaction of bodies in specific environments, a moment when rigid distinctions between subject and object become blurred. Werrett radicalises his intuition about pyrotechnics, arguing that the distinction between subject and object would have been alien to early modern people, who never conceived of 'objects' as passive entities. For them, the identities of people and objects were the result of a process of co-construction. In the case of fire management, the crucial passage in the reification of emoterial bodies was the conceptualization of human emotions in pyrotechnical terms and the insistence on the need to control them: moderating 'explosive' temperaments was seen as a step away from volatile subjectivity towards the objectivity of modern science.

Far from the background role that scholarship has traditionally assigned to fire, the essays in this volume demonstrate its centrality to understanding the entangled histories of science, technology, and society in the early modern

---

38 For a broader picture see Hogwood C., *Handel: Water Music and Music for the Royal Fireworks* (Cambridge – New York: 2005).

39 Mauskopf S.H., "Lavoisier and the Improvement of Gunpowder Production", *Revue d'histoire des sciences* 48.1–2 (1995) 95–122.

period. In the general conclusion, Liliane Hilaire-Pérez points out that these collected essays present a history of the construction of fire as an epistemic object, with an emphasis on practical and vernacular knowledge. In parallel with these scientific developments, in early modernity fire was also increasingly technicised: the seventeenth and eighteenth centuries marked the transition from a time when people – in the words of Daniel Roche, who in turn borrowed the expression from Gaston Bachelard – 'maintained an act of Prometheus' by caring for fire and maintaining a direct relationship with it, to a time of technological control of fire – the 'hidden fire' channelled into metal pipes, furnaces, ovens, etc. – which marginalised the direct experience of fire and thus the place of the emotions associated with the spectacle of the flame.[40] The chapters also underline, as Hilaire-Pérez notes, the non-linearity of this historical transition, if not the coexistence of these two 'worlds' of fire. Even today, in the age of triumphant technoscience, fire remains to some extent untamed – to quote a verse from William Blake's famous poem *The Tyger* (1794): 'What the hand dare seize the fire?'[41]

Bibliography

Primary Sources

Bevilacqua P., *Prometeo e l'aquila. Dialogo sul dono del fuoco e i suoi dilemmi* (Roma 2005).

Blake W., *Collected Poems*, ed. by W.B. Yeats (London – New York: [1905] 2002).

Gauger Nicolas, *La mécanique du feu, ou l'art d'en augmenter les effets, & d'en diminuer la dépense* (Paris, J. Estienne & J. Jombert: 1713).

Hesiod, *Theogony, Works and Days, Testimonia*, ed. by Glenn W. Most (Cambridge, MA – London: 2006).

Kessler Franz, *Holzsparkunst: Das ist, Ein solche new, zuvorn niemahln gemein, noch am Tag gewesen Invention etlicher unterschiedlichere Kunstofen* (Frankfurt a. M.: 1618).

Pepys S., *Diary*, vol. 7, ed. by Robert Latham – William Matthews (London: 1972).

Plato, *The Dialogues*, vol. 3: *Ion, Hippias Minor, Laches, Protagoras*, ed. by R.E. Allen (New Haven – London: 1996).

Secondary Sources

Il fuoco nell'Alto Medioevo: Spoleto, 12–17 aprile 2012 (Spoleto 2013).

Abad R., "L'Ancien Régime à la recherche d'une transition énergétique?", in Bouvier Y. – Laborie L. (eds.), *L'Europe en transitions. Energie, mobilité, communication, XVIIIe–XXIe siècles* (Paris: 2016) 23–84.

---

40   Roche D., *Le peuple de Paris: essai sur la culture populaire au XVIIIe siècle* (Paris: 1981) 141.

41   Blake W., *Collected Poems*, ed. by W.B. Yeats (London – New York: [1905] 2002) 74.

INTRODUCTION: THE EARLY MODERN FIRE

Anders G., *Die Antiquiertheit des Menschen*, vol. 1: *Über die Seele im Zeitalter der zweiten industriellen Revolution* (München 1956).

Boissière J., "La grande disette de bois à Paris des années 1783–1785", in Higounet C. (ed.), *L'approvisionnement des villes. De l'Europe occidentale au Moyen Âge et aux Temps Modernes* (Toulouse: 1985) 237–242.

Field J.F., *London, Londoners and the Great Fire of 1666: Disaster and Recovery* (London – New York: 2018).

Floris Cohen H., *How Modern Science Came into the World: Four Civilizations, One 17th-Century Breakthrough* (Amsterdam: 2012).

Garrioch D., "Towards a Fire History of European Cities (Late Middle Ages to Late Nineteenth Century)", *Urban History* 46.2 (2019) 202–224.

Goudsblom J., *Fire and Civilization* (London: 1992).

Gras A., *Le choix du feu. Aux origines de la crise climatique* (Paris: 2007).

Guedj M. – Mayrargue A., "Éclairages historiques sur l'émergence du concept d'énergie", *Recherches en didactique des sciences et des technologies* 10 (2014) 35–61.

Hilaire-Pérez L., "Technology as a Public Culture in the Eighteenth Century: The Artisans' Legacy", *History of Science* 45.2 (2007) 135–153.

Hilaire-Pérez L. – Garçon A.-F., "'Open Technique' between Community and Individuality in Eighteenth-Century France", in de Goey F. – Veluwenkamp J.W. (eds), *Entrepreneurs and Institutions in Europe and Asia, 1500–2000* (Amsterdam: 2002), 237–256.

Hogwood C., *Handel: Water Music and Music for the Royal Fireworks* (Cambridge – New York: 2005).

Jarrige F. – Vrignon A., *Face à la puissance. Une histoire des énergies alternatives à l'âge industriel* (Paris: 2020).

Klein U., "Hybrid Experts", in Valleriani M. (ed.), *The Structures of Practical Knowledge* (Cham: 2017) 287–306.

Landes D.S., *The Unbound Prometheus: Technological Change and Industrial Development in Western Europe from 1750 to the Present* (Cambridge: 1969).

Locqueneux R., *Sur la nature du feu aux siècles classiques. Réflexions des physiciens et des chimistes* (Paris: 2014).

Mathis C.F., *La civilisation du charbon. En Angleterre, du règne de Victoria à la Seconde Guerre mondiale* (Paris: 2021).

Mauskopf S.H., "Lavoisier and the Improvement of Gunpowder Production", *Revue d'histoire des sciences* 48.1–2 (1995) 95–122.

Mukherjee A., *Penury into Plenty: Dearth and the Making of Knowledge in Early Modern England* (Abingdon – New York: 2015).

Newman W.R. – Principe L.M., *Alchemy Tried in the Fire: Starkey, Boyle, and the Fate of Helmontain Chemistry* (Chicago: 2002).

Pyne S., *The Pyrocene: How We Created an Age of Fire, and What Happens Next* (Oakland, CA: 2021).

Rampling J.M., *The Experimental Fire: Inventing English Alchemy, 1300–1700* (Chicago: 2020).

Roche D., *Le peuple de Paris: essai sur la culture populaire au XVIIIᵉ siècle* (Paris: 1981).

Saelens W. – Blondé B. – Ryckbosch W. (eds.), *Energy in the Early Modern Home: Material Cultures of Domestic Energy Consumption in Europe, 1450–1850* (London – New York: 2023).

Sloterdijk P., *Die Reue des Prometheus. Von der Gabe des Feuers zur globalen Brandstiftung* (Berlin 2023).

Smil V., *Energy in World History* (Boulder, CO: 1994).

Smith P.H., *From Lived Experience to the Written Word: Reconstructing Practical Knowledge in the Early Modern World* (Chicago – London: 2022).

Spary E.C., *Eating the Enlightenment: Food and the Sciences in Paris, 1670–1760* (Chicago – London: 2012).

Valleriani M., "The Epistemology of Practical Knowledge", in Valleriani M. (ed.), *The Structures of Practical Knowledge* (Cham: 2017) 1–19.

Warde P., *Energy Consumption in England and Wales, 1560–2000* (Rome: 2007).

Werrett S., *Fireworks: Pyrotechnic Arts and Sciences in European History* (Chicago – London: 2010).

Werrett S., *Thrifty Science: Making the Most of Materials in the History of Experiment* (Chicago – London: 2019).

Zwierlein C., *Der gezähmte Prometheur: Feuer und Sicherheit zwischen früher Neuzeit und Moderne* (Göttingen: 2011).

PART 1

Fire in Early Modern Science

∴

CHAPTER 1

Purity, Purification, and Fire in the Seventeenth-Century Chymical Texts of Nicaise Le Febvre and Michael Sendivogius

Hannah Elmer

Abstract

Purity – be it of metals or medicines – was a central concern for early modern alchemists, and the purification of substances was a primary goal of alchemical processes. From distillation to calcination, the application of heat was often the key to transforming the substances at hand. To this end, fire was ubiquitous in early modern workshops. Yet fire was not the only source of heat (dung, for instance, could be used as a source of consistent, low-grade warmth) and it was often mediated (perhaps through water, in a *bain marie*). This paper examines the interplay between techniques of purification, especially those relying on heat, and broader ideas about the role of fire as a purifying agent, in the work of the seventeenth-century chymists Nicaise Le Febvre and Michael Sendivogius.

Keywords

chymistry – elements – pharmacy – principles (chymical) – purification – purity – separation (chymical) – transmutation

Early modern Europeans were worried about purity. They were worried about the purity of their communities, their religions, their bloodlines. Indeed, so important was purity and attendant acts of purification that historian Peter Burschel has recently argued these concepts could very well characterize the entire epoch.[1] His work, along with that of many other historians, has demonstrated the social, political, and religious manifestations and consequences of

---

1 Burschel P., *Die Erfindung der Reinheit: eine andere Geschichte der frühen Neuzeit* (Göttingen: 2014).

© KONINKLIJKE BRILL BV, LEIDEN, 2025 | DOI:10.1163/9789004521766_003

these concerns.[2] And when we shift our gaze from the broader social and religious contexts to consider the natural world and material practices, we indeed encounter more discussions of purity and purification.[3] Yet, crucially, in "applied" contexts, these ideas seem to function a bit differently. An examination of purity and purification through chymistry[4] brings into question some of the qualities assigned to them if they are only investigated from a social or religious perspective. Arguments like Valentin Groebner's of purity always being 'unmixed' no longer quite hold true[5] – but it also becomes difficult to pin the concepts to tight definitions. Nevertheless, Groebner's distinction between 'purity' and 'purification' provides a useful starting place: 'purity' is an (unachievable) condition, while 'purification' is the process by which something is urged towards the state of purity, perhaps through religious or technical means.[6] Yet the instability of the concepts alongside their regular evocation indicate this area to be one of particular importance: material purity – like social or religious purity – was being particularly intensely contested and constructed during the early modern period.[7]

One thing, however, that almost always appears in conjunction with purity and purification in chymical contexts and can also come up in social ones is fire. Burning things – and people – was often perceived to be a means of cleansing, and many chymists recognized fire to be the chief instrument by which they worked upon various substances. Yet fire was not merely an instrument, not simply a process that could be harnessed in various situations to

---

2 For instance, the significant amount of recent work on race and blood purity, especially in Iberia and the Americas. One of the most important theoretical foundations remains Douglas M., *Purity and Danger: An Analysis of the Concepts of Pollution and Taboo* (London – New York: 1994).

3 Art historians have been at the forefront of bringing material practices into conversation with other social and cultural dynamics. See, for instance, Jacobi L. – Zolli D. (eds.), *Contamination and Purity in Early Modern Art and Architecture* (Amsterdam: 2021).

4 To avoid the pseudoscientific/pejorative connotations of 'alchemy' as well as the anachronistic implications of 'chemistry', I follow William Newman and Lawrence Principe in using the early modern spelling 'chymistry'. Newman W.R. – Principe L.M., "Alchemy vs. Chemistry: The Etymological Origins of a Historiographic Mistake", *Early Science and Medicine* 3 (1998) 32–65.

5 Groebner V., *Wer redet von der Reinheit? Eine kleine Begriffsgeschichte* (Vienna: 2019).

6 *Ibidem*, 40.

7 Scholars of chymistry have also clearly shown that questions about purification contributed significantly to the overthrow of Aristotelian theories of matter that posited a formal alteration of substances as they combined to form a 'mixt'. See Newman W.R., *Newton the Alchemist: Science, Enigma, and the Quest for Nature's "Secret Fire"* (Princeton: 2019), and Principe L.M., *The Transmutations of Chymistry: Wilhelm Homberg and the Académie Royale des Sciences* (Chicago: 2020) 132ff.

effect certain kinds of change; for most of the early modern period, at least in chymical circles, fire was also widely recognized to be a type of substance in and of itself, be it as an element, a principle, or perhaps phlogiston. Both these identities – fire as instrument and fire as "element" – need to be viewed in order to grasp the complexity of fire, not to mention its relation to purity.

Although fire has been the subject of much historical study, this has mostly been from environmental or technological perspectives, with urban conflagrations and military developments being particularly common.[8] Historians of chymistry have addressed fire in both its instrumental and cosmological roles, but not as the main point of focus.[9] Ursula Klein is one of the few to have also put concepts of purity centre stage, demonstrating their importance for the history of chemistry, especially for the eighteenth and nineteenth centuries.[10] Bringing the study of fire and purity together thus seems to be generative for our understanding of both sets of concepts. Purity is intimately linked to fire and on a few different levels simultaneously: fire as an element and as an observable phenomenon embodied the highest degree of purity in the universe – its qualities of light, ethereal (in)corporeality, and ability to penetrate deeply into other substances became conflated with ideas of purity. In this way, fire provided a set of characteristics that shaped how people perceived the 'pure'.

This chapter seeks to map out some of the different relationships between fire-as-element, fire-as-instrument, purity, and purification in seventeenth-century chymical thought. As this is a wide and varied field, I have focused on the Paracelsian tradition, bringing the pragmatic medical perspective of apothecary Nicaise Le Febvre (1610–1669) into conversation with the metallurgical concerns of the 'adept' Michael Sendivogius (1566–1636). Le Febvre, who published his massive chymical textbook in 1660 after decades of practice, was one of the last to work with matter before the field became even more

---

8   For environmental history, Stephen Pyne has been the major forerunner with his *Vestal Fire: An Environmental History, Told through Fire, of Europe and Europe's Encounter with the World* (Seattle: 1997) and most recently *The Pyrocene: How We Created an Age of Fire and What Happens Next* (Oakland: 2021). From a technological angle, especially noteworthy for early modern Europe is Werrett S., *Fireworks: Pyrotechnic Arts and Sciences in European History* (Chicago: 2010).

9   It tends to come up as a feature of individual thinkers' works and thus a point of comparison of different cosmological claims, but not as a historical problem in itself. See, for instance, the various works by Ursula Klein, William Newman, Lawrence Principe, and Didier Kahn.

10 Klein U., "Objects of Inquiry in Classical Chemistry: Material Substances", *Foundations of Chemistry* 14 (2012) 7–23. This article also briefly, but helpfully, treats the seventeenth-century Paracelsian tradition (especially citing Le Febvre).

complicated with debates on corpuscular theory. He cultivated connections with courts, becoming a demonstrator in chymistry at the *Jardin Royal des Plantes médicinales* in Paris in 1652 and a royal apothecary and distiller two years later.[11] The *Jardin Royal* was connected to the royal court of Louis XIV, and was the major safe haven for Paracelsian medicine in an otherwise hostile Paris, which was still dominated by the traditional Galenic faculty of medicine at the university.[12] In 1661, just after publishing his textbook, *Traité de la chymie* (in English as *A Compendious Body of Chymistry*), Le Febvre moved to London to serve Charles II and only two years after that was inducted into the Royal Society. His textbook, quickly translated into several languages, was his main scholarly contribution, and is the focus of this study.[13]

While Le Febvre's text is part of a tradition stretching back to the early part of the century, and many of its ideas can indeed be found among its predecessors,[14] my purpose here is not to analyze this specific chymical tradition but rather to explore some of the range of relationships that fire and purity/purification could occupy in a broader chymical landscape of the earlier

---

11   While many scholars reference Le Febvre's work, few have engaged with it significantly, in part because it has been viewed as not particularly innovative. The most substantive treatment can be found in Klein U., *Verbindung und Affinität: die Grundlegung der neuzeitlichen Chemie an der Wende vom 17. zum 18. Jahrhundert* (Basel: 1994), and Klein U., "Nature and Art in Seventeenth-Century French Chemical Textbooks", in Debus A. – Walton M.T. (eds.) *Reading the Book of Nature. The Other Side of the Scientific Revolution* (Kirksville: 1998) 239–250, followed by Debus A.G., *The French Paracelsians: The Chemical Challenge to Medical and Scientific Tradition in Early Modern France* (Cambridge: 1991). Hélène Metzger's study has also been foundational for Le Febvre and his colleagues: Metzger H., *Les doctrines chimiques en France du début du XVIIᵉ à la fin du XVIIIᵉ siècle* (Paris: 1923). For a brief survey of the establishment of the *Jardin Royal* and the teaching tradition (including the fierce debates between the chymists and the medical faculty at the University of Paris) that directly preceded Le Febvre, see Kahn D., "The First Private and Public Courses of Chymistry in Paris (and Italy) from Jean Beguin to William Davisson", *Ambix* 68 (2021) 247–272.

12   For an overview of the conflicts see Kahn D., *Alchimie et paracelsisme en France à la fin de la Renaissance (1567–1625)* (Geneva: 2007) and Debus, *The French Paracelsians*.

13   It went through 6 French editions, 5 German, 3 English, and 1 Latin edition. I worked with the 1660 French edition and the first English edition (1662); I quote from the English edition. See Le Febvre Nicaise, *Traicté de la Chymie* (Paris, Thomas Jolly: 1660), and Le Febvre N., *A compendious body of chymistry, which will serve as a guide and introduction both for understanding the authors which have treated of the theory of this science in general: and for making the way plain and easie to perform, according to art and method, all operations, which teach the practise of this art, upon animals, vegetables, and minerals, without losing any of the essential vertues contained in them. By N. le Fèbure apothecary in ordinary, and chymical distiller to the King of France, and at present to his Majesty of Great-Britain* (London, Theo. Davies and Theo. Sadler: 1662).

14   For instance Jean Beguin, Joseph Du Chesne, William Davisson, and Étienne de Clave; see Kahn, "The First Private and Public Courses of Chemistry" for a helpful survey.

seventeenth century. Sendivogius's writings are thus particularly instructive as they are concerned with metallic transmutation and the production of the Philosophers' Stone, and they allow us to see how fire and purification could appear in the more enigmatic tradition of chymical adepts. Also circulating among several central European courts, Sendivogius published a relatively small number of treatises in the earlier decades of the seventeenth century.[15] These texts, however, circulated widely, and some of his core ideas have been shown to have been quite influential on later writers, not the least of whom being Isaac Newton.[16] Here I focus foremost on his *Treatise of Sulphur* but will also consider his *New Light of Alchymie*.[17] While overlap between the authors can be anticipated from their shared eclectic Paracelsianism – both also demonstrating clear Aristotelian and Neoplatonic influences – such shared moments are nonetheless instructive because of their otherwise different intellectual projects and textual genres.

1 Cosmological Fire

Both fire and purity feature prominently in *A Compendious Body*. Le Febvre begins his textbook by describing basic cosmology in order to indicate how

---

15   For an overview of Sendivogius's life and career, see Prinke R., "Beyond Patronage: Michael Sendivogius and the Meanings of Success in Alchemy", in López-Pérez M. – Kahn D. – Rey-Bueno M. (eds.), *Chymia. Science and Nature in Medieval and Early Modern Europe* (Newcastle upon Tyne: 2010) 175–231. Prinke R., "New Light on the Alchemical Writings of Michael Sendivogius (1566–1636)", *Ambix* 63 (2016) 217–243. For an important examination specifically of the *Treatise of Sulphur* that situates the text also among Sendivogius's other works, see Kahn D., "Le *Tractatus de sulphure* de Michael Sendivogius (1616), une alchimie entre philosophie naturelle et mystique", in Thomasset C. (ed.), *L'Écriture du texte scientifique au Moyen Âge* (Paris: 2006) 193–221.

16   See Newman *Newton the Alchemist* especially 66–69.

17   I worked with the Latin editions of both texts and the 1650 English edition and translation (which contains both texts). As the early modern English edition is faithful to the Latin in the sections analyzed here, I quote from it. Sendivogius Michael, *Tractatus De Sulphure Altero Naturae Principio* (Cologne, Ioannes Crithius: 1616); Sendivogius Michael, *Novum Lumen Chymicum e Naturae Fonte et Manuali Experientia deproptum, et in duodecim Tractatus diuisum, ac iam primum in Germania editum. Cui accessit Dialogus, mercurii, Alcymistae et Naturae, perquam utilis* (Cologne, Anton Bötzer: 1610) ; Sendivogius Michael, *A New Light of Alchymie: Taken out of the Fountain of Nature, and Manuall Experience. To Which is Added a Treatise of Sulphur / Written by Micheel Sandivogius ... Also Nine Books of the Nature of Things, Written by Paracelsus ... Also a Chymicall Dictionary Explaining Hard Places and Words Met Withall in the Writings of Paracelsus, and Other Obscure Authors. All of Which are Faithfully Translated out of the Latin into the English Tongue, by J.F.* (London, Richard Cotes for Thomas Williams: 1650).

and where humans can mimic and manipulate nature. Here his main distinction is between a system of four elements and a system of five principles. He proposes to 'reserve the appellation of Elements for those great and vast Bodyes which are the general Matrixes of natural things,' and to designate as 'principles' the 'constitutive parts of the Compound' that are encountered in everyday 'mixt bodies'.[18] Elements are ontologically prior to the principles, but this remove makes them difficult, if not impossible, for humans to access. Fire features prominently in both of these systems, but its relationship to purity varies somewhat between the two. We will turn first to the elements.

In Le Febvre's system fire is itself one of the four elements, alongside earth, water and air, but his system is also characterized by emanations and the circulation of spirits.[19] All elements remain in their respective spheres as places where substances are generated through the imprint of the universal spirit. In this capacity as element, fire remains in the superlunary sphere, where it is the first transmitter of the universal spirit:

> the Element of Fire is not enclosed under the sphere of the Moon, as we have already hinted above, and that consequently no other Fire can be admitted, then the *Aether* or Heaven it self, which hath its Matrixes and fruits, as the other Elements. For that great number of Stars which we see moving in that vast Element, are nothing else but particular Wombs or Matrixes, where the Universal Spirit takes a very perfect Idea, before it descends to incorporate it self in the Matrixes of other Elements.[20]

It is fire's subtle and ethereal nature that enables this element to receive the universal spirit and begin the process of informing the bodies with the spirit. It then moves down and passes on to the realm of air. The process of emanation continues, with the realms of water and earth being the most significant for the generation of bodies we encounter. Through later processes of separation, the 'corporified' fire returns to the heavens, where being consumed by stars is re-'spiritualized' and then begins the process of 'corporification' all over again.[21]

---

18 Le Febvre, *A Compendious Body* 20.

19 His system is an eclectic blend of Aristotelian, Stoic, and Neoplatonic elements. For a precise intellectual genealogy of Le Febvre that helps situate him among a number of his contemporaries, see Emerton N.E., *The Scientific Reinterpretation of Form* (Ithaca: 1984). Klein, *Verbindung und Affinität* is also foundational.

20 Le Febvre, *A Compendious Body* 36.

21 Le Febvre, *A Compendious Body* 37. 'The continual Influxes of Heaven and its Starres, do incessantly produce the Fire or Spiritual Light, which beginneth first to embody it self in the Air, where it takes the Idea of Hermaphroditical Salt, which thence falls in the waters

Thus, fire holds an exalted position in the system, but it is for us the most distant. Cycles of purification underlie the entire universe, with the motion of fire and its transformations through more or less pure states characterizing these cycles. But while such change characterizes fire (as well as the other elements), by framing it as changing states of purity, the element remains ontologically stable, a static ideal underneath accretions and reductions of the 'impurity' of other elemental properties. Thus, while change is inherent to the system, fire itself is not a process.

Sendivogius provides a more complicated picture, not least because *A New Light of Alchymie* and the *Treatise of Sulphur* emphasize different aspects of his philosophical system. For simplicity's sake we will focus on the *Treatise of Sulphur*, but even a brief look at *A New Light of Alchymie* reveals the central role of fire and important connections to purity for Sendivogius. In this earlier work, he introduces the generation of matter as a complex process involving two primary fire sources: a celestial fire and an earthly fire, the latter located in the centre of the earth. While he also draws on a kind of emanation and circulation of spirits to explain the introduction of form into matter, his focus in this work is rather on seminal forces, where substances are the product of seeds developing within specific matrixes. It is the purity of the matrix that greatly determines what substance will develop. The *Treatise of Sulphur* shifts emphasis somewhat from the seeds to elements and principles, thus echoing Le Febvre's organization.

As in the apothecary's work, elemental fire plays a complex role in connecting the sub- and superlunary realms in Sendivogius's texts, interacting with the other elements in the generation of natural substances. It is likewise characterized by purity. But Sendivogius also distinguishes different degrees of purity within each element, resulting in what might be thought of as different subspecies, each of which being fit to achieve specific ends. Fire's substance, we learn:

---

and upon the earth, where it takes a body, either Mineral, Vegetable, or Animal, by the character and efficacy of some particular Matrix, imprinted in it by the action of its ferment; And when this body comes to be dissolved by the means of some potent Agent, its Sulphur, fire or Light corporified is so depurated, that the Starrs attract it for their food; for the Stars (as we suppose)are nothing else but a Fire, a Sulphur, or some most pure Light actuated. Not unlike the link of a Lamp, which once being lighted, doth continually attract the Oyl to feed its flame: so that the Starrs in the same manner attract this fire, which is depurated by that action, and spiritualize it anew, to precipitate again by their kinde influency, and restore it to the Air, the Water, and the Earth, to corporifie it, or give it a body again'.

Is the purest of all, and its essence was first of all elevated in the Creation with the throne of divine Majesty […] out of the lesse pure part of its substance the Angells were created; out of that which was lesse pure then that, being mixed with the purest Aire, were the Sun, Moon and Stars created. That which is lesse pure then that is raised up to terminate, and hold up the Heavens: but the impure, and unctuous part of it is left, and included in the Center of the Earth by the wise and great Creator for to continue the operation of motion, and this wee call Hell. All these Fires are indeed divided, but they have a Naturall sympathy one towards another.[22]

Thus, within a general elemental system characterized by degrees of purity, fire holds the foremost position.[23] It also helps to join various parts of the universe, with an inherent 'sympathy' allowing the different types to remain vaguely connected. Divisions within the single element based on level of purity also characterize the other elements, making purity an essential ontological category for Sendivogius. Elemental fire itself thus plays a number of roles (including also being the source of the soul[24]), but it is specifically purity-impurity that enables this diversity.

Le Febvre does not locate different types of elemental fire in different places or assign it different purposes based on degree of purity, but he does invoke a concept of mixture to explain the relationship between elemental fire and the fire we encounter on earth:

When I have said, that Fire in its ascent doth forsake the commerce of other Elements, that I have so spoken, only because the visible Fire which we use in our Hearths, is nothing else but a Meteor or Body imperfectly mixt of some Elements or Principles, wherein Fire or Sulphur are predominant, and its flame an oleagineous and sulphureous smoak kindled; and when fire is spiritualized by that forsaking, it never ceaseth till it hath returned to its natural place, which of necessity must be above, and beyond the Air.[25]

---

22 Sendivogius, *Treatise of Sulphur* 99–100.
23 His opening words of the chapter are, 'Fire is the purest, and most worthy Element of all' (99).
24 Sendivogius, *Treatise of Sulphur* 100–101. 'This soule is of that most pure Elementary Fire, infused by God into the vitall spirit'.
25 Le Febvre, *A Compendious Body* 36.

Earthly fire differs from that in the heavens through a kind of temporary union with other elements, a union which is broken as the fire 'spiritualizes' and escapes back to the realm of the stars. This difference in fire is not, however, here explicitly framed as a difference in purity, as in the works of Sendivogius, though impurity is alluded to through the idea of mixture. While purity may provide an important category for differentiating between what we experience in our daily life and what is in the superlunary sphere, it does not result in the more static, ontological gradations within the element that Sendivogius describes.

But through all this, fire remains for both authors an essential building block in the natural world, endowing creation with its ethereal and volatile characteristics. It is this lightest, least-corporeal quality that enables it to be the matter of angels and the substance through which the universal spirit can enter ultimately into the sublunar realm. These qualities become interchangeable with those associated with purity.[26]

2   Principles

But fire as a constituent of the natural world does not remain solely in the category of element for either author. As mentioned above, both Sendivogius and Le Febvre seek to join the elements to a system of principles, so that by the latter, they can explain how humans can intervene in natural processes. Whereas the elements have clear Aristotelian connections (albeit with significant additions), the system of principles invokes the heritage of Paracelsus; by seeking to integrate the two systems, Sendivogius and Le Febvre do not threaten the Aristotelian theory of matter the way some of their contemporaries did.[27] Ultimately drawing on ideas from the medieval Islamicate alchemical tradition (among others), Paracelsus shifted emphasis away from the elements and towards a system of 'principles'. The Islamicate tradition, which had focused foremost on the transmutation of metals, conceived of substances (especially metals) as composed of the two principles of 'mercury' and 'sulphur', not to be conflated with the substances of the same names occurring naturally in the world and used in chymical workshops.[28] These principles actually functioned

---

26   Such an identification between purity and fire in Le Febvre's work is also recognized by Klein in "Objects of Inquiry" 14–15.

27   Kahn, "The First Private and Public Courses of Chymistry".

28   For more on these traditions and their transference into the Latin West, see Principe L.M., *The Secrets of Alchemy* (Chicago: 2013) and Burnett C. – Moureau S. (eds.). *A Cultural*

similarly to the Aristotelian elements in that they focused on and combined palpable qualities: mercury, cold and wet, lent the malleable, fluid properties to materials, while sulphur, hot and dry, provided the friable, flammable, and congealing characteristics. To these two, Paracelsus added a third – salt – to form his *'tria prima'*. Salt contributed stability, inflammability, and served to keep the other principles joined in whichever material body they currently composed. On the one hand, fire in this system of principles shifts somewhat from centre stage, now being subsumed under the idea of sulphur. However, the fiery aspect simultaneously achieves heightened importance as a principle: 'flammability' has now become one of the organizing categories. Not just hot-cold, substances comprised specifically flammable-inflammable qualities.

A number of thinkers adopted a system of principles, sometimes using them to replace the elements entirely, but sometimes trying to integrate them into a larger structure in which elements could still play a role.[29] Sendivogius and Le Febvre belong to the latter. They claim the principles to be both the means by which nature works and (therefore) the means by which humans can effect change. In the words of Sendivogius we can see this relationship between principles and elements clearly:

> Because Nature out of these Principles, which are Salt, Sulphur, and Mercury, doth produce Mineralls and Metalls and all kinds of things, and it doth not simply produce Metalls out of the Elements, but by Principles, which are the medium betwixt the Elements, and Metalls.[30]

Thus, in theory, to work with fire as a material component of natural substances, one would go through the sulphuric principle. Yet, interestingly, despite both writers asserting the same thing, Le Febvre follows through with it much more consequently. He quickly shifts from the elements to the principles in the textbook, spends longer explaining the workings of the principles, and draws much more on the language of principles than elements in the practical part of the book. Sendivogius, by contrast, reverts consistently to the language of elements, especially when describing meaningful transformation, despite his claim that principles are the medium between elements and metals; for instance:

History of Chemistry in the Middle Ages (London: 2022).

29   For instance, Jean Beguin, Joseph Du Chesne, William Davisson, and Étienne de Clave. Kahn, "The First Private and Public Courses of Chymistry" 252–254. See also Klein, *Verbindung und Affinität.*

30   Sendivogius, *Treatise of Sulphur* 144.

And whosoever desires to attaine to his desired end, let him understand the conversion of the Elements to make light things heavy, and to make spirits no [sic] spirits; then hee shall not worke in a strange thing. The fire is the Rule, whatsoever is done, is done by Fire.[31]

Although Sendivogius speaks often of sulphur and mercury (the treatise, after all, is 'of Sulphur'), these sections are especially vague, characterized by digressions about the general relationship between nature and the principles or about allegories that emphasize the hiddenness of sulphur in all things.

This difference in emphasis between the two writers connects to their larger chymical concerns. Not only does Sendivogius work primarily with metals, but he positions himself as an adept (or one in training – he is explicit about having "not yet" achieved the Philosophers' Stone): both *A New Light of Alchymie* and the *Treatise of Sulphur* are about generating the Philosophers' Stone, which is also explicitly connected to human prolongevity in the latter treatise. Thus his larger project of deep material transformation connects more directly to problems of material generation, and thus back to elements.

Le Febvre's concerns, by contrast, were more about the intermediate steps to the larger change: preparing specific medications to be applied for specific complaints. Although he does address universal medicines (more later), in practice he and the other apothecaries he was helping to train would be most often tasked with producing particular remedies. Such a focus on specific (and focused) medications comes through not only in the book's hundreds of medicinal recipes, but also in the privileging of principles over elements (or seeds). Le Febvre draws heavily on the Paracelsian *tria prima*, but following some of his predecessors at the *Jardin Royal*, he recognizes two additional principles: earth and water or phlegm.[32] These new principles – a watery and a more solid/dry substrate – were conceived of as 'passive' and as providing more of the material heft to the substance.[33] Fiery qualities were assigned still to sulphur, which was among the 'active' principles.

---

31   Sendivogius, *A New Light of Alchymie* 45–46.

32   Le Febvre, *A Compendious Body* 20: 'Three of these substances offer themselves to our sight, by the help of Chymical Operation, under the form of a Liquor, which are the Phlegm or Water, the Spirit, and the Oyl; the two other in a solid body, *viz.* Salt and Earth. The Water and Earth are commonly called passive Principles, material and of lesse efficacy than the other three; as contrary wise the Spirit or Mercury, Sulphur and Salt, are reckoned active and formal Principles, by reason of their penetrative and subtle virtue'.

33   The introduction of water and earth into the system of principles could take different forms. As Kahn shows, some thinkers held water and earth to be mere 'bodies' with the 'tria prima' being the proper principles, while others saw all five as principles, though

> That substance [...] is the third in order of those we extract by the artificial resolution of the Compound; [...] it is an Oleagineous substance, which easily takes fire, being a combustible nature, by whose means also the Mixts are rendred such. [...] This Substance is also considered two several wayes: For being loosened from the others, it swims above the Phlegm and Spirits, because it is lighter and more aetherial; but if it be not absolutely separated from the Salt and the Earth, it will sometimes precipitate it self to the bottom, or swim in the middle.[34]

Through this example of sulphur, we can see how the principles become visible through chymical operations and what kinds of characteristics are deemed most significant. Sulphur appears in terms of its flammability but also in terms of how it separates out from the other principles, what form it takes depending on how completely it is severed from the more solid salt and earth. If it is fully separated, it demonstrates the ethereal qualities we encountered above in the discussion of elemental fire.

The emergence of the extra principles, however, also points to the role fire played in people's discernment of material substances. As Klein has noted, the new principles can be connected to the different empirical practices of Paracelsus and the later apothecaries; Paracelsus worked primarily with combustion – burning substances outright – while the apothecaries worked especially through distillation.[35] Distillation processes allowed additional layers of the material to be discerned, and the apothecaries, in turn, sought to integrate these into the system of principles. Nevertheless, although the principles were shown to be much more accessible to human manipulation than the elements, the two remain similar insofar as they possess ideal forms that are in essence static, and hence, separable through chymical processes.

3 From Purity to Purification

As we have seen in the above discussions, the key characteristic of fire as element and principle was especially its sensory quality of ethereal lightness. This

---

differing in levels of passivity/activity. Others recognized yet different numbers and configurations. See Kahn, "The First Private and Public Courses of Chymistry" 252–254; also Klein, *Verbindung und Affinität* 36.

34   Le Febvre, *A Compendious Body* 26.

35   Klein U., "Nature and Art" 243. For a discussion of Paracelsus's discernment of principles during processes of combustion, see also Klein *Verbindung und Affinität* 44–45.

could and was often combined with the quality of heat and burning, but it was the lightness that enabled fire to be the substance of angels, and the lightness that most connected it to ideas of purity. The heat and potential to burn, however, help make this ethereal quality into something more active. While the ethereal quality might have enabled fire to penetrate other substances, it appears to be by the action of its heat that the other substances could be transformed. But most significantly, it is the properties of heat that humans can directly manipulate, endowing fire with its second identity of agent of change. It is in this capacity – what we today might describe as catalytic – that fire became linked with ideas not of purity but of purification.

With this in mind, let us turn now to examine more directly purity and purification as they appear in Le Febvre's and Sendivogius's works. Above we saw how the other elements and all principles could be identified as pure in the sense of 'unmixed'. Further, especially in Sendivogius's writings, we saw how degrees of purity could play an ontological role, determining different grades of each element. Such ontological significance appears consistently throughout his texts, above all with the purity of a place (as matrix) determining the specific substance that would develop from the seed once it inhered in that particular space; purity and generation are fundamentally linked.[36] Although issues of purity and purification permeate Sendivogius's texts, they are often in service to other concepts and thus not theorized directly. In one of the more direct moments, interestingly found in the section on fire in the *Treatise of Sulphur*, Sendivogius tells us,

> So now thou seest how corrupt Elements come to bee in a subject, and how they are separated when one exceeds the other, and because then putrefaction is made by the first separation, and by putrefaction is made a separation of the pure from the impure, if then there be a new conjunction of them by vertue of Fire, it doth acquire a form much more noble than the first was. For in its first state, corruption was by reason of grosse matter mixed with it, which is not purged away by putrefaction, the subject thereby being bettered, and this could not be but by the vertues of the foure Elements, which are in every compound thing.[37]

---

36 It is beyond the scope of this essay to explore Sendivogius's ideas about seeds and seminal reasons. See Newman, *Newton the Alchemist* 66–68. Hiro Hirai has also dealt extensively with this topic; see Hirai H., *Le concept de semence dans les théories de la matière à la renaissance de Marsile Ficin à Pierre Gassendi* (Turnhout: 2005).

37 Sendivogius, *Treatise of Sulphur* 109.

Here we encounter processes of purification that involve multiple stages of separation, first through putrefaction and then through fire, with the latter stage being what would also elevate the substance to its 'nobler' state. Purification relies on separation, but not only. We will find something similar later in Le Febvre's work.

In contrast, Le Febvre devotes the entire second part of his theoretical section to purity and impurity, where he seeks to explain what these are and how they function in the medicinal chymistry of the apothecaries. Nevertheless, it is still difficult to pin down precisely what purity means; Klein, for instance, identifies in Le Febvre's text four main concepts associated with the term: homogeneity, utility for human health, incorporeality, and singleness of gender and species.[38] Sometimes he addresses this multiplicity directly:

> To return then to our Operations, we have said above that by the ministery of fire the Artist did separate from each Compound five Substances or differing Principles, which though very pure in themselves, yet in several respects may have their impurities, either in respect to themselves or in respect to our intention. For if we have occasion to use only the Spirit extracted out of a Mixt and that this Spirit be joyned with some portion of the Phlegm of it, we shall say this Spirit to be impure in such a respect, and so of the other Principles.[39]

Here purity-impurity is located on the level of the principles, but purity-impurity can be relative terms. They can apply to principles, if they are not entirely separated from one another, but this does not mean that the mixt[40] from which they come is itself corrupt. In an earlier section Le Febvre proposes a kind of compromise between the two common views of purity-as-efficacy and purity-as-homogeneity, associating the pure with that which promotes human health, but leaving room for it to also characterize non-relational qualities of a substance. Similarly here, a combination of context-bound use – 'our intention' – and essential quality combine to inform the meaning of purity.

---

38   Klein, "Nature and Art", 246. In "Objects of Inquiry" she emphasizes above all purity as efficacy in Le Febvre's work. Klein, "Objects of Inquiry" 14–15.

39   Le Febvre, *A Compendious Body* 54.

40   Newman explains this concept in detail; in short, it is not our modern idea of 'mixture' because it entails an idea of homogeneity, bringing it more in line with what we would consider a 'compound'. See Newman, *Newton the Alchemist* 115–119. This also forms the core of Klein's work, *Verbindung und Affinität*.

PURITY, PURIFICATION, & FIRE IN THE 17TH-CENTURY CHYMICAL TEXTS   37

Things become even more complex when Le Febvre describes the relationship between purity, extraction, and the substances which result from chymical operations:

> Besides the five Substances or Principles, which we have formerly said may be extracted out of natural Compounds by the ministery of fire, there may be yet some Essences drawn by diversifying the Operations of Art, which exalts, and do ennoble the Principles of these Mixts and raise it to their *purity*. These Essences do not only differ in body from that of the Compound whence they were extracted, but are advanced also to nobler and more efficacious qualities and vertues than those which during it's intireness did adorn their bodies, and after its dissolution and artificial separation will possess more then any of the Principles of the Compound. But though these wonderfull Essences obtain several names in the Books of Philosophers, who call them Arcana's, Magisteries, Elixirs, Tinctures, Panacea's, Special Extracts, &c., they are comprehended nevertheless under the general notion of pure and purity, and it proceeds from hence that after those Essences are drawn out of their mixts, the remainder is commonly rejected as impure.[41]

Transported through some of these terms – 'elixirs' and 'panaceas' for instance – is a further complication: purity can derive from *composite* substances – and not only that, sometimes the compound actually enhances the degree of purity. This raises questions about the relationship between purity and the 'unmixed'. At what stage does a substance cease being a mixture, let alone a mixt?

It seems that what we also have here is a conflation of 'purifying' with 'purity'. Those things which are most efficacious in engendering health are associated with purity through their ability to consume or drive out impurities. But as recipes for universal medicines or the Philosophers' Stone reveal, universal purifying agents are themselves never derived from a single substance. Le Febvre presents this clearly when he describes how impurity can be expelled from a substance:

> There is two several ways of removing the impurity of things; the first universal, and the other particular: The first is an universal medicine, which is or may be extracted from several subjects, after they have been, as much as Art can perform it, reduced to their Universality, and devested

---

41   Le Febvre, *A Compendious Body* 55.

of their specification and natural fermentation, which caused them to be this or that determinate Mixt: For this Medicine being by a fit digestion, and requisite coction and maturation, reduced to the highest degree of its exaltation; it is sufficient to expel impurity out of all Bodies indifferently, because it insensibly consumes it, both by the help of Fixation and Volitalization. The second is a particular Medicine, which by its specifical vertue and faculty, may expel particular impurities.[42]

Crucial is not that the medicine be 'unmixed' but rather that each component be appropriately processed: each must be purified. Once a substance has attained its own state of purity, it can be mixed with other substances of the same status. And although Le Febvre speaks of reducing the substances 'to their universality', it is never an arbitrary substance that is chosen to be reduced, implying that principles extracted from mixts retain enough of their original identity that this must factor into their ultimate medicinal agency. Moreover, after everything has gone through the right procedures of spiritualizing and corporifying – here, 'digestion', 'concoction', and 'maturation' – the constituted whole can then, in turn, spiritualize and corporify ('fixation' and 'volitalization') other substances to bring them into a state of purity. Purification begets purification.

Sendivogius gives us a slightly different way of thinking about this. His brief, direct statement about the Philosophers' Stone reveals through the more traditional ideas of elemental balance how the ultimate purifying agent can indeed derive from mixture: 'For it is certain that when pure Elements are joined together equally in their vertues, such a subject must be incorrupted, and such must the Philosophers Stone bee.'[43] For both authors, then, purity (and purification) imply a certain type of efficacy that might be enhanced by the presence of multiple substances.

4　Separation and Purification

Operations of separation enable the chymist to isolate the principles, and these are what then can embody purity. While there is something seemingly intuitive about the connection between separation and purification – one might imagine the use of fire in processing metallic ores or in "freeing" gold or silver from a mercury amalgam – it is actually not self-evident that the action of fire

---

42　Le Febvre, *A Compendious Body* 54.
43　Sendivogius, *Treatise of Sulphur* 107.

PURITY, PURIFICATION, & FIRE IN THE 17TH-CENTURY CHYMICAL TEXTS   39

must result in such separation rather than some other type of transformation. Thus we find that such ideas of separating and purifying became central in defining chymical medicine against traditional Galenic pharmacy both for the chymists and the Galenists. This can be seen through the words of a Galenic apothecary – Jean de Renou – as he describes 'chymical extracts':

> There is no little difference betwixt the extractions of Apothecaries and those of Alcumists, for Apothecaryes extract onely a certain liquor, as Rosin, Gums, or such fluid matters, and separate them from the more grosse and solid substance; but Alcumists do not onley desert the grosse body, but exhale the thinnest substance till a very little portion, and that somewhat thicker be left, to which the vertue of the whole doth adhere as united to its subject, whence they call it an extract, as it were the essence extracted and separated from the body.[44]

Both Galenic apothecaries and their chymist counterparts were busy separating substances, and both were concerned specifically to separate the lighter from the grosser and to discard the latter. Yet a cursory examination of de Renou's text suggests that discourses of purity/impurity are far less common than in chymical texts,[45] and here we see that efforts to acquire the 'thinnest' are presented as something specific to the chymists. This amplifies the ideas about 'essences' that we see in Sendivogius's and Le Febvre's works, and the clear association of the essences with the 'thinnest' part evokes a similar sensory quality that we encountered in the descriptions of elemental fire: held up as the purest and most essential part of the substance is that which is subtle, rarefied, ethereal – the spirit that will fly away. What is also noteworthy here is how the emphasis on the thinnest essence expands the concepts of purity and

---

44   De Renou Jean, *A Medicinal Dispensatory, Containing The whole Body of Physick: Discovering The Natures, Properties, and Vertues of Vegetables, Minerals, and Animals: The manner of Compounding Medicaments and the way to administer them. Methodically digested in Five Books of Philosophical and Pharmaceutical Institutions; Three Books of Physical Materials Galenical and Chymical. Together with a most Perfect and Absolute Pharmacopoea or Apothecaries Shop. Accommodated with three useful Tables. Composed by the Illustrius Renodaeus, Chief Physician to the Monarch of France; And now Englished and Revised By Richard Tomlinson of London, Apothecary* (London, J. Streater and J. Cottrel: 1657) 79–80.

45   Andrew Wear's study suggests that for the Galenists language of purity/impurity might apply especially to counterfeits or adulterated medicine. He also shows language of purity/purification to be particularly central to van Helmont's programme and criticism. See Wear A., *Knowledge and Practice in Early Modern English Medicine, 1550–1680* (New York: 2000) 91, 100–102.

purification beyond being mere synonyms for homogeneity and the separation of substances, for the other parts resulting from chymical operations, which may also become more 'homogenous' as a consequence of the separation, are not necessarily considered pure because of it.

How one mixes pharmaceutical components constitutes one of the major lines of dispute between the traditional Galenists and the chymists. Joan Baptista van Helmont, for instance, criticized (Galenic) apothecaries for deviating from Galen's original pharmacopeia of simples and creating ever-more elaborate recipes of several or perhaps even dozens of ingredients.[46] But the clash here between the chymists and the Galenists is not about purity in the sense of a simple (which would support the idea of purity as 'unmixed'), because as we have seen, Le Febvre uses ideas of purity to *support* compound formulations (even if most of his recipes focus on single substances). The argument is about techniques of preparation, which Le Febvre presents in terms of purification:

> Why shall Chymists be debarred from the same Remedies, after they have prepared, corrected, and devested them from the malignity and venom which they did contain, by separating of Purity from Impurity? which is much a better way then the pretended correction of Galenists, who indeavor to mitigate the vices and malignity of Mixts, used by the Ancients and Moderns also, by the addition of some other Substance, which may have, and really hath, in it self its particular Vice and Impurity: as is obvious in Hellebore, Spurge, Scamony, [...] which they pretend to correct by a simple addition of Mastick, Cynamon [...] But to show the manifest difference between this Correction and that of Chymists we may use the vulgar comparison of an unskilful Cook, which to dress a savoury and delicate dish of Tripes, should think it enough to boyl them with odoriferous Herbs and good sents without washing and cleansing them from their inward filth.[47]

The purification that matters here is not that which the completed medicine will be able to accomplish in the body (though, naturally this is what is ultimately at stake), but rather the purification necessary to create an effective medicine in the first place. The impurity found in mixts is linked to 'vice and malignity' and this must be 'corrected', echoing the language found in the

---

46   Wear, *Knowledge and Practice* 95ff.

47   Le Febvre, *A Compendious Body* 110.

Galenic texts.[48] Yet where the Galenists see balance as the means of 'correcting' and achieving efficacy and see mixing substances as a further means of attaining that balance, Le Febvre argues that substances must first be purified before being mixed (and *if* being mixed at all). The final analogy of animal excrement amplifies this connection between purity and preparation. But, interestingly, from the Galenic side we also encounter criticism directed at the chymical processes of purification that focuses precisely on the use of fire; some people were convinced that the fire corrupted, rather than corrected, the substances.[49]

These discussions of proper preparation for achieving purity bring us back now to fire, this time in its 'material' form, to borrow Le Febvre's term. Interestingly, Galenists used plenty of fire in preparing their remedies, but as we have seen, the chymists associated their own use of fire with processes of purification, and used language of purification to distinguish their work from their more traditional contemporaries.[50] It was how the fire was understood to transform or alter the substance that divided the two groups, not the application of fire in general.

5 Material Fire

We saw above that it was not merely the occupational context of the practitioners (chymical apothecaries versus metallurgists, for instance) that led to a different discernment of materials – five principles instead of three – but that the different uses of 'material' fire stood behind this. The modulation of heat, perhaps through water or sand, afforded new ways of understanding the basic components of material bodies. But what exactly was going on during these chymical operations? The principles discerned through processes of distillation raised questions about the identities of these separated substances: were they actually components, separated out, or were they rather first generated through these chymical processes?[51] Behind these questions is a larger

---

48 For a discussion of 'correcting' and compounds in Galenic medicine, see Wear, *Knowledge and Practice* 100ff.

49 Debus presents a few examples, some of which specifically locate the corruption in the smell of fire that remains in the substance. Debus, *The French Paracelsians* 81–88.

50 Wear describes the points of debate between the Galenists and chymists (especially Helmontians), see Wear, *Knowledge and Practice* 100–102 and passim.

51 Debus, *The French Paracelsians* 56, 110 and passim; Debus A.G., "Fire Analysis and the Elements in the Sixteenth and the Seventeenth Centuries", *Annals of Science* 23 (1967) 127–147; Klein, *Verbindung und Affinität* 36.

one about what the material fire – the fire of the workshops – was actually doing, which in turn inspired questions about what such earthly fire actually *was*.[52] Thus, how one understood the agency of fire contributed to how one understood what happened to substances when they underwent chymical procedures, and vice versa, that how one saw the chymical substances influenced how one interpreted the identity of material fire.

Both Sendivogius and Le Febvre clearly see fire processes as separating constituents of a mixt rather than creating new substances, though they do not explicitly theorize fire as an agent of change. Before Le Febvre details the nine degrees of heat and six main types of furnaces (not to mention numerous vessels and tools) that a chymical apothecary must master, he describes the material requirements of fire, that 'most potent Agent that Nature hath furnished us,' explaining that 'to feed and maintain it self, [fire] doth need first a combustible oily and sulphureous matter, either Mineral as Sea-coal, or Vegetable as Chark-coal, and the Oyls of Vegetables; or finally, Animals Fats, Suets and Oyls of Animals'.[53] Unsurprisingly the fuels all suggest a significant proportion of the sulphuric principle. But, he continues, fire also 'needs a continual Air, that may by its action drive away the excrements and fuliginous emanations of the substances which are burned, and that may animate the Fire, to make it more or less act upon its subject'.[54] Proper ventilation is critical for clearing away the components of the fuels that remain after the sulphur has been separated out.

Fire's action is then distinguished between mediate and immediate: 'Immediate, when without opposition Fire acts upon the matter, or Vessel that contains it, whether it be a Crucible, Retort, or any other thing; and this is commonly called Open Fire; fire of Calcination and suppression. Mediate is, when any thing is interposed between the Fire and the matter, which doth hinder its destructive action'.[55] The art is then in governing the fire, for although Le Febvre distinguishes nine classes of heat, 'the ingenuous Artist,' he tells us, 'may yet vary [these] in an infinite number of manners, according to his intention, and that the quality of the Mixt upon which he works, doth require it'.[56] Moderating heat is foremost achieved through distance or the interposition of another body between the fire and the matter being processed. Different ways

---

52 Debus shows how over the seventeenth century the status of fire as one of the elements was contested by a number of people (such as Du Chesne and van Helmont), and this often occurred through arguments derived from scripture. Debus, *The French Paracelsians* 51–54 and passim.

53 Le Febvre, *A Compendious Body* 81.

54 Ibidem.

55 Le Febvre, *A Compendious Body* 82.

56 Ibidem.

of achieving this are presented in detail in the following sections on degrees of heat, furnaces, vessels, and lutes.

Although Sendivogius also claims to work practically, his work remains far less precise. Thus, when discussing fire as a tool at the start of *A New Light of Alchymie*, in passing, he distinguishes only two degrees of fire: 'The first fire, or of the first operation is a fire of one degree, continuall, which goes round the matter; the second is a naturall fire, which digests and fixeth the matter'.[57] In the subsequent texts only rarely do specific forms of heat or degrees come up and even then, not in particularly applicable form. One of the most concrete passages, which is tellingly situated in a dream allegory, describes successful chymical operations with impossible vagueness:

> Hee [Saturn] then took two Mercuries of a differing substance, but of one originall, which Saturn washed in his owne urine, and called them Sulphurs of Sulphurs, and mixed the fixed with the volatile, and the composition being made, hee put it into its proper vessel, and lest the Sulphur should fly away, hee set a keeper over him, and then put him into a bath of a most gentle heat.[58]

Although Sendivogius assumes his readers will already know proper fire-handling techniques and thus provides no practical instruction, he does provide a helpful distinction when presenting fire as the element by which substances are to be purified and separated:

> This stirring up of the fire is done by the will of Nature, sometimes by the will of the skilfull Artificer disposing of Nature. For naturally all impurities, and pollutions of things are purged by Fire: All things that are compounded, are dissolved by Fire: as water washeth, and purgeth all things imperfect, which are not fixed, so the Fire purgeth all things that are fixed, and by Fire they are perfected: As Water doth conjoine all things that are dissolved; so fire separates all things that are conjoined; and what is naturall, and of affinity with it, it doth very wel purge, and augment it, not in quantity but in vertue.[59]

While we do not learn about degrees of heat, we do learn about fire's particular role in separating substances. Both water and fire purify, but fire removes that

---

57   Sendivogius, *A New Light of Alchymie* 31–32.
58   Sendivogius, *Treatise of Sulphur* 139.
59   Ibidem, 105.

which is 'fixed', meaning that it has the capacity to penetrate and break up a substance. Interestingly, although water too is used to purify, by washing away the loose impurities, it is also given the capacity to join substances. Fire only separates, but through this it also perfects.

This is echoed in Le Febvre's descriptions of chymical operations, where operations such as filtering, sifting, washing are designated 'material' separations, while those that tend to rely on heat/fire are labelled 'formal':

> Purity first must be separated from Impurity, which is performed materially or formally: Materially, by cribration or sifting, ablution or washing, edulcoration or sweetening, detersion or cleansing, effusion or powering, colation and philtration, or running through a bag, and despumation: Formally, by distillation, sublimation, digestion, and several other reiterated Operations, wereof hereafter.[60]

While all separation may contribute to purity, separation itself occurs on functionally different levels. The 'material' separation prepares the way for the 'formal,' and it is the formal – which very likely involves heat[61] – that enables one 'to obtain a perfect exaltation of the Mixt'.[62]

But, interestingly, once we get to the practical part of Le Febvre's textbook, with its concrete instructions, the language shifts from 'purity' and 'impurity' to 'extracting vertue' and managing/promoting volatilization, subtilizing; the adjectives 'gross' and 'heavy' also are invoked. This we can see clearly in the description of the process of 'spiritualization', which would seem to be a prime candidate for discussions of purity and purification:

---

60   Le Febvre, *A Compendious Body* 74.

61   Of the long list of procedures that follows, the majority of operations uses heat, and those that don't tend to be used in combination with those that do, making the application of fire highly likely for most pharmaceutical preparations. The operations are 'Limation or filing, rasion or scraping, pulverisation or reducing to powerder, alkoholisation or reducing to atomical parts, incision or cutting, granulation or reducing into small grains by melting, lamination or converting into thin Plates, putrefaction, fermentation, maceration, fumigation, which is either dry or moyst, cohobation, precipitation, amalgamation, distillation, rectification, sublimation, calcination either actual or potential, vitrification, projection, lapidification, extinction, fusion, liquation, cementation, stratification, reverberation, fulmination or detonation, extraction, expression, inceration, digestion, evaporation, desiccation, exhalation, circulation, congelation, crystallization, fixation, volatilisation, spiritualisation, corporification, mortification and revivification'. Le Febvre, *A Compendious Body* 74–75.

62   Le Febvre, *A Compendious Body* 74.

Spiritualization doth change the whole body into Spirit, so that it becomes no more palpable nor sensible to us: And by Corporification, the Spirit re-assumes its Body, and manifests it self again to our senses; but the body so qualified is an exalted body, very different in vertue from that from which it hath been extracted, since this body so glorified contains in itself the mystery of its own Mixt.[63]

Only under three techniques, 'cementation,' 'detonation,' and 'vitrification', is purification explicitly indicated. This suggests that 'purification' and by extension 'purity' is achieved through a series of steps, each of which contributes a particular aspect, but not the whole. Purification emerges from an aggregate of operations. But in this aggregate, fire is the central agent of change.

And yet, as mentioned above, it is also exactly in these moments that fire's agency is particularly disputed. While Le Febvre and Sendivogius see it 'exalting' the substances by separating out their fixed constituents, others see it generating the new form of these substances, while yet others see it corrupting the substances. What people observe happening through the application of fire is explained and construed in very different ways, revealing how far from straightforward even fire-as-instrument is (let alone fire-as-element and fire-as-principle).

6 Conclusion

This examination of how fire was understood to work in a chymical context prompts further questions about purity and purification. The purity embodied by fire-as-element might, for instance, allow us to expand Groebner's claims of purity being unattainable. In Groebner's discussion of the social, political, and religious contexts, purity tends to be located in the past – in the Garden of Eden, for instance – and therefore out of reach. While both Sendivogius and Le Febvre also refer to this, each mentioning the Fall and the corruption it brought into the natural world, the purity of fire, visible in the night sky as stars, was spatially, not temporally, removed. Indeed, the presence of that perfection, even if seemingly impossibly removed, had for centuries inspired ideas about the quintessence: those highly distilled spirits might just be that celestial purity freed from coarse, earthly bodies.[64] And the production of purified

---

63   Le Febvre, *A Compendious Body* 80.
64   DeVun, L., *Prophecy, Alchemy, and the End of Time: John of Rupescissa in the Late Middle Ages* (New York: 2009). While John of Rupescissa likely did not believe the alcohol he

substances that could, in turn, purify other substances similarly raises questions about identifying purity too closely with the 'unmixed', for as we have seen, purity may also be something that can be heightened *through mixing*. A universal medicine, the Philosophers' Stone here included, was a highly complex substance, comprised of numerous ingredients that had to be processed in very specific ways. Itself a product of processes of spiritualizing and corporifying, it made balance/equilibrium something active – this perfectly exalted substance acquired the capacity to 'consume' impurity, but not be tainted in the process.

Here we can detect the importance of fire in its conceptualization – flames are not 'contaminated' by the grossness of the body from which they might come. Fire escaping from the mixts and returning to the stars provides an image of purity where the purity itself has a kind of substance, rather than simply being the absence of 'impure' qualities. Where washing carries away the impure, leaving the 'pure' behind and the water dirty, burning releases the pure, leaving the dross behind but the flame clear.

Bibliography

Primary Sources

De Renou Jean, *A Medicinal Dispensatory, Containing The whole Body of Physick: Discovering The Natures, Properties, and Vertues of Vegetables, Minerals, and Animals: The manner of Compounding Medicaments and the way to administer them. Methodically digested in Five Books of Philosophical and Pharmaceutical Institutions; Three Books of Physical Materials Galenical and Chymical. Together with a most Perfect and Absolute Pharmacopoea or Apothecaries Shop. Accommodated with three useful Tables. Composed by the Illustrius Renodaeus, Chief Physician to the Monarch of France; And now Englished and Revised By Richard Tomlinson of London, Apothecary.* (London, J. Streater and J. Cottrel: 1657).

Le Febvre Nicaise, *A compendious body of chymistry, which will serve as a guide and introduction both for understanding the authors which have treated of the theory of this science in general: and for making the way plain and easie to perform, according to art and method, all operations, which teach the practise of this art, upon animals, vegetables, and minerals, without losing any of the essential vertues contained in them. By N. le Fèbure apothecary in ordinary, and chymical distiller to the King of France, and at present to his Majesty of Great-Britain* (London, Theo. Davies and Theo. Sadler: 1662).

distilled to actually be the substance of stars, later authors began to take the metaphoric connections more literally, facilitated by Neoplatonic ideas of emanation.

Le Febvre Nicaise, *Traicté de la Chymie* (Paris, Thomas Jolly: 1660).

Sendivogius Michael, *A New Light of Alchymie: Taken out of the Fountain of Nature, and Manuall Experience. To Which is Added a Treatise of Sulphur / Written by Micheel Sandivogius ... Also Nine Books of the Nature of Things, Written by Paracelsus ... Also a Chymicall Dictionary Explaining Hard Places and Words Met Withall in the Writings of Paracelsus, and Other Obscure Authors. All of Which are Faithfully Translated out of the Latin into the English Tongue, by J.F.* (London, Richard Cotes for Thomas Williams: 1650).

Sendivogius Michael, *Novum Lumen Chymicum e Naturae Fonte et Manuali Experientia deproptum, et in duodecim Tractatus diuisum, ac iam primum in Germania editum. Cui accessit Dialogus, mercurii, Alcymistae et Naturae, perquam utilis* (Cologne, Anton Bötzer: 1610).

Sendivogius Michael, *Tractatus De Sulphure Altero Naturae Principio* (Cologne, Ioannes Crithius: 1616).

Secondary Sources

Burschel P., *Die Erfindung der Reinheit: eine andere Geschichte der frühen Neuzeit* (Göttingen: 2014).

Burnett C. – Moureau S. (eds.), *A Cultural History of Chemistry in the Middle Ages* (London: 2022).

Debus A.G., "Fire Analysis and the Elements in the Sixteenth and the Seventeenth Centuries", *Annals of Science* 23 (1967) 127–147.

Debus A.G., *The French Paracelsians: The Chemical Challenge to Medical and Scientific Tradition in Early Modern France* (Cambridge: 1991).

Douglas M., *Purity and Danger: An Analysis of the Concepts of Pollution and Taboo* (London – New York: 1994).

DeVun, L., *Prophecy, Alchemy, and the End of Time: John of Rupescissa in the Late Middle Ages* (New York: 2009).

Emerton N.E., *The Scientific Reinterpretation of Form* (Ithaca: 1984).

Groebner V., *Wer redet von der Reinheit? Eine kleine Begriffsgeschichte* (Vienna: 2019).

Hirai H., *Le concept de semence dans les théories de la matière à la renaissance de Marsile Ficin à Pierre Gassendi* (Turnhout: 2005).

Jacobi L. – Zolli D. (eds.), *Contamination and Purity in Early Modern Art and Architecture* (Amsterdam: 2021).

Kahn D., *Alchimie et paracelsisme en France à la fin de la Renaissance (1567–1625)* (Geneva: 2007).

Kahn D., "Le *Tractatus de sulphure* de Michael Sendivogius (1616), une alchimie entre philosophie naturelle et mystique", in Thomasset C. (ed.) *L'Écriture du texte scientifique au Moyen Âge* (Paris: 2006) 193–221.

Kahn D., "The First Private and Public Courses of Chymistry in Paris (and Italy) from Jean Beguin to William Davisson", *Ambix* 68 (2021) 247–272.

Klein U., "Nature and Art in Seventeenth-Century French Chemical Textbooks", in Debus A. – Walton M.T. (eds.), *Reading the Book of Nature. The Other Side of the Scientific Revolution* (Kirksville: 1998) 239–250.

Klein U., "Objects of Inquiry in Classical Chemistry: Material Substances", *Foundations of Chemistry* 14 (2012) 7–23.

Klein U., *Verbindung und Affinität: die Grundlegung der neuzeitlichen Chemie an der Wende vom 17. zum 18. Jahrhundert* (Basel: 1994).

Metzger H., *Les doctrines chimiques en France du début du XVIIe à la fin du XVIIIe siècle* (Paris: 1923).

Newman W.R., *Newton the Alchemist: Science, Enigma, and the Quest for Nature's "Secret Fire"* (Princeton: 2019).

Newman W.R. – Principe L.M., "Alchemy vs. Chemistry: The Etymological Origins of a Historiographic Mistake", *Early Science and Medicine* 3 (1998) 32–65.

Principe L.M., *The Secrets of Alchemy* (Chicago: 2013).

Principe L.M., *The Transmutations of Chymistry: Wilhelm Homberg and the Académie Royale des Sciences* (Chicago: 2020).

Prinke R., "Beyond Patronage: Michael Sendivogius and the Meanings of Success in Alchemy", in López-Pérez M. – Kahn D. – Rey-Bueno M. (eds.), *Chymia. Science and Nature in Medieval and Early Modern Europe* (Newcastle upon Tyne: 2010) 175–231.

Prinke R., "New Light on the Alchemical Writings of Michael Sendivogius (1566–1636)", *Ambix* 63 (2016) 217–243.

Pyne S., *The Pyrocene: How We Created an Age of Fire and What Happens Next* (Oakland: 2021).

Pyne S., *Vestal Fire: An Environmental History, Told through Fire, of Europe and Europe's Encounter with the World* (Seattle: 1997).

Wear A., *Knowledge and Practice in Early Modern English Medicine, 1550–1680* (New York: 2000).

Werrett, S. *Fireworks: Pyrotechnic Arts and Sciences in European History* (Chicago: 2010).

CHAPTER 2

'Atoms of Fire': Galileo's Unachieved Theory of Heat and the Beginnings of Thermometry (*c.*1603–1638)

Stefano Salvia

Abstract

Galileo never developed a systematic theory of heat in his works, nor of the atomic structure of matter. All we know about it can be derived from Galileo's correspondence with Giovan Francesco Sagredo (1612–1615), from *Il Saggiatore* (1623), from Benedetto Castelli's letter to Ferdinando Cesarini (1638), and from Vincenzo Viviani's *Racconto istorico* (1654). The limits and unresolved issues of Galilean atomism have been thoroughly discussed in the literature. My paper is focused on Galileo's mechanistic-corpuscular account of heat and related phenomena (water ebullition, air expansion, the glowing of hot charcoal and metals, etc.) as effects produced by subtle, mobile, acute, and indivisible *minima* of fire penetrating the inter-atomic *vacua* of material substances. The hypothesis was not original, as the connection between heat and motion at microscopic level already belonged to the tradition of ancient and early modern corpuscularianism. Galileo's specific contribution is to be found in his indirect quantification of the heat released or acquired by a material body, given the impossibility of a direct method to measure the amount of fire corpuscles emitted or absorbed. Such an attempt exposed Galileo to challenging objections from his contemporaries: if hot bodies did not seem to be heavier than cold ones, heat must not be identified with fire itself, but with the effects of igneous atoms streaming through larger particles (friction, agitation, disruption, and transport). The conceptual shift towards studying the action of fire on material bodies in terms of measuring their temperature, by means of the already known phenomenon of thermal dilatation, was the key step leading to Galileo's thermoscope (ca. 1603) and to its first medical application as graduated (proto-)thermometer by Sanctorius in the 1610s.

Keywords

atoms – fire – Galileo – theory of heat – thermometry

© KONINKLIJKE BRILL BV, LEIDEN, 2025 | DOI:10.1163/9789004521766_004

1   Outlines of an Open Research Programme

The emergence of a new science of temperature (thermometry) and heat (calorimetry) from the domain of the experimental philosophy of nature between the 17th and 18th centuries was the result of a complex, multi-actor process of both conceptual and practical domestication of the traditional element 'fire' as an epistemic object, whose allegedly intrinsic qualities (heat, dryness, lightness, luminosity, subtleness, mobility, pervasiveness, etc.) were gradually replaced by its direct or indirect action and effects on material bodies (heating, drying, rarefying, calcining, dilating, etc.). This fundamental shift in focus 'from substance to function' – to quote the title of Ernst Cassirer's 1910 essay[1] – would gradually lead to a clear distinction between 'fire' as physical-chemical reaction and 'heat' as associated but distinct physical phenomenon, as well as between 'temperature' as its objective measure and the subjective, physiological sensation of 'feeling hot'.[2] The same process would be later at work in replacing the *phlogiston* theory with Antoine-Laurent Lavoisier's notion of oxidation to account for fire as combustion, and in rethinking the concept of a 'caloric fluid' in terms of thermal energy.[3]

In this paper, I will retrace the beginnings of this historical path, dating back to Galileo Galilei's first inquiries in Padua around 1603, and then in Florence in 1612–1615, about heat and how to quantify it. If not directly, by weighting the amount of fire acquired or released by a given body, Galileo quantified it indirectly, by isolating, studying, and comparing its effects on different substances in terms of rarefaction or dilatation, fusion or ebullition, trying to find a common criterion and a kind of graduated scale, to which different degrees of heat could be related. Following what might be defined as an open (and unachieved) mechanistic-corpuscular research agenda, I will discuss Galileo's main treatment of heat both as a physical and a physiological phenomenon in the *Assayer* (1623), together with Benedetto Castelli's later account of Galileo's thermoscopic researches in his letter to Ferdinando Cesarini (30 September 1638).

In the last section, I will provide an excursus on further Galilean developments of the new 'science of heat,' in the context of the Accademia del

---

1   Cassirer E., *Substanzbegriff und Funktionsbegriff: Untersuchungen über die Grundfragen der Erkenntniskritik* (Hamburg: 1910; reprint, Hamburg: 2000).

2   Bachelard G., *La psychanalyse du feu* (Paris: 1938).

3   See, e.g., Morris R.J., "Lavoisier and the Caloric Theory", *The British Journal for the History of Science* 6.1 (1972) 1–38; Best N.W., "Lavoisier's 'Reflections on Phlogiston' II: On the Nature of Heat", *Foundations of Chemistry* 18 (2016) 3–13.

Cimento (1657–1667), and on the late 17th-century gradual transition from semi-quantitative thermoscopy to quantitative thermometry, that also marks an advance toward modern thermology as a branch of classical physics.[4]

The choice of focusing on Galileo's studies on heat and thermoscopy – which are no less known to scholars than the later developments of thermometry, calorimetry, and thermology – rests on the idea that the first half of the 17th century was crucial, from a historical-epistemological perspective, in the slow evolution from fire as one of the four classical elements, with heat as one of its inherent properties, to fire as a physical-chemical process affecting any kind of flammable material. Heat production and diffusion, as a conceptually separated phenomenon implied in the action of 'burning' (i.e., combustion), became one of its main manifestations, together with visible, lighting flames and other effects, like smoke and ash production.

Galileo's Paduan and Florentine investigations on the heating effects of fire, with the essential distinction between fire (still regarded as elemental substance) and its mechanical action upon other bodies by causing, among other outcomes, what we perceive through our sense of touch as 'being hot,' represented a research programme, so to speak, that was integral part of his attitude as scientist-engineer, practitioner, and skilled instrument-maker.[5] As I will argue, Galileo's attitude was not far from that of some contemporary 'chymists' – a hybrid, transformational figure on the road from alchemy to modern chemistry – whose work required technical expertise in fire management and heat gradation for their everyday laboratory operations.[6]

2   Water, Air, and Fire: the Paduan Thermoscope

The only direct and explicit reference to a Galilean theory of heat generation, as of the observable and measurable effects of heat propagation through solid or fluid materials, can be found in Galileo's *Assayer*. This is not accidental, insofar as heat (or rather what is perceived as such by our sense of touch) is mechanically reduced to a special kind of motion and attrition induced by streams of igneous particles at the microscopic level of matter constituents. The context is that of a harsh controversy between Galileo and the Jesuit

---

4 Fenby D.V., "Heat: Its measurement from Galileo to Lavoisier", *Pure and Applied Chemistry* 59.1 (1987) 91–100.

5 Valleriani M., *Galileo Engineer* (Berlin: 2010).

6 Newman W.R., *Atoms and Alchemy: Chymistry and the Experimental Origins of the Scientific Revolution* (Chicago: 2006).

father Orazio Grassi on various interconnected subjects, ultimately linked to the nature and proper motion of comets; starting from what is the object of true and reliable knowledge in natural philosophy: according to Galileo, the measurable 'primary qualities' of the physical world, the only ones which are written in mathematical language.[7]

Apart from such an attempt in the *Assayer*, where Galileo keeps, however, a cautious attitude towards his own hypotheses, no other reference to systematic research on measuring the 'quantity of heat' acquired or released by material bodies can be found in his published works. All we know beyond the *Assayer* can be derived from the correspondence with his Venetian friend and collaborator Giovanni Francesco Sagredo.

The other two sources on the topic are Castelli's already-mentioned 1638 letter to Cesarini and Vincenzo Viviani's *Historical Account of the Life of Galileo Galilei* (1654), where Viviani makes a short, dense, and *a posteriori* reference to his master's early investigations on heat, 'thermoscopy,' and temperature.

> During this same time, Galileo invented thermometers, that is, those glass instruments filled with air and water that are used to measure the changes of heat and cold, and the variation of temperatures in different places. With his sublime mind, the great Ferdinando II, our Most Serene Sovereign, developed and improved this amazing invention with curious and subtle new effects, which, cleverly hidden under ingenuous appearances, are deemed prestigious by those who do not know their causes.[8]

Castelli's letter contains the first detailed description of the instrument (and, at the same time, experimental apparatus) that Galileo, according to Castelli's report, showed him around 1603.

Castelli wrote his letter to Ferdinando Cesarini more than thirty-five years later, essentially agreeing with Viviani; the latter, indeed, dates the construction of the first 'thermoscope' to *c.*1593, shortly after Galileo's arrival in Padua. This was also the period in which Galileo had just abandoned the early views on motion and free fall contained in his unachieved Pisan treatise *De motu* (*c.*1590) and was reconsidering the same problems from a different perspective.[9]

---

7   Gorham G. – Hill B. – Slowik E. – Waters C.K. (eds.), *The Language of Nature: Reassessing the Mathematization of Natural Philosophy in the Seventeenth Century* (Minneapolis: 2016).

8   Viviani V., *Racconto istorico della vita del Signor Galileo Galilei*, in Favaro – Del Lungo – Marchesini (eds.), *Opere di Galileo Galilei* XIX 607. English translation in Gattei S. (ed.), *On the Life of Galileo* 17.

9   Salvia S., "From Archimedean Hydrostatics to Post-Aristotelian Mechanics: Galileo's Early Manuscripts *De motu antiquiora* (ca. 1590)", *Physics in Perspective* 19.2 (2017) 105–150.

I recall an experiment shown to me more than thirty-five years ago by our master Galileo; who, taking a glass flask the size of a small hen's egg, with a neck around two palms in length and as thin as a barley-stalk, heated the flask with the palms of his hands and then, turning the mouth of it over into a vessel placed under it, in which there was a little water, when he left the vessel free of the heat of his hands, suddenly the water began to climb up the neck and rose to above the level of the water in the vase by more than a palm. Galileo made use of this effect to fabricate an instrument to estimate the degrees of heat and cold.[10]

What Castelli describes is actually an open-air thermoscope, although the word *thermoscopium* was coined by the Jesuit astronomer and mathematician Giuseppe Biancani only in 1620. Viviani reports of 'thermometers' being invented by his master already in the Paduan years and later improved by Grand Duke Ferdinand II de' Medici himself, but this is quite anachronistic and should instead be referred to the 1650s, when Viviani was writing his biography of Galileo. As Castelli says, the device realized in Padua (more probably around 1603) was essentially an apparatus that could be used to detect and estimate a certain increase or decrease in the temperature of the bulb, though without the possibility of quantifying such a variation:

> One might say a lot about this instrument. However, for what is relevant to our purposes, it is sufficient to notice that the colder the air surrounding the flask, the higher the level of water; the warmer the air is, the lower the level of water inside the neck of the flask. [...] I was led to say that because, in the flask put in colder air, the air locked inside the flask condenses and reduces in a smaller volume. Since nothing else than the water in the lower vessel can enter to fill the space left, therefore it climbs up.[11]

This first model of open-air thermoscope was affected by fluctuations of atmospheric pressure and temperature, being a thermoscope and a baroscope at the same time, therefore very imperfect. It was also a kind of small air pump, inasmuch as air thermal expansion or condensation in the bulb caused an increase or decrease in its volume, pushing down or pulling up the water contained in the tube and in the lower vessel. If the tube was filled with a solid material,

---

10   Benedetto Castelli to Ferdinando Cesarini (20 September 1638), in Favaro – Del Lungo – Marchesini (eds.), *Opere di Galileo Galilei* XVII (1906) 377. English translation available on: catalogue.museogalileo.it/object/Thermoscope.html (accessed on 21 February 2023).

11   Ibidem, 377–378.

or just sealed at its mouth, so that resistance would be opposed to air expansion, heating the bulb would make the tube crack beyond a certain breakpoint. Given the constant volume of the air confined in it, its internal pressure, being higher and higher than the external atmospheric pressure, would overcome the resistance of the glass.

Less intuitively at the time when Castelli wrote his letter, even cooling the bulb, without allowing water to enter the tube and replace the volume previously occupied by air, would finally lead to the same result, this time because the pressure inside the tube would be increasingly lower than outside. In this case, instead of exploding, the tube would implode, because of the partial *vacuum* produced inside.

> So let A be the flask, BCD its neck, with the mouth D in the vessel, where let water be up to the level EF: it is manifest that, once the flask is heated up and then its mouth dipped into the water of the vessel, as the flask is free from heat the air A will decrease in volume, making water in the vessel climb through the neck up to C, then to B and higher, since no other body can enter the small tube. However, if other matter could enter, water would not go up to fill the space left by condensed air; and if no other external body could enter the flask, in this case either the air in A would stay rarefied or the flask would break.[12] [Figs. 2.1, 2.2]

Castelli's further remarks on glass vessels that are too much and too rapidly heated or cooled are significant, since they indicate that, around 1638, various observations and experiments had already been done on the subject within Galileo's entourage. If the glass is not properly tempered, with the glassmaker keeping it warm, once moulded outside the furnace, in order to make it more homogeneous, compact, and so resistant, the little bubbles of air encapsulated in the glass will behave in the same way as a thermoscope bulb with a sealed tube.

> Maybe, from this accident, one can solve the question why glasses, and other bodies too, sometimes break by themselves by immersing them either in very cold or very hot water. I would say that this occurs because, when glass is immersed in cold water, the air locked in many small bubbles scattered in the substance of glass shrinks and condenses. Since

---

12   Castelli, *Opere di Galileo Galilei* XVII 378.

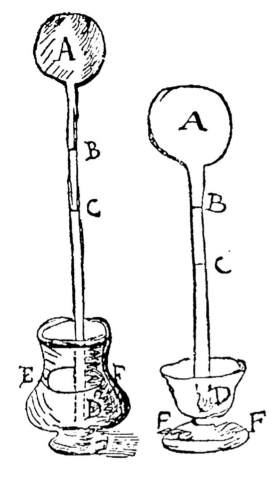

FIGURE 2.1
The thermoscope, as illustrated by Benedetto Castelli in his letter to Ferdinando Cesarini in Rome (September 20, 1638). From: Favaro, Del Lungo, Marchesini (eds.), *Opere di Galileo Galilei* XVII (1906) 37
COPYRIGHT: LIBRARY OF THE MUSEO GALILEO, FLORENCE

no other body can enter the bubbles, they are forced to break, and this causes the whole vessel to break apart. The same occurs by immersing the glass in very hot water: then the air locked in the bubbles expands becoming more rarefied and forces them to open up, so the glass breaks. Notice that this effect, considered by us when the vessel is put into hot or cold water, will also occur when the air surrounding the vessel changes suddenly, passing from hot to cold, or from cold to hot.[13]

The passage quoted above is interesting, because we also know that Evangelista Torricelli was a disciple and collaborator of Benedetto Castelli in Rome from

---

13   Castelli, *Opere di Galileo Galilei* XVII 378.

FIGURE 2.2   Nineteenth century replica of "Galileo's" thermoscope.
Museo Galileo (Florence), Room VII. Unknown maker.
Inventory: 2444. Materials: glass, cork. Height: 440 mm
COPYRIGHT: MUSEO GALILEO (FLORENCE)

1632 (the year of Torricelli's first letter to Galileo) to 1641, when Torricelli moved to Arcetri to assist Galileo in his last researches, together with Vincenzo Viviani.[14] It is highly probable that Castelli is referring to his collaboration with Torricelli, who might have developed the idea of his 1644 mercury 'barometer'

---

14   Toscano F., *L'erede di Galileo. Vita breve e mirabile di Evangelista Torricelli* (Milan: 2008).

'ATOMS OF FIRE': GALILEO'S UNACHIEVED THEORY OF HEAT 57

just by working with the open-air thermoscope realized by Galileo in Padua almost four decades before.[15] As we will see, the history of Torricelli's barometer is entwined with that of the first bulb thermometers, whose concept is that of a 'reversed thermoscope'.

Castelli shows a particular interest in the composition, features, and properties of glassware, especially in relation to the effects of heating and cooling in natural philosophy experiments dealing with incompressible fluids, compressible air, pressure, controlled fire, ebullition, freezing, and the possibility of producing local, artificial *vacua*. There is also more at stake than simple curiosity, considering Castelli's acquaintance with the practical knowledge of glassmaking: the research interests of a 17th-century chymist, beyond those of a mathematician and hydraulic engineer, seem to emerge between the lines of his letter to Cesarini.

3 Quantifying Heat: Sanctorius's 'Clinical' Thermoscope

The fascinating figure of Sanctorius (latinization of Santorio Santori), Istrian physician and natural philosopher living in Padua and Venice, sometimes considered the founder of 'iatro-mechanics' (although 'medical statics' would be a better definition of his approach), is well-known as a pioneer of the introduction of quantitative methods in physiology, beyond the traditional reference framework of Hippocratic-Galenic medicine.[16]

We know from Sanctorius's works, in particular his *Commentaria* on Galen's *Ars medica* (1612)[17] and the *Commentaria* on Avicenna's *Canon medicinae* (1625),[18] that instruments like the *statera*, the *pulsilogium*, the hygrometer, but also what would later be called thermoscope or thermometer, were designed to study the combined effects on body weight (i.e., the result of feeding, transpiration, and evacuation), heartbeat, sweating, respiration, and body temperature produced by fever as a symptom of disease. The question whether Galileo in Padua, between 1593 and 1603, developed 'his own' thermoscope, inspiring Sanctorius's apparatus, or rather the opposite, was already a matter

---

15 Shank J.B., "What exactly was Torricelli's barometer?", in Gal O. – Chen-Morris R. (eds.), *Science in the Age of Baroque* (Dordrecht: 2013) 161–195.

16 Bigotti F., *Physiology of the Soul: Mind, Body and Matter in the Galenic Tradition of the Late Renaissance (1550–1630)* (Turnhout: 2021).

17 Sanctorius Santorio, *Commentaria in artem medicinalem Galeni, libri tres*, 2 vols. (Venice, Giacomo Antonio Somasco: 1612).

18 Sanctorius Santorio, *Commentaria in primam Fen primi libri Canonis Avicennae* (Venice, Giacomo Sarcina: 1625).

of contention in the correspondence between Sagredo and Galileo on the subject (1612–1615).[19] Sagredo announced in his first letter to Galileo (June 30, 1612) the publication of Sanctorius's *Commentaria* on Galen's *Ars medica*, where the author described his 'clinical' thermoscope for the first time. Unfortunately, we do not have Galileo's replies, but from Sagredo's subsequent letters (9 May 1613, 7 February and 15 March 1615) we can infer that Galileo, at least in private, claimed priority in the invention of the thermoscope, which Sagredo maintained he believed without doubt. The same is stated by Galileo himself in his letter to Cesare Marsili (25 April 1626), by Castelli in his 1638 letter to Cesarini, and in an even more assertive way by Viviani in his 1654 *Racconto istorico*, where Galileo is credited as inventor of the 'thermometer,' thus establishing a direct connection between the original open-air device and the later sealed-bulb instrument.[20]

Interestingly enough, a letter from Sanctorius to Galileo (9 February 1615), the only one we have, seems to 'enter the discussion' between Sagredo and Galileo, although it does not mention the thermoscope. Galileo must have insisted on receiving a copy of Sanctorius's recently published *Ars de statica medicina* (1614),[21] whose delayed delivery to Florence was apparently due to the Venetian bookseller.[22] Sanctorius proudly explained to Galileo his innovative ideas on transpiration and quantitative medicine on a corpuscular basis, as well as the invention of his 'physical-medical' instruments in the last part of the letter. The impression is that Galileo felt an urge to be updated on Sanctorius's work, while Sagredo was conducting experiments with the thermoscope on his own, reporting to Galileo his technical improvements to upgrade its performance.

Rather than establishing the priority of Sanctorius or Galileo in the invention, it is safe to assume mutual influences between the two (with Sagredo as mediator). As a matter of fact, they had shared almost the same 'research agenda' (together with Paolo Sarpi) since the very beginning of their relationship in Padua and Venice during the 1590s, not to mention their common

---

19 Giovanni Francesco Sagredo to Galileo Galilei in Florence (Venice, June 30, 1612; May 9, 1613), in Favaro – Del Lungo – Marchesini (eds.), *Opere di Galileo Galilei*, XI (1901) 349–351, 505–506; Giovanni Francesco Sagredo to Galileo Galilei in Florence (Venice, February 7, 1615; March 15, 1615), in Favaro – Del Lungo – Marchesini (eds.), *Opere di Galileo Galilei* XII (1902) 138–140, 156–158.

20 Galileo Galilei to Cesare Marsili in Bologna (Florence, April 25, 1626), in Favaro – Del Lungo – Marchesini (eds.), *Opere di Galileo Galilei* XIII (1903) 319–320.

21 Sanctorius Santorio, *Ars de statica medicina et de responsione ad staticomasticem, aphorismorum sectionibus septem comprehensa* (Venice, Niccolò Polo: 1614).

22 Santorio Sanctorius to Galileo Galilei in Florence (Venice, February 9, 1615), in Favaro – Del Lungo – Marchesini (eds.), *Opere di Galileo Galilei* XII 141–142.

interest in Archimedes's hydrostatics, Hero's pneumatics and hydraulics, and more generally in the Alexandrian physical-mathematical school.[23]

From Sanctorius's works it emerges that the initial configuration of his thermoscope was essentially the same as described in Castelli's letter: two flasks of tempered glass, one inside the other, the lower partially filled with water, the upper dipping upside down in water by means of a long and narrow tube. To measure, or better to estimate by comparison, the 'degrees of heat' (as defined by Castelli, 'temperaments' in Sagredo's words) of feverish patients, Sanctorius let their hand hold the upper egg-like bulb for a given amount of time, in order to make the air contained in it expand along the tube, pushing water in the lower vessel. As an alternative, he put the bulb directly inside the patient's mouth for a more accurate detection in order to minimize the cooling effect of the environment. He had understood the importance of using always the same anatomical part for measurements, since the temperature was not uniform all over the body.

It seems, however, that since the 1610s both Sanctorius and Sagredo (if we believe what he reports to Galileo) started improving the instrument, whose main fault as an open-air thermoscope was being easily affected by atmospheric pressure and environmental temperature at the same time. Sanctorius was indeed aware that air had a 'weight' and *vacuum* was physically possible, coherently with his atomistic-corpuscular views and contrary to the Peripatetic opinion: like water or any other fluid, air exerted a pressure on material bodies immersed in it. A full explanation, though, of the omnidirectional force exerted by fluidostatic pressure and of the relation between pressure and the height of the fluid column would be provided after Torricelli's and Blaise Pascal's researches.[24]

Most of the information we have on these improvements is found in the 1625 *Commentaria* on Avicenna's *Canon medicinae*, which are also of high interest for their theoretical implications and, most probably, for their direct influence on Galileo's views on matter, light, heat propagation, and magnetism.[25] First

---

23   Boas M., "Hero's *Pneumatica*. A Study of its Transmission and Influence", *Isis* 40 (1949) 38–48. Hero endorsed a form of atomism, according to which discontinuous voids between discrete material particles were necessary to make motion in space possible.

24   Bigotti F., "The Weight of the Air: Santorio's Thermometers and the Early History of Medical Quantification Reconsidered", *Early Modern Studies* 7.1 (2018) 73–103.

25   Grmek M.D., "L'énigme des relations entre Galilée et Santorio", in Gruppo Italiano di Storia delle Scienze (ed.), *Atti del simposio internazionale di storia metodologia, logica e filosofia della scienza "Galileo nella storia e nella filosofia della scienza"* (Florence: 1967) 155–162; Bigotti F. – Barry J., "Introduction", in Bigotti – Barry (eds.), *Santorio Santori and the Emergence of Quantified Medicine, 1614–1790. Corpuscularianism, Technology and Experimentation* (Cham: 2022) 26–34.

of all, Sanctorius used colored water to make its variations in height more visible inside the tube. More importantly, he had the idea of closing the lower vessel, with no other opening left than that for the tube passing through. He could thus isolate the air and the water contained in the thermoscope from the effects of external pressure, also preventing the evaporation of the water and thus the frequent need to refill the lower flask [Figs. 2.3, 2.4].

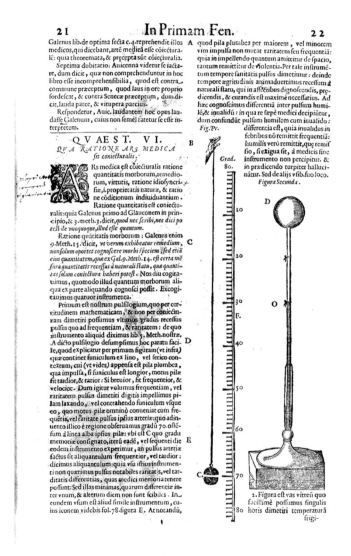

FIGURE 2.3 Sanctorius's *pulsilogium* and closed-air thermoscope, from his *Commentaria in primam Fen primi libri Canonis Avicennae* (Venice, Giacomo Sarcina: 1625) 22
COPYRIGHT: WELLCOME LIBRARY, LONDON

'ATOMS OF FIRE': GALILEO'S UNACHIEVED THEORY OF HEAT

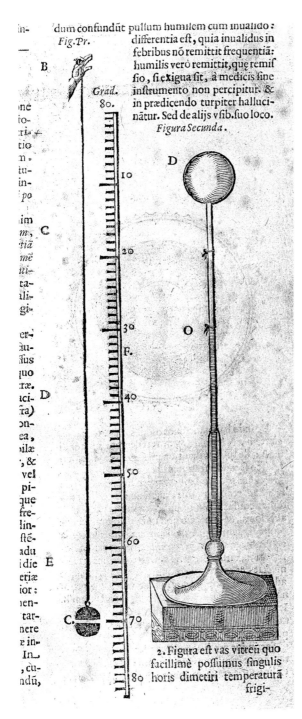

FIGURE 2.4
Sanctorius's *pulsilogium* and closed-air thermoscope (detail), from his *Commentaria in primam Fen primi libri Canonis Avicennae* (Venice, Giacomo Sarcina: 1625) 22
COPYRIGHT: WELLCOME LIBRARY, LONDON

This version of Sanctorius's thermoscope has notches on the tube, but no graduated scale.

The immediate consequence was that, in the new version of the apparatus, the water in the tube had a finite range of height variations, from a minimum to a maximum, corresponding to the degree of sensitivity of the instrument. Sanctorius provided his clinical and 'closed-air' thermoscope with graduated scales, as far as we can deduce from his published works and from their illustrations. Moreover, given Sagredo's frequent references to 100, 150, or even 360 'degrees' measured by his improved thermoscopes, while reporting his experiments in the 1612–1615 correspondence with Galileo, we may infer that graduated thermoscopes were already common in the 1610s, although without standard scales yet.

4   Fire Particles in Motion: a Mechanical-Corpuscular View

Sanctorius's quantitative approach to physiology was a direct consequence of his very peculiar kind of atomistic-corpuscular theory of matter, with its inherent features (*situs*, *figura*, *numerus*) and transformation processes, its causal interactions with and inside living bodies, and its effects on the soul, which is always treated in terms of *anima vegetativa*, *sensitiva*, and *intellectiva* according to the Aristotelian definition as *in actu* form and ἐντελέχεια (*entelécheia*) of the organic body.

Sanctorius developed his views on matter and its ultimate constituents over more than thirty years, especially in the *Methodi vitandum errorum omnium* (1603), in the *Commentaria* on Galen's *Ars medica*, in the *Ars de statica medicina*, and more extensively in his *opus magnum*, the *Commentaria* on Avicenna's *Canon medicinae*. The recent discovery of Sanctorius's own copy of the book in the Sloane Collection of the British Library (with autograph marginal notes on the properties of mixtures in relation to elementary corpuscles under the action of fire) has revived scholarly interest in atomism, corpuscularianism, 'chymistry,' observation and experimentation in post-Aristotelian natural philosophy, as in theoretical-practical medicine, highly specialized craftsmanship, and practitioner knowledge in Padua from the 1590s to the 1630s.[26] [Fig. 2.5]

---

26   Bigotti F., "A Previously Unknown Path to Corpuscularianism in the Seventeenth Century: Santorio's Marginalia to the *Commentaria in Primam Fen Primi Libri Canonis Avicennae* (1625)", *Ambix* 64.1 (2017) 29–42.

cocto ex violis, & lactuca, quod in ventum conuertitur, quo referto cubiculo, aer frigidus & humidus fummo cū beneficio infpiratur: vbi eſt C eſt foramen perexiguum per quod vas prius igne calefactum, mox frigido decocto immerſū repletur: poſtea verò per idem foramen, fi vas paulo magis igneſcat, decoctum inclufum in magnū ventum faceſſit, qui cubiculi aerem refrigerat & humectat, mox infpiratus maximè iuuat.

Præterea notandum per febrem hecticam nō exficcari humidum natiuum, & radicale, fed folum alimentale, quod prouenit à fanguine: quia fecundæ humiditates nihil aliud funt, quam fanguis benè præparatus ad nutritionem: humidum enim natiuum, & radicale non poteſt per alterantia humectari, quia eſt fubſtantia, & in fubſtantia non funt alterationes. Hinc Galenus 1. de diff. feb. 8. & 9. vult hecticam febrem primi gradus fieri, dum calor febrilis exficcat humidū alimentale folidarum partium: hecticam verò

FIGURE 2.5 Sanctorius's *marginalia* to col. 406C–D of his own copy of the *Commentaria in primam Fen primi libri Canonis Avicennae* (Venice, Giacomo Sarcina: 1625) 406. British Library (Sloane Collection), 542.h.11. Image from Bigotti F., "A Previously Unknown Path to Corpuscularianism in the Seventeenth Century: Santorio's Marginalia to the *Commentaria in Primam Fen Primi Libri Canonis Avicennae* (1625)", *Ambix* 64.1 (2017) 33
COPYRIGHT: BRITISH LIBRARY, LONDON

Typical of the 'hybrid' model adopted by Sanctorius (essentially appropriated by Galileo as a conjectural hypothesis) is the use of the term *minimum* instead of *atomus*, with two fundamental implications. First, *minimum* is a relational, therefore relative notion (as in Giordano Bruno's late works), which can be applied to different levels and scales of organization, from microcosm to macrocosm. In the case of material substances, the minimal quantity of a

64 SALVIA

compound is not the same thing as the minimal quantity (*minimi quanti*, in Galileo's words) of a simple substance, which does not coincide in turn with an absolute *minimum* of matter.[27]

The *minima* Sanctorius refers to are conceptually more similar to Anaxagoras's ὁμοιομέρειαι (*homoioméreiai*), i.e., the minimal units of a given substance (element or compound), whose further division or 'resolution' is impossible; not because they are ἄτομοι (*átomoi*) in Democritus's sense, but since their further division would cause the substance to lose its essential and distinctive features with respect to other substances (Descartes would introduce the terms *molecula* and *molécule* at the exact time, namely the 1620s, when Sanctorius was working on his *Commentaria*). This leads to the second implication: the four traditional elements of natural philosophy (earth, water, air, and fire) are not absolutely elementary, they are rather the four simplest substances in nature and the basic principles that enter the composition of all other derived substances (sulphur, mercury, and salt as constituents of all metals, in alchemy; the four Hippocratic-Galenic humors with their respective temperaments, in medicine).[28]

As in Plato's *Timaeus*, what distinguishes a *minimum* of earth from a *minimum* of fire is not due to intrinsic qualitative differences, but to differences in their respective shape and configuration. Hence, the specific properties of each element 'emerge' from and can be explained (at least in principle) by the specific geometric configuration of its *minima*. Unlike Plato, however, Sanctorius did not go into details about the four fundamental shapes that such *minima* should have in order to account for the different but interrelated and complementary qualities that they are supposed to produce in the material world (actually, the 'qualities' they produce in human sense organs, so that the soul perceives them as such). The only explicit consideration he makes, in this respect, is that igneous particles must be the smallest and sharpest ones (presumably pyramids, tetrahedra, or some other 'spiky' shape), the most subtle, mobile, and penetrating, compared with those of the other elements.

Another important consequence: if the four elemental substances have a corpuscular nature, if they can also transform into one another by means of heating/rarefaction or cooling/condensation (thanks to the voids among their

---

27   Lüthy C., "Atomism in the Renaissance", in Sgarbi M. – Valleriani M. (ed.), *Encyclopedia of Renaissance Philosophy* (Cham: 2018) 1–14.

28   See, e.g., Gregory T., "Studi sull'atomismo del Seicento", *Giornale critico della filosofia italiana* 20 (1966) 38–65; Baldini U et al. (eds.), *Ricerche sull'atomismo del Seicento* (Florence: 1977); Salem J. (ed.), *L'atomisme aux XVIIᵉ et XVIIIᵉ siècles* (Paris: 1999).

particles), and the same elements can enter the composition of very different compounds, according to various mixing proportions and 'constitutions' (*complexiones*, i.e., particle dispositions and arrangements), there must be a more fundamental level and state of matter; some kind of absolute and irreducible *minimum* (a true ἄτομος [*átomos*]), whose primary aggregations are directly responsible for the different shapes of the four elemental *minima* and, indirectly, for the innumerable configurations their higher-level aggregates can assume.

Reappraising Bruno's original foundation of all reality upon ultimate physical *minima* interspersed with 'active' *vacua* (i.e., empty spaces that are positionally significant in relation to their dimensions and form), Sanctorius concluded that such 'atomic' *minima* should be the smallest, simplest, discrete, and homogeneous unities of universal *materia prima* that can exist (under which corporeality itself would dissolve into non-existence). These unities are all undifferentiated and identical to one another, with no other properties except their extension (minimal but finite), shape (spherical), mutual position, and motion.[29]

However, the atomic-corpuscular model that I have briefly outlined above leaves the way open to a new sort of 'quintessence' or *aether* that is not supposed to be separated from all other substances, with its own special properties, but should be the underlying common substrate of *materia prima*, made of free absolute *minima*. All elements, compounds, and aggregates come from and return to it by constantly transforming into one another via aggregation, disaggregation, and recombination.

Sanctorius appeals in his works, particularly in the *Commentaria* of 1625, to a 'vortex theory' derived from Democritus, but curiously destined to have a great influence on Cartesian anti-atomistic *plenum* physics. According to this, the spherical *minima* of prime matter have an intrinsic, chaotic motion, so that free atoms cannot be at rest. Change is therefore structural to matter, although, as bound units in corpuscles and higher-level aggregations, the absolute *minima* are somehow confined; at least until microscopic incessant motion prevails on macroscopic apparent stability.[30]

---

29   Grant E., *Much Ado About Nothing: Theories of Space and Vacuum from the Middle Ages to the Scientific Revolution* (Cambridge: 1981); Lüthy C., "Bruno's *Area Democriti* and the Origins of Atomist Imagery", *Bruniana & Campanelliana* 4.1 (1998) 59–92.

30   Bigotti F., "'Gears of an Inner Clock': Santorio's Theory of Matter and Its Applications", in Bigotti – Barry (eds.), *Santorio Santori and the Emergence of Quantified Medicine* 75–77.

The same multi-level combinatory mechanisms (whose Llullian inspiration can be traced back to Bruno) are at work in Sanctorius's distinction between mixtures and temperaments. Mixtures are compound substances in which the particles of the four elements are decomposed into their constituent *minima*, losing their elemental features to give birth to new particles (i.e., new substances) with completely different properties. Temperaments are instead blends of the four elements, or of more complex substances, in various proportions, where each component preserves its original properties, although their mutual interactions may give rise to new effects.[31]

What kind of matter is made of absolute *minima* which are just free to move in space without forming any corpuscle or aggregate of corpuscles? Is such a 'zero-level' configuration possible? One distinctive characteristic of Bruno's solution in his Latin poems, significantly shared by Sanctorius as well as by Galileo in their hypotheses on physical *minima* and particles, is the identification of this basic state of matter with *aether* as the subtlest, ubiquitous, and omnipervasive substance in the universe. As such, it should be the true and ultimate medium which all material bodies and their aggregates, from corpuscles to stars, are immersed in, penetrated by, and moved through.[32]

Furthermore, 'light' (*lux* as formal principle and action, distinct from *lumen* as its material and efficient cause) is treated as a phenomenon pertaining to *aether*, although there is an ambiguity between two possible interpretations. In the first, light (*lux*) is the result of absolute *minima* being emitted or projected by a light source, streaming like corpuscles (*lumen*) in all directions along rectilinear trajectories (with infinite or finite velocity, this is also an issue): in this case, the *minima* themselves, or very simple particles composed of a few of them, act as 'luminiferous' corpuscles. In the second, light (*lux*) coincides with a particular state of the *aether* being moved in all directions (*lumen*) by a light source, like sound propagating in the air or in any other fluid medium moved by a vibrating object: in this case, the *aether* itself acts as the 'luminiferous' principle. The two versions are alternative but not incompatible, inasmuch as

---

31 Moreau E., "Atoms, Mixture, and Temperament in Early Modern Medicine: The Alchemical and Mechanical Views of Sennert and Beeckman", in Bigotti – Barry (eds.), *Santorio Santori and the Emergence of Quantified Medicine* 137–164.

32 Clericuzio A., *Elements, Principles and Corpuscles: A Study of Atomism and Chemistry in the Seventeenth Century* (Dordrecht: 2000); Lüthy C. – Nicoli E. (eds.), *Atoms, Corpuscles and Minima in the Renaissance* (Leiden: 2022).

'ATOMS OF FIRE': GALILEO'S UNACHIEVED THEORY OF HEAT

the free *minima* under the form of *aether* are not 'emitted,' but rather conveyed and streamed by light sources.[33]

Light, therefore, seems to be the final 'resolution' of matter in this model, immediately below fire as 'the most elementary element' constituted by simple, small, and mobile particles (*ignicula*), before coming to the ultimate *minima* of *materia prima*. This level of proximity between fire and light would also explain why the two are almost always associated, although Sanctorius is well aware that light in nature can be 'cold' (as in the case of fireflies), whereas heat, conversely, can be 'dark' (such as hot smoke). On the one hand, light is formally distinguished from fire: the latter 'contains' the former, so to speak, as more fundamental, but the contrary is not always true. On the other hand, heat still seems to be considered an intrinsic or at least a distinctive property of fire, as the specific element responsible for heating, drying, and rarefaction.

5 *Motus est causa caloris*: Galileo on Heat, Fire, and Light

In Galileo's writings (including his correspondence) there is nothing like an achieved, coherent, and detailed account of the intimate structure of matter, of the different features of all substances entering the composition of material bodies, why they are subject to transformation (if not transmutation, as for metals in the alchemic tradition), and how these features, emerging from atomic-corpuscular structures, affect in various ways our *sensorium commune* to produce the sensations and cogitations that allow us to have knowledge of them. Unlike Sanctorius, Galileo was not interested in (re-)elaborating such a comprehensive physical-philosophical and physiological system from a theoretical perspective.

Astronomy (with optics supporting telescopic observations) and mechanics (with acoustics derived from mechanical principles) were at the top of Galileo's 'research agenda'. The internal constitution of matter was a secondary issue to him, except for three particular cases, in which it was inevitably called into question. First, the resistance, fracture, and elasticity of materials

---

33   Gómez López S., "Galileo y la naturaleza de la luz", in Montesinos J. – Solís C. (eds.), *'Largo campo di filosofare': Eurosymposium Galileo 2001* (La Orotava: 2001) 403–418; and "The Mechanization of Light in Galilean Science", *Galilaeana* 5 (2008): 207–244. As for Newton's corpuscular theory of light, any commitment with luminous particles being produced/emitted by a source implied a mass loss through time.

(also involving the physical-geometrical notion of 'indivisible').[34] Second, heat, its measure, and the effects of heating/cooling on mechanics, hydrostatics, hydraulics, and pneumatics. Third, magnetism, with its apparently 'occult' action on ferromagnetic objects that needed to be explained in mechanical terms, without any reference to 'affinities' or 'sympathies'.[35]

In all three cases, however, unprovable conjectures about atoms, corpuscles, aggregates, and mixtures or temperaments, no matter how powerful and suggestive they might be, were largely unavoidable at the time, given the lack of a robust empirical and experimental basis. This is the main reason why Galileo usually refrained from speculating on remote causes and ultimate principles in natural philosophy, confining himself to observable phenomena and their proximate causes, described and explained in physical-mathematical terms.[36]

Nonetheless, such methodological caution, expressly claimed in the *Assayer* against Orazio Grassi's arguments, did not prevent the author from giving the formal recipient of his published 'letter-treatise,' the noble Lyncean academic Virginio Cesarini in Rome, a short but dense essay on an atomic-corpuscular theory of heat, fire, and light.[37] The general framework is clearly inspired by Sanctorius's model, although pragmatically outlined in the text just as a hypothesis ('some thoughts of mine'). This is the sole place in his works where a quite extensive account of heat and light is provided.

Galileo's core argument reconsiders the traditional maxim of Scholastic natural philosophy – *motus est causa caloris* – in the light of the previously discussed 'hybrid' theory, prudently presented in a conjectural form. Actually, this is not only a cautious attitude towards controversial and potentially heterodox thoughts on atomism and corpuscularianism, although perfectly understandable in a publication intended to be a scientific-philosophical 'manifesto'.[38]

---

34   Baldini U., "La struttura della materia nel pensiero di Galileo", *De Homine* 57 (1976) 91–164; Le Grand H.E., "Galileo's Matter Theory", in Butts R.E. – Pitt J.C. (eds.), *New Perspectives on Galileo* (Dordrecht-Boston: 1978) 197–208; Bucciantini M. – Torrini M. (eds.), *Geometria e atomismo nella scuola galileiana*, (Florence: 1992).

35   Biener Z., "Galileo's First New Science: The Science of Matter", *Perspectives on Science* 12.3 (2004): 262–287; Galluzzi P., *Tra atomi e indivisibili. La materia ambigua di Galileo* (Florence: 2011); Baldin G., *Hobbes and Galileo: Method, Matter and the Science of Motion* (Cham: 2021).

36   Meinel C., "Early Seventeenth-Century Atomism. Theory, Epistemology, and the Insufficiency of Experiment", *Isis* 79 (1988) 68–103.

37   Galilei G., *Il Saggiatore*, in Favaro – Del Lungo – Marchesini (eds.), *Opere di Galileo Galilei* VI 347–352. English translation in Drake S. (ed.), *The Assayer* 22–25.

38   See Shea W.R., "Galileo's atomic hypothesis", *Ambix* 17 (1970) 13–27, and "Galileo e l'atomismo", *Acta Philosophica* 10.2 (2001) 257–272; Redondi P., "Atomi, indivisibili e dogma", *Quaderni storici*, 20.2 (1985) 529–575; Nonnoi G., *Saggi Galileiani: Atomi, immagini e ideologia* (Cagliari: 2000).

Galileo's reluctance to go into detail, besides his rhetorical strategies, is also due to his personal doubts and oscillations on such a speculative matter.

First of all, however, he needs to refute the common-sense and Peripatetic idea that sensible phenomena and qualities like heat are really to be found 'out there' in the physical world, even that 'being hot and dry' are essential properties of the element fire, instead of being the effect produced by the *numerus, figura et situs* (to quote Sanctorius) of material particles and interstitial voids on our sense organs. For example, in the case of touch, the whole skin and the innervated flesh, responsible for what we call 'heat' and for its apparently opposite quality, 'cold' (i.e., a lack, or better a comparatively lower degree of heat as perceived by our body). This is one important instantiation of the famous distinction, formalized by John Locke but already traceable in ancient atomism, between 'primary' and 'secondary' qualities (as later codified in the historiography and philosophy of science), that would later make the *Assayer* a milestone of the 'Scientific Revolution'.[39]

> It now remains for me to tell Your Excellency, as I promised, some thoughts of mine about the proposition 'motion is the cause of heat,' and to show in what sense this may be true. But first I must consider what it is that we call heat, as I suspect that people in general have a concept of this which is very remote from the truth. For they believe that heat is a real phenomenon or property, or quality, which actually resides in the material by which we feel ourselves warmed.
>
> Now I say that whenever I conceive any material or corporeal substance, I immediately feel the need to think of it as bounded, and as having this or that shape; as being large or small in relation to other things, and in some specific place at any given time; as being in motion or at rest; as touching or not touching some other body; and as being one in number, or few, or many. From these conditions I cannot separate such a substance by any stretch of my imagination. But that it must be white or red, bitter or sweet, noisy or silent, and of sweet or foul odor, my mind does not feel compelled to bring in as necessary accompaniments. Without the senses as our guides, reason or imagination unaided would probably never arrive at qualities like these. Hence I think that tastes, odors, colors, and so on are no more than mere names so far as the object in which we place them is concerned, and that they reside only in the consciousness.

---

39 See Nolan L. (ed.), *Primary and Secondary Qualities: The Historical and Ongoing Debate* (Oxford: 2011).

Hence if the living creature were removed, all these qualities would be wiped away and annihilated.[40]

What follows is a corpuscular treatment of heat and all related phenomena, both on the side of material bodies, under the action of fire (in all its possible forms) and on that of living organic bodies, like ours, whose interaction with fire produces certain sensations and associated feelings we usually refer to as being hot/cold, heating/cooling, scalding/chilling, burning/freezing, etc. Similarly to Sanctorius, Galileo writes of igneous minute particles (*ignicoli*, Italian calque of the Latin *ignicula*): a whole class of corpuscles under the general name of 'fire', with different dimensions and shapes, but all sharing the common features of being sharp and able to penetrate the grosser particles of matter in large numbers at a high speed through their voids, and thus enter the pores of our body.

Dimension, shape, concentration, and velocity of the different species of *ignicoli* (so that we should rather speak of 'fires' plural) are responsible for the various manifestations of heat, in the physical as well as in the physiological sense. A high concentration of penetrating and fast-travelling fire particles will cause disaggregation, disruption, 'resolution' (i.e., decomposition) of matter into smaller and simpler corpuscles, making us experience burns and pain, while a moderate quantity of less sharp and/or slower particles will cause gentle heating and a pleasant sensation of warmth. Up to this point, Galileo's account is not original, looking more or less like a contemporary reappraisal of ancient Greek atomism. However, in this reappraisal, atoms are substituted with corpuscles that can in principle be further divided into more fundamental particles, according to the Late Medieval re-elaboration of Aristotle's *minima naturalia* (i.e., the smallest parts into which a homogeneous natural substance could be divided while still retaining its essential properties).[41]

Having shown that many sensations which are supposed to be qualities residing in external objects have no real existence save in us, and outside ourselves are mere names, I now say that I am inclined to believe heat to be of this character. Those materials which produce heat in us and make us feel warmth, which are known by the general name of 'fire',

---

40 Galilei G., *Il Saggiatore*, in Favaro – Del Lungo – Marchesini (eds.), *Opere di Galileo Galilei* VI 347–348. English translation in Drake S. (ed.), *The Assayer* 22–23.

41 Van Melsen A.G., *From Atomos to Atom: The History of the Concept 'Atom'* (New York: 1960); Murdoch J.E., "The Medieval and Renaissance Tradition of *Minima Naturalia*", in Lüthy – Murdoch – Newman (eds.), *Late Medieval and Early Modern Corpuscular Matter Theories* 91–133.

would then be a multitude of minute particles having certain shapes and moving with certain velocities. Meeting with our bodies, they penetrate by means of their extreme subtlety, and their touch as felt by us when they pass through our substance is the sensation we call 'heat'. This is pleasant or unpleasant according to the greater or smaller speed of these particles as they go pricking and penetrating; pleasant when this assists our necessary transpiration [*insensibile traspirazione*], and obnoxious when it causes too great a separation and dissolution of our substance. The operation of fire by means of its particles is merely that in moving it penetrates all bodies, causing their speedy or slow dissolution in proportion to the number and velocity of the fire-corpuscles and the density or tenuity of the bodies. Many materials are such that in their decomposition the greater part of them passes over into additional tiny corpuscles, and this dissolution continues so long as these continue to meet with further matter capable of being so resolved.[42]

The key issue, apart from a direct, though anonymous, reference to Sanctorius's *perspiratio insensibilis*, is Galileo's distinction between the matter of fire(s), on the one hand, as a group of similar elementary particles sharing some distinctive properties, and their action on material bodies, on the other hand, as main cause and principle of heat(ing), with its psycho-physical correlates. None of the different types of fire (burning flames, roasting embers, glowing metals, bright sparks, as well as lightless variants like hot smoke, boiling water, expanding air, or bodies that are simply hot, i.e., with a high temperature) would produce any sort of heat by themselves, if they were at rest, just contained and confined in the innumerable voids disseminated through the corpuscular aggregates that constitute ordinary matter. At most, they would represent a potential, 'hidden' or 'latent' heat that could be released in given circumstances, as in the case of flints when they are rubbed.

To generate heat, the *ignicoli* need to move, moving other particles around them with their own motion according to their shape, number, and speed. Heat is nothing else but the result of this motion or agitation at a microscopic level. At the same time, the sensation of heat results from the action that this motion or agitation has on our sense organs, when they enter in contact with *effluvia* of fire particles and other moving corpuscles. This is Galileo's re-interpretation

---

42   Galilei G., *Il Saggiatore*, in Favaro – Del Lungo – Marchesini (eds.), *Opere di Galileo Galilei* VI 350–351. English translation in Drake S. (ed.), *The Assayer* 24–25. Drake omits the adjective 'insensibile' and thus conceals Galileo's implicit reference to Sanctorius's *perspiratio insensibilis*. Significantly enough, Galileo never mentions Sanctorius in the *Assayer*.

of the traditional *motus est causa caloris*, originally referring to friction, the dynamics of which is extended to the deep structure of matter.

> Since the presence of fire-corpuscles alone does not suffice to excite heat, but their motion is needed also, it seems to me that one may very reasonably say that motion is the cause of heat ... But I hold it to be silly to accept that proposition in the ordinary way, as if a stone or piece of iron or a stick must heat up when moved. The rubbing together and friction of two hard bodies, either by resolving their parts into very subtle flying particles or by opening an exit for the tiny fire-corpuscles within, ultimately sets these in motion; and when they meet our bodies and penetrate them, our conscious mind feels those pleasant or unpleasant sensations which we have named heat, burning, and scalding. And perhaps when such attrition stops at or is confined to the smallest quanta, their motion is temporal and their action calorific only; but when their ultimate and highest resolution into truly indivisible atoms is arrived at, light is created. This may have an instantaneous motion, or rather an instantaneous expansion and diffusion rendering it capable of occupying immense spaces by its – I know not whether to say its subtlety, its rarity, its immateriality, or some other property which differs from all these and is nameless.[43]

The presence of fire particles is therefore a necessary but insufficient condition to produce heat, if they do not stream with enough impetus through matter, causing motion, displacement, agitation, and even disruption of its diverse aggregations of corpuscles, as in combustion or in the most spectacular cases of explosion and conflagration. As a consequence, heat (and 'being hot') is not an intrinsic property of fire, but rather a combined effect of both its presence and motion affecting material bodies.

On combustion of flammable substances, in particular, Galileo sketches an interesting solution, that hints at a further decomposition of igneous particles, as of any other kind of corpuscles. When compounds or aggregates of compounds susceptible to burning are 'activated' by some fiery or just hot piece of matter, the resting *ignicoli* trapped inside them interact with the moving ones coming from the 'triggering' substance. By opening their way through the pores and cavities of the flammable compound, the fire particles start a

---

43   Galilei G., *Il Saggiatore*, in Favaro – Del Lungo – Marchesini (eds.), *Opere di Galileo Galilei* VI 351–352. English translation in Drake S. (ed.), *The Assayer* 25.

'ATOMS OF FIRE': GALILEO'S UNACHIEVED THEORY OF HEAT

'chain reaction,' in which those already free to move unravel and cause others to move in turn.

Bare displacement and circulation of *ignicula* through matter only causes its heating; decomposition of complex aggregations into simpler corpuscles, due to their motion, causes its melting; finally, further 'resolution' of larger corpuscles into smaller fire particles generate new fire particles that feed the reaction, whose visible result is flames coming out of the burning compound, which goes through a process of transformation, being reduced to thinner and lighter aggregates: smoke, vapor, soot, and ashes.[44]

A reduction and recombination mechanism like this would be impossible, if the *minima* of fire, the simplest among all elemental *minima*, were not 'resolvable' into more fundamental components, actually the ultimate, universal, and irreducible ones ('*atomi realmente indivisibili*', truly indivisible atoms). Like Sanctorius, Galileo imagines that, when no further division is possible, 'light is created', so that light as *lumen* would be at the same time the most elementary, almost immaterial state of matter and the foundation of any form of materiality. The illuminating property and action of light as *lux* would be somehow due to *atomi* of prime matter being free to travel or to expand through space ('*una sottilissima sostanza eterea*', a very subtle ethereal substance) instantaneously or rather at a finite speed, although extremely high, since an infinite velocity of expansion would imply no propagation of light, i.e., its absolute immobility.[45] In his *Discorsi*, Galileo would discuss an experiment (however unsuccessful) to measure this elusive value.[46]

---

44   As for most 'chymical' theories of combustion before the 1770s, the intuitive idea that fire (then '*phlogiston*') must be issued of burning substances, which should hence lose mass (being reduced to lighter compounds), was never questioned; also because the uncontrolled loss of gaseous products in the atmosphere prevented experimenters from 'weighing' combustion by means of an appropriate experimental apparatus, with the same quantifying attitude displayed by Sanctorius and later acknowledged by Lavoisier. See Kohler R.E., "The Origin of Lavoisier's First Experiments on Combustion", *Isis* 63.3 (1972) 349–355; Guerlac H., *Lavoisier – The Crucial Year: The Background and Origin of His First Experiments on Combustion in 1772* (Ithaca, NY: 1961).

45   Galilei G., *Il Saggiatore*, in Favaro – Del Lungo – Marchesini (eds.), *Opere di Galileo Galilei* VI 317. See Benedetto Castelli to Galileo Galilei in Florence (Rome, May 8, 1612), in Favaro – Del Lungo – Marchesini (eds.), *Opere di Galileo Galilei* XI 294–295; Galileo Galilei to Piero Dini in Rome (Florence, March 23, 1615), in Favaro – Del Lungo – Marchesini (eds.), *Opere di Galileo Galilei* V (1895) 297–305. See also Shea W.R., "Galileo e l'atomismo", *Acta Philosophica* 263–266.

46   Galilei G., *Discorsi e dimostrazioni matematiche intorno a due nuove scienze*, in Favaro – Del Lungo – Marchesini (eds.), *Opere di Galileo Galilei* VIII 86–89.

6   A Paradigm Shift: Fire, Heat, Temperature

Only in a few places in Galileo's works, as well as in his correspondence, can one find an atomic-corpuscular account of heat through the action of fire, in relation to the problem of quantifying the physical phenomenon (once separated from its psycho-physiological correlates) by means of a graduated thermoscope. The above-quoted passages from the *Assayer* are in fact the main places where such a theory is most extensively outlined. The overall implication is that Galileo had to deal urgently with ancient and contemporary controversies on atomism and geometry, as with physical-philosophical objections from contemporaries, which made him change his mind over decades.

Since the publication of his 1612 *Discourse on floating bodies*, Galileo was concerned with the atomic-corpuscular model itself, against Peripatetic *continuum* physics. To him, only discrete particles and aggregates of particles as constituents of matter, with innumerable interstitial *vacua* among them (like grains of sand), could provide a satisfying explanation for motion of solid bodies through a fluid medium, as for Archimedes's principle. Namely, that the sole specific weight (i.e., density) of a body with respect to that of the medium was relevant and responsible for its fluidostatic behavior (later distinguished from the fluidodynamic effects): floating on the surface of the fluid, if less dense; being immersed in the fluid and in equilibrium, when of equal density; sinking into the fluid, if denser.[47]

Compared with the general problems of understanding how a 'hybrid', multi-level, and hierarchical model (made both of divisible corpuscles and indivisible atoms as their ultimate 'resolution') should actually work to account for all observable phenomena, as with the question of whether and how inter-particle voids played an active role in giving matter its various properties (not to speak of how physical discreteness can match with mathematical *continuum*), the treatment of heat and light occupies a special position in the evolution of Galileo's peculiar form of 'atomistic corpuscularianism'.

Heat entered and influenced the 1612–1615 debate on floating bodies, insofar as Galileo had to reply to the criticisms of his Aristotelian adversaries about two main issues, among many others. First, the counter-intuitive expansion and rarefaction of solid ice with respect to liquid water, on which it floats. Second, the fact that hot water expands in its container and is therefore less dense than

---

47   Galilei G., *Discorso al Serenissimo Don Cosimo II Gran Duca di Toscana intorno alle cose, che stanno in su l'acqua, o che in quella si muovono* (Florence, Cosimo Giunti: 1612), in Favaro – Del Lungo – Marchesini (eds.), *Opere di Galileo Galilei* IV (1894) 17–142.

cold water, combined with Democritus's theory according to which hot water (i.e., fire heating water) is able to make bodies which are slightly denser than water itself rise up to its boiling surface and float (at least for a time), as if fire pushed them upwards with a force additional to Archimedes's thrust.[48]

About the first issue, the fact that ice was less dense than liquid water, so that the same amount of water occupied a larger volume if solidified, was an authentic paradox for the Peripatetics, because to them solidification was always equivalent to condensation. Something unexpected and against common sense should occur when water froze into ice, something that could be explained only supposing a discrete constitution of both liquid water and solid ice, with water corpuscles both separated and united by empty spaces.

Already around 1612, when Sanctorius was publishing his *Commentaria* on Galen's *Ars medica* and Johannes Kepler had published his 1611 little treatise *De nive sexangula*,[49] Galileo had come to the conclusion that, differently from other substances, there should be many more and larger voids among water particles when they were bound together to form solid ice (causing their paradoxical rarefaction), than when they were free to move one over the other as liquid water. In other words, 'solid' did not necessarily mean 'condensed', in the sense of being denser or more compact, i.e., having less and/or smaller inner voids.

This conclusion led Galileo to think of a possible role for voids (their number, dimensions, and shape) and of the *horror vacui* in inter-particle cohesion, also with respect to the action of fire corpuscles in the process of heating. This became a decade-long reflection, entwined with the study of the resistance of materials, as with the development of a 'geometry of indivisibles', that would never come to a definitive solution, being reported in the First Day of the *Discorsi* actually as an open question, if not an open 'research programme'.[50]

---

48   Shea W.R., "Galileo e l'atomismo" 258–260. The expansion of fluids under the action of heat would later be the working principle of the so-called "Galilean thermometer", developed by the Accademia del Cimento as an evolution of the aerometer: a sealed tube filled with alcohol, in which different little weights were attached to small glass spheres containing air. The instrument would be used to measure the external temperature in relation to the density variations of air and alcohol inside the flask. The Museo Galileo in Florence hosts the most important collection of original 17th-century thermometers of the Cimento: catalogo.museogalileo.it/sezione/TermometriDellAccademiaVetriDarteScienza.html (accessed on 21 February 2023).

49   Kepler J., *Strena seu de nive sexangula* (Frankfurt, Gottfried Tampach: 1611).

50   Galilei G., *Discorsi e dimostrazioni matematiche intorno a due nuove scienze*, in Favaro – Del Lungo – Marchesini (eds.), *Opere di Galileo Galilei VIII* 56–78; Shea, W.R., "Galileo e l'atomismo" 269–270.

Here Salviati-Galileo, reflecting upon the phenomenon of the dilatation, glowing, and finally fusion of heated metals (apparently the most dense, compact, and least breakable among all solid substances except diamonds), asks himself and his companions, Sagredo and Simplicio, whether the combined effect of innumerable, very minute voids among many compressed and 'packed' metal particles could paradoxically have a much stronger cohesive action, compared with less dense, more porous, and therefore more fragile solids (exactly like water ice), where there are altogether far fewer but much larger empty interstices.

In the case of melting metals, in particular, the subtle and sharp fire *particole* would be the only ones able to enter the otherwise impenetrable microscopic voids among the dense aggregates of metal particles, streaming through them and causing their dilatation, until the whole metallic substance becomes fluid. Once the *ignicoli* are gone and the metal cools down, it would solidify because of the micro-voids reforming into its 'compactified' particles; so strongly tied together by so many micro-voids that no combustion could ever develop.

> Considering sometimes how fire, going through the minimal particles of this and of that metal, which are so strongly conjunct, finally separate and disjoin them; and how then they come back to join to one another with the same tenacity as before, without any decrease in the amount of gold, and with a very little in other metals, even if they stay melted for a long time; I thought that this could happen because the very subtle particles of fire, by penetrating through the narrow pores of the metal (through which, due to their narrowness, neither the minima of air nor those of other fluids could pass) and filling the minimal voids interposed among them, liberated its minimal particles from the violence with which the same voids attracted them to one another, preventing them from separating; thus, once they could freely move, their mass became fluid, maintaining this state until the igneous particles remained among them; then, when the latter went away, leaving the previous voids, their usual attraction came back, and consequently the cohesion of the parts. Coming to sir Simplicio's question, I think that one can answer that, although such voids would be very minute, so that each of them would be easy to be overcome, anyway their innumerable multitude innumerably (so to speak) multiplies the resistances.[51]

---

51 Galilei G., *Discorsi e dimostrazioni matematiche intorno a due nuove scienze*, in Favaro – Del Lungo – Marchesini (eds.), *Opere di Galileo Galilei* VIII 66–67.

'ATOMS OF FIRE': GALILEO'S UNACHIEVED THEORY OF HEAT

Concerning the second issue, the 1612 *Discourse on floating bodies* and Castelli's subsequent responses (with Galileo's approval) to the objections raised by the Aristotelian philosophers Ludovico Delle Colombe and Vincenzo Di Grazia already contains the essential idea that heat is no intrinsic quality of fire. Rather it is a phenomenon due to the motion of material particles and multiplication and/or expansion of inter-particle empty spaces, induced by pervasive igneous *minima* streaming through them, hitting them, pushing them apart, and, when sufficiently numerous and impetuous, taking corpuscles or entire portions of aggregates away, or even breaking them into smaller particles.[52] This should account for the level of boiling water rising up in vessels, because of its expansion/rarefaction, so that a solid body with little more density than cold water could be pushed upwards, thanks to the lower density of hot water and to the action of fire.

Ebullition was quite easily explained with such a mechanism, but, unlike Delle Colombe, Castelli-Galileo did not identify the creation of bubbles with vapor forming in the liquid and ascending to the surface. Rather he identified them with 'clusters' of fire particles coming from the heating source (typically an underlying fire), penetrating the walls of the container through their imagined microscopic pores, and finding their way out through water into the air. The transformation of water into vapor was, so to speak, a side effect generated by a constant and intense stream of fire corpuscles causing more and more disruptive turbulence inside the liquid, with the aggregates of *ignicoli* taking larger amounts of water particles away.

> Take a flask of glass with a long and very narrow neck, [...] and fill it with water up to the middle of the neck, and mark accurately the level reached by the water; then put such a vessel on some lit coals, and observe that, as soon as the fire hits the glass, the water starts to rise up [...]: then if you want to understand what this expansion is due to, observe diligently, and you will see that, as the atoms of fire are multiplying and many of them are aggregating together through the water, they form some small globules, which in great number are ascending through the water and going out of its surface, and the more many they will enter the water, the more it will rise up in the neck of the vessel [...]. These, sir Colombo, are

---

52 Castelli B., *Risposta alle opposizioni del Signor Lodovico delle Colombe e del Signor Vincenzio di Grazia contro al Trattato del Signor Galileo Galilei, delle cose che stanno su l'Acqua, o che in quella si muovono* (Florence, Cosimo Giunti: 1615), in Favaro – Del Lungo – Marchesini (eds.), *Opere di Galileo Galilei* IV 449–788.

not, as you believe, vapors generated by some parts of the water, that, by means of the hot quality of fire, is going to dissolve and transform into those vapors: which is manifest because, if you remove the coals [...], you will see the water going down slowly, and finally coming back to the same level you marked on the neck of the flask, without decreasing a single drop; and if you repeat such an operation one thousand times, you will see millions of those littles spheres of fire passing through the water, with the latter never being diminished a single hair.[53]

The passage is very revealing of how controversies can be shaped by mental models and their epistemic contexts. Castelli-Galileo's primary intention was to refute Delle Colombe's orthodox Aristotelian idea that, during ebullition, water was qualitatively transformed in its very essence into something completely different by the rarefying action of fire, which was in turn intrinsically hot/heating and dry(ing); vapor being traditionally conceived of as a kind of hot and humid 'air' into which the element water, liquid by definition, was transmuted. Castelli-Galileo needed to demonstrate, against Delle Colombe, that vapor was still made of water, even better that it was nothing but (hot) water, just in a more rarefied form where water particles were dispersed in and mixed with air. He could not accept the fact (although compatible with his mechanical-corpuscular account of heat) that vapor was produced in the liquid phase and then rose to its surface as gaseous bubbles (what actually happens, indeed). This led Castelli-Galileo to the deceptive conclusion that the globules increasingly observable in heating and then boiling water were made of aggregating 'atoms of fire' (here Galileo's uncertainties in terminology overlap with Castelli's own convictions) finding their way through water particles and their interstitial voids.

On the contrary, if to be more sure you make close the mouth of the same vessel, after you have put water into it, you can leave it on the coals for entire months, and you will always see the small globules of fire ascending and, passing through the glass of the other side, going to the air; not even an ounce of the enclosed water will ever be consumed in one hundred years, while, until fire is mixed with water, the latter will rise up to leave place to the former, and once the fire is gone, it will return to its original, immutable state. However, if you take large and open vessels, and you heat the water very much, then the great abundance of fire, that you can see coming up, will aggregate in very large globes, which

---

53   Castelli, *Risposta alle opposizioni del Signor Lodovico* 654.

will ascend with more impetus, causing that effect we call boiling, and by running out they will raise and take away many water atoms with them, like when strong winds raise the dust bringing away its most subtle parts: as dust, while transported, does neither convert into air nor into vapors, so water atoms, brought away by those of fire, still remain water, they do not transform into another thing.[54]

Castelli-Galileo's argument is quite problematic, if we also consider that he rejected the idea of a superficial tension in liquids, while both a resistance of water to let less dense matter leave its surface and an equal resistance of water to let it enter its surface should be taken into account. To him, instead, the above-standing air (i.e., atmospheric pressure) opposed a resistance to fire particles coming out of the surface of water, while no such resistance prevented them (as any other sort of particles) from penetrating the liquid.

Another striking aspect is that, always in contrast to Ludovico Delle Colombe, Castelli-Galileo was forced to underestimate the common phenomenon of room-temperature evaporation (well known to Sanctorius and an integral part of his theory of *perspiratio insensibilis*). He also overlooked the fact that heated water started evaporating well before coming to ebullition. Once again, a notion of superficial tension and vapor pressure was needed, with respect to environmental temperature and atmospheric pressure. Therefore, in a sealed vessel saturated with vapor, with water maintained at a constant temperature, either vapor would condense on the internal walls of the vessel, in a state of equilibrium between liquid and vapor at a constant pressure, or the vessel would break, its internal temperature and pressure being too high and exceeding the resistance of the glass.

> The same igneous atoms, which, by running out of the coals, where so many were stacked and compressed, moved at a high speed and with such an impetus that many passed quickly through the most narrow pores of the glass, once they have come into water, they move slower deep inside it, since they have lost that first impetus they had received from their compression; and if they find in the water some flat layer, of less gravity because of the subtlety or quality of its matter, they aggregate in minute globules, which appear to be almost like dew; this aggregate of innumerable vesicles of light matter slowly raises the layer, bringing it back to the surface of the water: since the reason for all these effects is

---

54   Castelli, *Risposta alle opposizioni del Signor Lodovico* 655.

always amenable to the same principle, i.e., that bodies less dense than water go upwards through it.[55]

According to Castelli-Galileo, flammable fuel like charcoal contained a great amount of 'igneous atoms' that were deeply 'stacked' and 'compressed' inside its substance, among many other material particles and their interstices. Already in 1615, it was clear to him that an immobilized reservoir of fire corpuscles trapped in potentially ignitable matter could not heat anything, not even the fire-rich fuel itself (which could not be properly said to be 'calorific' until it was lit), precisely because of its immobility. Only when fuel was (literally) ignited by some external fire, the opening of its inner cavities and the breaking of its aggregates could mobilize the fire particles, starting the above-mentioned chain reaction of combustion that generated heat(ing) and light(ing) as its main dynamic effects.

Interestingly, in the last quoted passage, one finds the idea that glass must be a porous substance (almost like clay, ceramic, or porcelain) by virtue of its atomic-corpuscular structure, despite its apparently smooth and impenetrable constitution; with 'the narrowest pores', so subtle that no other kind of corpuscles except fire and light particles (as later magnetic, electric, caloric, and other ethereal *effluvia*) could get through them. This idea, together with the hypothesis, also discussed by Galileo and Orazio Grassi, of invisible exhalations of lightless fire corpuscles being released by the breaking of glass moulded in furnaces (sometimes made visible by grosser particles of glass powder), would last until the early 19th century;[56] with Robert Boyle as one of the most prominent and influential advocates of the 'porousness' and 'perviousness' of glass.[57]

7   Measuring Temperature, Managing Fire

Benedetto Castelli's reference to the properties of vitreous substances, in the 1638 letter to Ferdinando Cesarini about Galileo's thermoscope, appears to be something more than just a curious digression. The presence of minute air bubbles in glass, very common at that time because of the technical limitations of

---

55   Castelli, *Risposta alle opposizioni del Signor Lodovico* 656.

56   Boyle R., *Essays of the Strange Subtilty, Determinate Nature, Great Efficacy of Effluviums: to Which are annext New Experiments to make Fire and Flame Ponderable: Together with A Discovery of the Perviousness of Glass to Ponderable Parts of Flame* (London, William Godbid for Moses Pitt: 1673). See in particular 57–85.

57   Shea W.R., "Galileo e l'atomismo" 262–263.

artisanal glassmaking, reinforced the shared conviction that glass was a porous material, whose quality was higher inasmuch as its imagined pores were narrower. From a Galilean perspective, a glass vessel dotted with air bubbles was not so different, in principle, from a piece of water ice, compared with a piece of metal: more rarified and fragile because of larger interstitial spaces in its substance, furthermore filled with air particles that would expand under the action of fire, cracking the bubbles and so breaking the whole glass.

An unquestioned protagonist of our story, since the very beginning, was the round or egg-like flask with a long and narrow neck, straight to form a *gozzo* (the future reaction flask) or curved to form a *storta* (the retort, a key piece of chemical glassware in alembics used for distillation). As already seen, the first prototype of a thermoscope was nothing but a composition of two laboratory vessels: a lower flask with a larger neck and mouth, partially filled with water, and an upper, small flask with a very long and narrow neck, dipped upside down in the water. The long way that saw within a century the transition from the Sanctorius-Galileo open-air thermoscope of the 1600s, filled with water, to Daniel Gabriel Fahrenheit's 1709–1714 closed-loop and graduated bulb thermometers, filled with alcohol and then mercury, may be regarded as a series of transformations, implementations, and variations of the same, fundamental epistemic object: the egg-like flask with a long and narrow neck, that already came from a century-long alchemical tradition.

A major turning point in this transition was certainly the passage to a 'reversed' thermoscope, where the liquid contained in the bulb, instead of the air, became the substance expanding and rising along the glass tube. Initially, the liquid was water itself, as in the bulb thermoscope described by the French physician and 'chymist' Jean Rey in a letter to Marin Mersenne dated 1 January 1632.[58] However, the upper end of Rey's instrument was still open, so that it shared the same limitations affecting the open-air thermoscope, being sensitive to changes in atmospheric pressure, environmental temperature, and humidity, and to the constant evaporation of the water it contained. Sealing the mouth of the long neck, in order to isolate the system, was the decisive step that allowed for a systematic application of graduated scales to the tube.

If one goes back for a while to Castelli-Galileo's 1615 *Risposta* to Delle Colombe's and Di Grazia's objections to the Archimedean-Galilean theory of floating bodies, published in the same year that Sagredo wrote to Galileo on his thermoscopic experiments and when Sanctorius wrote to Galileo about his

---

58 Rey Jean, *Essais de Jean Rey docteur en médecine sur la recherche de la cause pour laquelle l'estain et le plomb augmentent de poids quand on les calcines* [1630] (Paris, Nicolas Ruault: 1777); McKie D. (ed.), *The Essays of Jean Rey* (London: 1951).

Ars de statica medicina, one further consideration may be added. Taken as an experimental set, the glass flask with a long and narrow neck, filled with water up to a given reference level (marked on the neck) and heated by lit coals, was already the concept of a 'reversed' open-bulb thermoscope, with water expansion as its working principle – not so different indeed from the instrument described by Rey in his 1632 letter to Mersenne.[59]

Moreover, when Castelli-Galileo imagined closing the mouth of the vessel to prevent water from evaporating (but not fire particles from passing unaffected through the flask), he made one step further in the direction of a sealed-bulb and closed-loop instrument to measure the variations of heat/temperature – the two notions still overlapped at the time, in terms of 'degrees of heat.' Independently of the dispute with Delle Colombe on the very nature of the bubbles visible during ebullition (globules of vapor or fire aggregates), sealing the flask/bulb necessarily implied fixing a minimum and maximum level for the water inside. The former was its level 'at rest,' the latter the critical level and breaking point of the vessel.

In the famous experiments performed by the Accademia del Cimento between 1657 and 1667, then selected and published by the secretary Lorenzo Magalotti as the *Saggi di naturali esperienze* (1667),[60] various types of sealed-bulb thermometers were realized – exactly the kind of instruments that Viviani refers to in his *Racconto istorico*. They were filled with wine spirit (*acquarzente*), since alcohol was a better indicator (having in today's words a larger coefficient of thermal dilatation) and did not leave traces in the glass tube.

Being also pieces of artwork made by specialized craftsmen working for the Medici court, the instruments used by the academicians of the Cimento were actually thermometers, not just thermoscopes, because they were fully equipped with graduated scales made of enamel dots arranged at regular intervals (degrees) along their tubes. They had two main disadvantages, one material and technical, the other conceptual. First, they were impractical for use outside the small circle of the Academy, due to their size and fragility (water and, even more so, alcohol thermometers required very long tubes), as they were made of delicate crystal glass. Second, and most importantly, they did not have standardized scales, so that each of them had its own scale and the

---

59   See also Sanctorius's very similar apparatus, illustrated in his *Commentaria* on Avicenna's *Canon medicinae* (1625) (fig. 2.5).

60   Magalotti L., *Saggi di naturali esperienze fatte nell'Accademia del Cimento* (Florence, Giuseppe Cocchini: 1667).

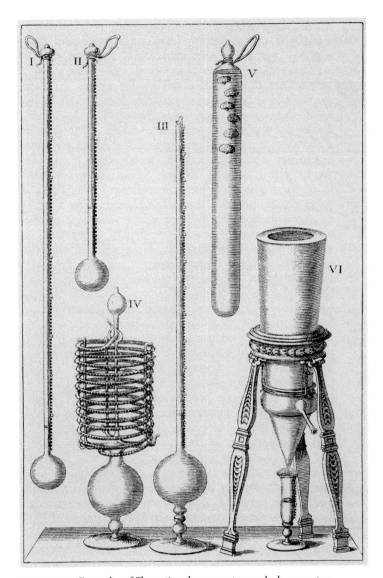

FIGURE 2.6 Examples of Florentine thermometers and a hygrometer.
Image from Magalotti L., *Saggi di naturali esperienze fatte nell'Accademia del Cimento* VIII
COPYRIGHT: LIBRARY OF THE MUSEO GALILEO, FLORENCE

same measure of temperature could not be replicated by other experimenters in other places [Figs. 2.6, 2.7].

Replacing water or alcohol with mercury as a more reliable indicator in the sealed-bulb thermometer, at the beginning of the 18th century, would

FIGURE 2.7 Mid-seventeenth-century spiral thermometers of the Accademia del Cimento. Museo Galileo (Florence), Room VIII. Unknown maker. Inventory: 193, 194/a, 194/b. Materials: glass, enamel, alcohol. Height: min 300 mm, max 340 mm
COPYRIGHT: MUSEO GALILEO, FLORENCE

be after all more a question of technological improvement than of 'scientific change,' although such an innovation would imply a sort of 'knowledge transfer' between the two complementary domains of thermometry and barometry. Torricelli's first *vacuum* tube and barometer was actually an open-air 'Galilean' thermoscope with a very long glass tube dipped in the lower vessel, filled with mercury instead of water and with an empty space left in the upper side of the tube instead of air. Different from a thermoscope and in a way complementary to it, a barometer needed to be an open-air instrument, at least in its earlier versions, because it must be affected by the atmospheric pressure it was made to measure. The problem was exactly the opposite to thermometers: it, too, was sensitive to the variations of environmental temperature and humidity, but this limitation would later become a potential advantage, with proper instrumental implementations, since it would make short-term weather forecasts possible.[61]

Daniel Gabriel Fahrenheit's mercury thermometer, from the perspective of its 'material biography', would have been, with respect to Torricelli's barometer, analogous to the sealed-bulb thermometer with respect to Galileo's thermoscope: a 'reversed' closed-loop barometer, with the key advantage of portability as a more compact and less fragile instrument, besides that of being far more sensitive, precise, and reliable. It was therefore suitable for use in physics cabinets, chemical laboratories, and for medical applications as a clinical thermometer.

The crucial point, the third fundamental turn in the history of thermometry, would be the establishment of standard, interconvertible scales, whose different degrees would be all derived from the uniform subdivision of the same temperature interval, ranging from the freezing point to the ebullition point of water as their common reference. As is well known, Fahrenheit was a former student and collaborator of the *communis totius Europae praeceptor*, the physician, chemist, botanist, and natural philosopher Hermann Boerhaave, who, among other things, introduced the systematic use of Fahrenheit's thermometer both in medical practice and in the study of physical-chemical processes, in which the action of fire was involved. Fire was here no longer an element in the traditional sense, but a transformative agent and operational instrument of chemistry.[62] After Lavoisier's experiments on calcination and combustion in the 1770s, the idea of a caloric fluid (possibly of the same 'ethereal' nature of gravitation, magnetism, electricity and light) would replace that of *phlogiston* as causing heat by the microscopic motion of matter, with temperature as its indirect measure – one could say: on the path already traced by Galileo and Sanctorius.

---

61   Knowles Middleton W.E., *A History of the Thermometer and Its Use in Meteorology* (Baltimore: 1966).

62   Powers J.C., "Measuring Fire: Herman Boerhaave and the Introduction of Thermometry into Chemistry", *Osiris* 29 (2014) 158–177.

Acknowledgements

Well beyond the limits of the present contribution, my sincere gratitude and affectionate memory goes to Paolo Brenni, who unfortunately is no longer with us. I had the privilege to know him personally, although we met but a few times, on the occasion of conferences and seminars. His passion for the material and experimental history of science was simply contagious and inspiring, like his personality. Thanks to the pioneering works of Paolo Brenni, to whom this paper is dedicated, the history of scientific instruments as products and generators of practical knowledge (as of knowledge practice) in its specific context, has definitely become a historical science of instruments.

Bibliography

Primary Sources

Boyle Robert, *Essays of the Strange Subtilty, Determinate Nature, Great Efficacy of Effluviums: to Which are annext New Experiments to make Fire and Flame Ponderable: Together with A Discovery of the Perviousness of Glass to Ponderable Parts of Flame* (London, William Godbid for Moses Pitt: 1673).

Castelli Benedetto, *Risposta alle opposizioni del Signor Lodovico delle Colombe e del Signor Vincenzio di Grazia contro al Trattato del Signor Galileo Galilei, delle cose che stanno su l'Acqua, o che in quella si muovono* (Florence, Cosimo Giunti: 1615), in Favaro – Del Lungo – Marchesini (eds.), *Opere di Galileo Galilei* IV 449–788.

Galilei Galileo, *Discorso al Serenissimo Don Cosimo II Gran Duca di Toscana intorno alle cose, che stanno in su l'acqua, o che in quella si muovono* (Florence, Cosimo Giunti: 1612), in Favaro – Del Lungo – Marchesini (eds.), *Opere di Galileo Galilei* IV (1894) 17–142.

Galilei Galileo, *Il Saggiatore. Nel quale con bilancia esquisita e giusta si ponderano le cose contenute nella Libra Astronomica e Filosofica di Lotario Sarsi Sigensano* (Rome, Giacomo Mascardi: 1623), in Favaro – Del Lungo – Marchesini (eds.), *Opere di Galileo Galilei* VI (1896) 197–372.

Galilei Galileo, *Discorsi e dimostrazioni matematiche intorno a due nuove scienze attenenti alla meccanica e i movimenti locali* (Leiden, Lodewijk Elzevier: 1638), in Favaro – Del Lungo – Marchesini (eds.), *Opere di Galileo Galilei* VIII (1898) 39–466.

Kepler Johannes, *Strena seu de nive sexangula* (Frankfurt, Gottfried Tampach: 1611).

Magalotti Lorenzo, *Saggi di naturali esperienze fatte nell'Accademia del Cimento* (Florence, Giuseppe Cocchini: 1667).

Rey Jean, *Essais de Jean Rey docteur en médecine sur la recherche de la cause pour laquelle l'estain et le plomb augmentent de poids quand on les calcines* [1630] (Paris, Nicolas Ruault: 1777).

Sanctorius Santorio, *Commentaria in artem medicinalem Galeni, libri tres*, 2 vols. (Venice, Giacomo Antonio Somasco: 1612).

Sanctorius Santorio, *Ars de statica medicina et de responsione ad staticomasticem, aphorismorum sectionibus septem comprehensa* (Venice, Niccolò Polo: 1614).

Sanctorius Santorio, *Commentaria in primam Fen primi libri Canonis Avicennae* (Venice, Giacomo Sarcina: 1625).

Secondary Sources

Bachelard G., *La psychanalyse du feu* (Paris: 1938).

Baldin G., *Hobbes and Galileo: Method, Matter and the Science of Motion* (Cham: 2021).

Baldini U., "La struttura della materia nel pensiero di Galileo", *De Homine* 57 (1976) 91–164.

Baldini U et al. (eds.), *Ricerche sull'atomismo del Seicento* (Florence: 1977).

Best N.W., "Lavoisier's 'Reflections on Phlogiston' II: On the Nature of Heat", *Foundations of Chemistry* 18 (2016) 3–13.

Biener Z., "Galileo's First New Science: The Science of Matter", *Perspectives on Science* 12.3 (2004) 262–287.

Bigotti F., "A Previously Unknown Path to Corpuscularianism in the Seventeenth Century: Santorio's Marginalia to the *Commentaria in Primam Fen Primi Libri Canonis Avicennae* (1625)", *Ambix* 64.1 (2017) 29–42.

Bigotti F., "The Weight of the Air: Santorio's Thermometers and the Early History of Medical Quantification Reconsidered", *Early Modern Studies* 7.1 (2018) 73–103.

Bigotti F., *Physiology of the Soul: Mind, Body and Matter in the Galenic Tradition of the Late Renaissance (1550–1630)* (Turnhout: 2021).

Bigotti F. – Barry J. (eds.), *Santorio Santori and the Emergence of Quantified Medicine, 1614–1790. Corpuscularianism, Technology and Experimentation* (Cham: 2022).

Bigotti F. – Barry J., "Introduction", in Bigotti – Barry (eds.), *Santorio Santori and the Emergence of Quantified Medicine* 26–34.

Bigotti F., "'Gears of an Inner Clock': Santorio's Theory of Matter and Its Applications', in Bigotti – Barry (eds.), *Santorio Santori and the Emergence of Quantified Medicine* 75–77.

Boas M., "Hero's *Pneumatica*. A Study of its Transmission and Influence", *Isis* 40 (1949) 38–48.

Bucciantini M. – Torrini M. (eds.), *Geometria e atomismo nella scuola galileiana*, (Florence: 1992).

Cassirer E., *Substanzbegriff und Funktionsbegriff: Untersuchungen über die Grundfragen der Erkenntniskritik* (Hamburg: 1910; reprint, Hamburg: 2000).

Clericuzio A., *Elements, Principles and Corpuscles: A Study of Atomism and Chemistry in the Seventeenth Century* (Dordrecht: 2000).

Drake S. (ed.), *Discoveries and Opinions of Galileo*. Translated with an Introduction and Notes by S. Drake (New York: 1957).

Favaro A. – Del Lungo I. – Marchesini U. (eds.), *Opere di Galileo Galilei*. Edizione Nazionale sotto gli auspici di Sua Maestà il Re d'Italia, 20 vols. (Florence: 1890–1909).

Fenby D.V., "Heat: Its measurement from Galileo to Lavoisier", *Pure and Applied Chemistry* 59.1 (1987) 91–100.

Galilei G., *The Assayer*, in Drake S. (ed.), *Discoveries and Opinions of Galileo* (1957) 229–280.

Galilei G., *Two New Sciences, Including Centers of Gravity and the Force of Percussion*. Edited and translated by S. Drake (Madison, WI: 1974; second edition, Ontario: 1989).

Galluzzi P., *Tra atomi e indivisibili. La materia ambigua di Galileo* (Florence: 2011).

Gattei S. (ed.), *On the Life of Galileo: Viviani's Historical Account and Other Early Biographies* (Princeton: 2019).

Gómez López S., "Galileo y la naturaleza de la luz", in Montesinos J. – Solís C. (eds.), *'Largo campo di filosofare': Eurosymposium Galileo 2001* (La Orotava: 2001) 403–418.

Gómez López S., "The Mechanization of Light in Galilean Science", *Galilaeana* 5 (2008) 207–244.

Gorham G. – Hill B. – Slowik E. – Waters C.K. (eds.), *The Language of Nature: Reassessing the Mathematization of Natural Philosophy in the Seventeenth Century* (Minneapolis: 2016).

Grant E., *Much Ado About Nothing: Theories of Space and Vacuum from the Middle Ages to the Scientific Revolution* (Cambridge: 1981).

Gregory T., "Studi sull'atomismo del Seicento", *Giornale critico della filosofia italiana* 20 (1966) 38–65.

Grmek M.D., "L'énigme des relations entre Galilée et Santorio", in Gruppo Italiano di Storia delle Scienze (ed.), *Atti del simposio internazionale di storia metodologia, logica e filosofia della scienza 'Galileo nella storia e nella filosofia della scienza'* (Florence: 1967) 155–162.

Guerlac H., *Lavoisier – The Crucial Year: The Background and Origin of His First Experiments on Combustion in 1772* (Ithaca, NY: 1961).

Knowles Middleton W.E., *A History of the Thermometer and Its Use in Meteorology* (Baltimore: 1966).

Kohler R.E., "The Origin of Lavoisier's First Experiments on Combustion", *Isis* 63.3 (1972) 349–355.

Le Grand H.E., "Galileo's Matter Theory", in Butts R.E. – Pitt J.C. (eds.), *New Perspectives on Galileo* (Dordrecht-Boston: 1978) 197–208.

Lüthy C., "Bruno's *Area Democriti* and the Origins of Atomist Imagery", *Bruniana & Campanelliana* 4.1 (1998), 59–92.

Lüthy C., "Atomism in the Renaissance", in Sgarbi M. – Valleriani M. (ed.), *Encyclopedia of Renaissance Philosophy* (Cham: 2018) 1–14.

Lüthy C. – Nicoli E. (eds.), *Atoms, Corpuscles and Minima in the Renaissance* (Leiden: 2022).

McKie D. (ed.), *The Essays of Jean Rey* (London: 1951).

Meinel C., "Early Seventeenth-Century Atomism. Theory, Epistemology, and the Insufficiency of Experiment", *Isis* 79 (1988) 68–103.

Moreau E., "Atoms, Mixture, and Temperament in Early Modern Medicine: The Alchemical and Mechanical Views of Sennert and Beeckman", in Bigotti – Barry (eds.), *Santorio Santori and the Emergence of Quantified Medicine* 137–164.

Morris R.J., "Lavoisier and the Caloric Theory", *The British Journal for the History of Science* 6.1 (1972) 1–38.

Murdoch J.E., "The Medieval and Renaissance Tradition of *Minima Naturalia*", in Lüthy – Murdoch – Newman (eds.), *Late Medieval and Early Modern Corpuscular Matter Theories* 91–133.

Newman W.R., *Atoms and Alchemy: Chymistry and the Experimental Origins of the Scientific Revolution* (Chicago: 2006).

Nolan L. (ed.), *Primary and Secondary Qualities: The Historical and Ongoing Debate* (Oxford: 2011).

Nonnoi G., *Saggi Galileiani: Atomi, immagini e ideologia* (Cagliari: 2000).

Powers J.C., "Measuring Fire: Herman Boerhaave and the Introduction of Thermometry into Chemistry", *Osiris* 29 (2014) 158–177.

Redondi P., "Atomi, indivisibili e dogma", *Quaderni storici*, 20.2 (1985) 529–575.

Salvia Stefano, "From Archimedean Hydrostatics to Post-Aristotelian Mechanics: Galileo's Early Manuscripts *De motu antiquiora* (ca. 1590)", *Physics in Perspective* 19.2 (2017) 105–150.

Salem J. (ed.), *L'atomisme aux XVIIe et XVIIIe siècles* (Paris: 1999).

Sgarbi M. – Valleriani M. (ed.), *Encyclopedia of Renaissance Philosophy* (Cham: 2018).

Shank J.B., "What exactly was Torricelli's barometer?", in Gal O. – Chen-Morris R. (eds.), *Science in the Age of Baroque* (Dordrecht: 2013) 161–195.

Shea W.R., "Galileo's atomic hypothesis", *Ambix* 17 (1970) 13–27.

Shea W.R., "Galileo e l'atomismo", *Acta Philosophica* 10.2 (2001) 257–272.

Toscano F., *L'erede di Galileo. Vita breve e mirabile di Evangelista Torricelli* (Milan: 2008).

Valleriani M., *Galileo Engineer* (Berlin: 2010).

Van Melsen A.G., *From Atomos to Atom: The History of the Concept 'Atom'* (New York: 1960).

Viviani V., *Racconto istorico della vita del Signor Galileo Galilei*, in Favaro – Del Lungo – Marchesini (eds.), *Opere di Galileo Galilei* XIX (1907) 599–632.

CHAPTER 3

Without a Thermometer: the Technical Knowledge of Heat in the Early Modern Age

Marco Storni

Abstract

The history of the scientific study of heat has been mostly reduced to the history of thermometry and calorimetry. Recently, historians of chemistry have suggested a more nuanced view of the evolution of the scientific study of heat in the early modernity by emphasizing the interplay of sensory evidence and thermometric measurement in laboratory practice and science teaching. However, historians mostly endorse the view by which the knowledge of heat based on sensory evidence has progressively been regarded insufficient per se and supplanted with thermometer-based quantification. In this paper, I reject the idea that the modern science of heat, grounded on a systematic employ of the thermometer, merely came to replace a knowledge of heat based on unmediated sense evidence. I rather claim that the advent of thermometry collided with a much more articulated and better-defined body of knowledge than the immediate and unthinking relationship to heat implied by the idea of "reliance on the senses." Until the late eighteenth century, the evolution of the thermometer, and more generally the emergence of a modern scientific study of heat, faced the persistence of a "technology of heat" – whereby I mean a codified body of knowledge that emerged from artisanal work and other experimental activities (chemical laboratory trials, cooking, etc.). The point is proved by analysing the emergence and codification of a technical vocabulary of heat, as well as the elaboration of non-thermometric methods to measure heat, in the period 1600–1750.

Keywords

heat – fire – thermometer – artisanal knowledge – codification – measurement – quantification – technical vocabulary

© KONINKLIJKE BRILL BV, LEIDEN, 2025 | DOI:10.1163/9789004521766_005

WITHOUT A THERMOMETER: THE TECHNICAL KNOWLEDGE OF HEAT 91

The history of the scientific study of heat has been – and still is – mostly reduced to the history of thermometry and calorimetry. This is the legacy of nineteenth-century retrospective narratives of the genesis of thermodynamics. In the preface of his *Theory of Heat* (1852), the father of classical electromagnetism James Clerk Maxwell observed that the first step 'by which our knowledge of the phenomena of heat has been extended [...] is the invention of the thermometer'.[1] The second step was 'the measurement of quantities of heat, or calorimetry', leading in turn to the last step, namely the formulation of thermodynamics.[2]

> The whole science of heat is founded on thermometry and calorimetry, and when these operations are understood we may proceed to the third step, which is the investigation of those relations between the thermal and the mechanical properties of substances which form the subject of thermodynamics.[3]

In one of the best recent accounts of the history of thermometry, Hasok Chang defines the idea that 'the scientific study of heat started with the invention of the thermometer' as a 'well-worn cliché'; nonetheless, Chang believes that this cliché 'contains enough truth to serve as the starting point of our inquiry'.[4] This assumption is translated into a historiographical analysis that, at least when dealing with thermometry in the early modern period, focuses on the problem that scientists faced of finding the fixed points to define a thermometric scale. The development of the early modern science of heat is thus contextualised in the research of a universally valid system of measures aiming to fulfil the promise of objectivity, achieved through accurate quantification. As Chang puts it:

> The positive motivation for change is an *imperative of progress*. Progress can mean any number of things [...], but when it comes to the improvement of standards there are a few obvious aspects we desire: the consistency of judgments reached by means of the standard under consideration,

---

1 Maxwell J.C., *Theory of Heat* (London: 1871) v.
2 Ibidem.
3 Ibidem.
4 Chang H., *Inventing Temperature: Measurement and Scientific Progress* (Oxford – New York: 2004) 8.

the precision and confidence with which the judgments can be made, and the scope of the phenomena to which the standard can be applied.[5]

In Chang's view, the progress of the early modern understanding of heat can be described as follows. The 'first stage in our iterative chain of temperature standards' are 'the bodily sensation of hot and cold'. Starting from there, thermoscopes were created as 'initially grounded in sensations', but then improved 'the quality of observations by allowing a more assured and more consistent ordering of a larger range of phenomena by temperature'. The next step was the creation of 'numerical thermometers', that 'constituted an improvement upon thermoscopes because they allowed a true quantification of temperature'. This represented a major advancement as 'thermometric observations became possible subjects for mathematical theorizing'. Finally, 'reasonably successful strategies for stabilizing the boiling point' were implemented, particularly with the establishment of 'the new numerical thermometer using the "steam point" as the upper fixed point constituted an improved temperature standard'.[6] The evolution of the scientific understanding of heat is thus characterised by Chang as a linear path, that started with sensations of hot and cold and pursued with an exact quantification of temperature, embodying the research for a universal standard.

In the last two decades, other historians – particularly in the field of the history of chemistry – have suggested a more nuanced view of the evolution of a science of heat in early modernity. The interpretative framework is that provided by studies of experimental practice and craftsmanship. In her seminal study *The Body of the Artisan* (2004), Pamela H. Smith has proposed the notion of 'vernacular (or artisanal) epistemology' to indicate the ability of sixteenth and seventeenth-century artisans – including artisans-chemists – to expand their knowledge about the natural world through 'bodily labour', namely the work in which craftsmen engaged materially.[7] In the eighteenth century, as Lissa Roberts has shown, the acquisition of bodily skills was still essential to

---

5   Ibidem, 44.

6   Ibidem, 47–48.

7   Smith P.H., *The Body of the Artisan: Art and Experience in the Scientific Revolution* (Chicago – London: 2004) 110. See also the analysis provided in Smith P.H., "Epistemology, Artisanal", in Sgarbi M. (ed.), *Encyclopedia of Renaissance Philosophy* (Cham: 2018) 1a: 'The ability of craftspeople to produce material things rests upon experientially derived bodies of knowledge that can be employed rigorously and methodically to extend, categorize, innovate, and accumulate new knowledge'. On vernacular epistemology and the artisans-chemists, see Powers J.C., "Measuring Fire: Herman Boerhaave and the Introduction of Thermometry into Chemistry", *Osiris* 29.1 (2014) 160.

WITHOUT A THERMOMETER: THE TECHNICAL KNOWLEDGE OF HEAT

chemical training, insofar as such skills were considered complementary to the production of quantified measures: 'Aspiring chemists were trained at mid-century to monitor the progress of chemical processes and make chemical determinations by sensibly examining qualitative characteristics'.[8] In Roberts's view, it is only toward the end of the century that 'chemists increasingly subordinated their bodies to the material technology of their laboratories'; this, however, did not imply 'that chemists stopped smelling, tasting, touching, or listening in the service of their analytical activities'.[9] As far as the study of heat is concerned, the interplay of sensory evidence and thermometric measurement has been recently re-evaluated as a key element to understand its early modern development. In his article *Measuring Fire*, John C. Powers claims that the early teachings of Leiden professor Herman Boerhaave aligned with the tradition of analysing chemical substances, particularly the effects of heat on them, 'in terms of their tastes, smells, and colours'.[10] In later years, however, Boerhaave endorsed a view by which the knowledge of heat based on sensory evidence was considered insufficient *per se* and was to be integrated (if not replaced) with 'philosophical knowledge', meaning 'a theory of how the thermometer works, how to make sense of its readings, and how to deploy that knowledge in various experimental and operational situations'.[11] Although not unanimously accepted by eighteenth-century chemists, in Powers's view the trend inaugurated by Boerhaave fostered the historical process leading 'from a reliance on the senses to the mediation of thermometers to understand and control heat'.[12]

In this essay, I pursue the line of inquiry inaugurated by the latter group of scholars, while criticising an assumption that most of them share. I reject the idea that 'philosophical knowledge' (in the sense defined above) coexisted and progressively replaced a knowledge of heat merely based on 'reliance on the senses', namely, a 'sensuous technology' intended as a set of bodily skills developed by individual practitioners.[13] I argue for the existence of an intermediary dimension between sensuous technology and the philosophical knowledge, which I categorise as the 'technical knowledge of heat'.

---

8   Roberts L., "The Death of the Sensuous Chemist: The 'New' Chemistry and the Transformation of Sensuous Technology", *Studies in History and Philosophy of Science Part A* 26.4 (1995) 506.
9   Ibidem, 507.
10   Powers, "Measuring Fire" 161.
11   Ibidem, 175.
12   Ibidem, 160.
13   See Ibidem and Roberts, "The Death of the Sensuous Chemist" 507.

In early modernity, the knowledge of the operations one could perform with fire (distilling, heating, burning, etc.) was progressively codified in the literature, and became widely shared across communities of experimenters, artisans, and household practitioners. The technical knowledge of heat involved the body but was not exclusively relatable to sensuous technology: it was in fact acquired through experiments and operations performed with specific technical apparatuses, the knowledge of which was progressively externalized and codified. This codification is well displayed by the evolution of a technical vocabulary of heat, which I examine in section 1. The technical vocabulary distinguished the types of fire used in laboratories and workshops by experimenters, artisans, and other practitioners, identifying each fire through the technical method by which heat was produced, or by the effects of heat on various materials.

Despite the process of codification epitomised by the development of a technical vocabulary, the technical knowledge of heat was never framed in a unified and consistent natural-philosophical picture, and remained closer to artisanal practice than to mathematised physics. In this sense, although norms to carry out measurements of the intensity of heat were part of this body of knowledge, the measurements produced in this context were not based on accurate tools nor universal scales of reference. Rather, as we shall see in section 2, one is confronted with a 'quasi-quantification', namely a form of measurement based on a minor level of abstraction than actual quantification.[14] The reference to thermometric scales was avoided in favour of other measuring techniques, closer to the everyday experience of practitioners, such as the number of coals used to produce fire, the action on materials, or the velocity in performing certain operations (e.g., distillation). The choice of quasi-quantification was also supported by *ad hoc* epistemological arguments – such as those formulated by the chemist Gabriel François Venel in the entry *Feu* of Diderot and d'Alembert's *Encyclopédie* – which will be discussed in section 3.

1 The Technical Vocabulary of Heat

A first crucial aspect in the development of a technical knowledge of heat in the early modern age is the codification of a vocabulary of fire. Practitioners of chemistry and related arts often had a tacit knowledge of the principles and

---

14 The lemma "quasi-quantification" has previously been used, although in a different context, by Richard H. Scryock in his essay on "The History of Quantification in Medical Science", *Isis* 52.2 (1961) 85.

WITHOUT A THERMOMETER: THE TECHNICAL KNOWLEDGE OF HEAT

tools of their discipline, including the different strategies for managing fire. This experimental know-how was progressively translated into taxonomies that made explicit the distinctions between 'types of fire', based on the widely shared knowledge of how heat was produced (materials, techniques). Such a codification was not only functional to ease the circulation of chemical knowledge across communities of practitioners, but also helped the promotion of chemistry among a larger public by fostering its teaching.

In the seventeenth and eighteenth centuries, one can observe two reverse tendencies characterizing the technical vocabulary of heat. On the one hand, in prescriptive texts such as introductions to or courses of chemistry, the classifications of types of fire increasingly became synthetic and simple, omitting the explanation of the categories as if unnecessary. On the other hand, in descriptive texts such as dictionaries and encyclopaedias, there was a proliferation of nomenclatures of 'fires' not only relevant to chemical practice, but also to cognate practices, such as glassmaking and cookery.

In his famous *Cours de chymie*, originally published in 1675 and reprinted multiple times during the eighteenth century, the preeminent French chemist Nicolas Lémery listed names commonly given to different types of fire. Lémery's distinctions were based on the method by which heat is produced; all the fires he mentioned therefore had a specific heating potential:

> The fire of sand, of the filings of iron, and of ashes, is made, when the vessel that contains the matter that is to be heated is covered underneath and on all sides with sand, or the filings of iron, or with ashes; this is done to heat the vessel the more gently. [...] The reverberatory fire is made in a close furnace, so that the heat or flame which always tends upwards, may reverberate or return upon the vessel which is placed on two iron bars. This fire hath its degrees, but may be raised to a greater violence than the rest.[15]

Similar distinctions were presented in many early modern companions to the study and practice of chemistry. Interestingly, taxonomies were sometimes more elaborated and organised more systematically. This is the case of the influential *Introduction a la chymie, ou à la vraye physique* by E.R. Arnaud, first published in 1650. Arnaud began his discussion with the traditional alchemical distinction between natural and artificial fire. While natural fire is merely that

---

15   Lémery Nicolas, *Cours de chymie, contenant la manière de faire les opérations qui sont en usage dans la médecine par une méthode facile, avec des raisonnements sur chaque opération* (Paris, chez l'Auteur: 1675) 25–26.

of sunbeams which 'warm up or burn matter by themselves', artificial fire can be further distinguished into material and essential.[16] Artificial material fire is the fire produced by an artisan. This fire can be distinguished into simple and composed, where the simple is that which 'serves only one operation', either digestion or separation, and the composed serves to both digestion (i.e., the dissolution of a material substance) and separation (of the subtler and thicker elements of a composed substance).[17] Throughout this discussion, Arnaud mentioned several types of fire which – as the consonance with Lémery's discussion indicates – became part of a standardised technical nomenclature during the seventeenth century:

> That [the heat] of ashes is halfway between the heat of bain[-marie] and the heat of sand; [...] a free [fire] is when the fire hits immediately the vessel, and separates even the most resistant liquids [...] through the fire of coal, or the fire of flame, which is called reverberatory fire.[18]

The picture is completed by artificial essential fire, defined as the heat 'which acts as fire, although it is not fire'.[19] This category includes materials that can produce the same effects as heat (burning, melting, etc.), without necessarily being inflammable. Examples of this type of fire are 'the cauteries, the butter of antimony [antimony trichloride], the oil of sulphur, the oil of vitriol [...]'.[20]

Another classic of seventeenth-century chemistry, namely Christophe Glaser's *Traité de la chimie* of 1663, which will be reprinted almost forty times over the following decades, presented the reader with a systematic taxonomy of the types of fire.

> We shall therefore say that the mildest fire is the steam bath [...]. The fire that is higher in intensity is the bain-marie [...]. After comes the fire of ashes, which is improperly called a bath [...]. After that in growing intensity is the fire of sand, which is also improperly called a bath [...]. The fire of iron filings comes after, as it is even more intense than the fire of sand. The closed reverberatory fire is the following, which is the one used to

---

16   Arnaud E.R., *Introduction a la chymie, ou à la vraye physique* (Lyon, chez Claude Proste: 1650) 55.
17   Ibidem, 56.
18   Ibidem, 57.
19   Ibidem.
20   Ibidem.

WITHOUT A THERMOMETER: THE TECHNICAL KNOWLEDGE OF HEAT 97

extract the spirits, and which is made with coal. The fire of flame or of fusion comes after: it is the most violent of all, and is made with wood.[21]

Glaser listed seven types of fire, but he also stressed that his list was far from complete. As a matter of fact, he granted that 'in addition to those, there are other types of fire, such as the fire of lamp, of manure, of burning glass, and others'.[22] The choice of excluding these from the list was motivated by their rarity, namely by the fact that a chemist would encounter such types of fire only in very specific circumstances. Since Glaser's book was a treatise exposing the fundamentals of chemistry, none of the experiments presented in it required a fire different from the main seven: the other typologies could thus be ignored 'not to bother the mind with useless studies'.[23]

In the same years as Glaser, an alternative taxonomy of types of fire was proposed by Nicaise Le Febvre in the first tome of his *Traicté de la chymie* (1660). As the apothecary Arthur Du Monstier – the eighteenth-century editor of Le Febvre's *Cours de chymie* – noticed, although Le Febvre and Glaser 'come to diverge on the degrees of fire', the difference between the two is after all minor:

> Le Febvre, who establishes nine degrees of fire, puts in the first place the strongest and most active one, namely that of flame, whereas Glaser starts from the weakest, that which is more natural [...]; and Glaser only identifies seven degrees. However, it comes down to the same thing as the latter [Glaser] adds other types of fire, such as those of lamp, of manure and of the sun, as subsidiary.[24]

The two types of fire added by Le Febvre in the *Traicté* are the 'great fire', namely that 'of iron blade heated to the point it becomes red' – which is an intermediate step between what Glaser called 'fire of iron filings' and 'closed reverberatory fire' (this latter is called by Le Febvre 'coal fire', 'fire of wheel' or 'suppression fire') – and the 'lamp fire', classified as the ninth degree (and therefore the weakest), 'which can be graduated by distancing or approaching the lamp [...] depending on whether we want more or less heating power'.[25]

---

21 Glaser Christophe, *Traité de la chymie, enseignant par une brève et facile méthode toutes ses plus nécessaires préparations* (Paris, chez l'Auteur: 1667) 58–59.

22 Ibidem, 60.

23 Ibidem.

24 Du Monstier Arthur, "Différences qui se rencontrent entre les chymies de Le Febvre et de Glaser", in Le Febvre Nicaise, *Cours de chymie, pour servir d'introduction à cette science*, tome 5 (Paris, Rollin: 1751) 296.

25 Le Febvre Nicaise, *Traicté de la chymie*, tome 1 (Paris, Thomas Iolly: 1660) 127–129.

Taxonomies such as Glaser's and Le Febvre's were most successful in the chemical literature of the following decades, especially in introductory courses expounding the basic principles of the discipline. In his *Compleat Course of Chymistry* (1699), the chemist George Wilson listed the same types of fire as Glaser, without providing a detailed account of any of them.[26] The same was done by William Lewis, in his *Course of Practical Chemistry*.[27] The listing of fires without explanation might be the consequence of the widespread use of these concepts, which had become common knowledge among practitioners by the early eighteenth century. Some other authors, especially while talking to a large audience, preferred to eschew the complexity of long classifications, and reduced the types of fire to four. This is the case of the physician Peter Shaw, who delivered a series of *Chemical Lectures* in London between 1731 and 1732. In lecture two, entirely dedicated to fire, Shaw stressed that 'there are in chemistry as many kinds of heat, as there are mediums thro' which it may be conveyed; or fuels that afford it'.[28] Shaw clarified that heat can be conveyed through water, ashes, or sand; 'when thro' no intermediate substance at all, [it is called] a naked fire'.[29]

As we see, the introductory courses to chemistry tried to simplify progressively the vocabulary of fire, not only to ease the learning process, but also to emphasise the detachment from the alchemical tradition, within which the taxonomies of fire had first been elaborated. A different trend is displayed by dictionaries and lexicons, which in general were no prescriptive texts, but rather recorded the actual uses of words and concepts. In chemical and, more generally, scientific dictionaries, one can observe the repetition of the classical distinctions, with the addition of several other categories – those which Glaser considered useless to recall. In the second volume of his *Medicinal Dictionary* (1745), Robert James stressed that 'the chymists, in order to perform their operations, use fires of various kinds'; he then listed nine, which he considered commonplace, with the addition of another four, lesser known.[30]

---

26   Wilson George, *A Compleat Course of Chymistry, containing near Three Hundred Operations* (London, at the Author's House and by Walter Kettilby: 1699) 3.

27   Lewis William, *A Course of Practical Chemistry: In Which Are Contained All the Operations Described in Wilson's Complete Course of Chemistry* (London, J. Nourse: 1746) 4.

28   Shaw Peter, *Chemical Lectures, Publickly Read at London, in the Years 1731 and 1732; and since at Scarborough, in 1733; for the Improvement of Arts, Trades, and Natural Philosophy* (London, J. Shuckburgh: 1734) 35.

29   Ibidem, 35.

30   James Robert, *A Medicinal Dictionary, including Physic, Surgery, Anatomy, Chymistry and Botany, in All Their Branches Relative to Medicine*, volume 2 (London, T. Osborne: 1745) unnumbered page [*Ignis*].

those of sand, filings of iron, and ashes, the reverberatory fire, the *ignis rotae*, or fire for fusion, the lamp fire, the *balneum mariae*, the vapour bath, and the fire of suppression. They [the chemists], also, use several other kinds of heats, which may be classed among the fires; such as insolation, a bath of horse-dung, a bath of the skins of grapes, and the heat of quick-lime.[31]

Not only did James list these typologies, but he also described each in detail. Some were also characterised by their uses in practice:

Insolation is, when any matter, design'd either to be put into fermentation, or dried, is exposed to the rays of the sun. The bath of horse-dung, called also the horse's belly, is, when a vessel, containing any matter, to be either digested or distill'd, is placed in a large heap of horse-dung. A bath of the skins of grapes, collected in large quantities after the vintage, may [...] serve for digestions or distillations. But the principal use made of these skins, in warm climates, where they become hotter than in such as are temperate, is, to penetrate and produce a rust on copper, for the production of verdegrise [verdigris].[32]

Published in the same years as James's dictionary, Gianfrancesco Pivati's *Nuovo dizionario scientifico e curioso sacro-profano* (1746–1751) included a rich entry on fire, that is a clear testimony to the proliferation of heat classifications in mid-eighteenth-century vocabularies. Pivati started by defining ten types of heat, which mostly coincided with the traditional ones. The most significant differences were, first, to place under the same heading ('*bagno maria*') several kinds of heat, formerly distinguished: '[fire] of bath of ashes, of bath of sand, of bath of iron filings, and others'.[33] Second, the inclusion in the nomenclature – as the tenth degree – of the 'Olympic fire', namely 'that of sun rays, which is made with burning mirrors', that dictionaries and encyclopaedias often mentioned as the highest degree of heat.[34] Unlike Glaser's reticence to enrich the list of fires with less common typologies, Pivati recognised a multitude of other names, of which this passage provides an example:

---

31   Ibidem.

32   Ibidem.

33   Pivati Gianfrancesco, *Nuovo dizionario scientifico e curioso sacro-profano*, tome 4 [F] (Venezia, B. Milocco, 1747) 623b–624a.

34   Ibidem, 624a. For another source on Olympic fire see Furetière Antoine, *Dictionnaire universel, contenant tous les mots français tant vieux que modernes*, tome 2 (The Hague, Arnout and Reinier Leers: 1690) 40a.

> Other chemical authors [...] call tartar the vegetal fire, and call material fire the fire of ashes; infernal fire is a moderately warm environment; nitrous fire is the fire of suppression, which I have already mentioned above; celestial or ethereal fire conveyed by some liquid is the philosophers' quicksilver; secret fire or fire of generation is the fire of lamp; natural fire is that of quicksilver, or of sunlight, or of sulfur; some call humid and natural fire the fire of lamp, of manure, of a bath, or of quicksilver; a dry fire is the fire of flame, or another type of violent fire.[35]

It is noteworthy that several denominations could refer to different types of fire, according to the manifold uses of these categories in different communities of chemical practitioners. Most likely, Pivati collected information on fire from different sources as he wished to provide his readers with a panoramic of the chemical knowledge of fire, which was far from homogeneous in the 1740s.

The proliferation characterizing Pivati's entry on fire is also a feature of the corresponding article in the *Dictionnaire de Trévoux* (in the seventh edition of 1771, tome 4). After listing the ten types of fire, which are the same as those described by Pivati, the Jesuit author listed several other expressions relating to fire which concerned practices loosely related to chemistry.

> The slow fire (*petit feu*), in the common use and in the arts, indicates the preparation or cooking of certain things which is carried out with a fire that is not violent, has very little intensity, is in small quantity, and acts slowly. [Examples:] Cook something with a slow fire; the tyrants have sometimes burnt martyrs with a slow fire; *lento igne torrere, comburere, coquere* [to dry out, to burn, to cook with a slow fire]. [...] We call infernal fire any great fire. During the assault, they made an infernal fire. In glassmaking they always use an infernal fire. In cooking, it is said: to use an infernal fire, to grill food with an infernal fire, meaning with a very strong and powerful fire. The dyers say to give the first or second fire to a fabric that is to be dyed, or to give the first or second flame (*réchaud*), by which they mean to pass it for the first or second time in a dye heated with a furnace (*teinture bouillante de la chaudière*).[36]

As is clear, the definition of a technical vocabulary of heat did not only concern chemistry, but also artisanal practices where fire played a central role, such

---

35   Pivati, *Nuovo dizionario* 624a–b.
36   *Dictionnaire universel français et latin, vulgairement appelé Dictionnaire de Trévoux*, tome 4 [F=JAM] (Paris, Compagnie des Libraires Associés: 1771) 118a.

WITHOUT A THERMOMETER: THE TECHNICAL KNOWLEDGE OF HEAT   101

as glassworks, dyeing and cookery. In cooking books, for instance, the codification of a technical vocabulary of heat throughout the early modern age can be observed neatly. François Marin's recipe book *Les Dons de Comus, ou l'art de la cuisine, réduite en pratique* (1739) displays the extent and articulation that the technical vocabulary of heat had reached in cooking practice at the mid-eighteenth century. Throughout the text, Marin consistently used the lemmas 'great fire', 'moderate fire', 'slow fire', 'very slow fire', 'clear fire', 'lively fire', 'hot ashes', 'mild fire', and other such expressions.[37] Likewise, in Joseph Menon's four-volume treatise *Les Soupers de la Cour, ou l'Art de travailler toutes sortes d'alimens pour servir les meilleures tables* (1755), different types of heat were associated with specific cooking techniques and time measures. The recipes for sauces (sauce *petite italienne, à la mariniere, à la royale, à la flamande*, etc.), for instance, included indications like 'to boil with a slow fire' for 'half an hour' or 'an hour'; the technical indications, however, could also be finer-grained: there were references to a type of heat called 'with a very slow fire', accompanied by measures of shorter time-lapses such as 'a quarter of an hour' or even 'half a quarter of an hour'.[38]

These examples indicate the emergence of a culture of 'technical precision', which is different from the precision characterizing the hard sciences. Unlike the natural-philosophical culture of precision – where accuracy was pursued through quantification, namely through abstract mathematical models – the technical knowledge was rather codified through qualitative distinctions, expressed by a specific nomenclature, whose accuracy lied in the capacity to generalise the practitioners' experience (the perception of heat) as well as their technical gestures (the ways in which heat could be produced, and its potential uses).

2   Non-thermometric Measures of Heat

The intermediary dimension between the sensuous technology and the thermometric analysis of heat cannot be reduced to the elaboration of a technical vocabulary, but also encompassed the multifarious strategies adopted to measure the intensity of heat. Experimenters elaborated different scales to

---

37   [Marin François], *Les Dons de Comus, ou les délices de la table*, 3 tomes (Paris, Prault: 1739). On fire management in cookery, see Bernasconi G., "Cuisine et cultures du feu. De l'âtre au 'feu enveloppé' (XVIIIe–début du XIXe siècle)", *Food and History* 20.2 (2022) 109–128.

38   [Menon Joseph], *Les soupers de la Cour, ou l'art de travailler toutes sortes d'alimens pour servir les meilleures tables, suivant les quatre saisons*, tome 1 (Paris, L. Cellot, 1778) 113–140.

codify the degrees of heat, most often without the help of a thermometer. As noticed by the Venetian apothecary Antonio De Sgobbis in his successful treatise *Nuovo et universale theatro farmaceutico* (1667), a scale of degrees had by then already become traditional: 'The degrees of heat are generally described as fourfold'.[39] As he noticed, there is no direct correspondence between the types of fire listed in the nomenclatures and the four degrees of intensity:

> The heat produced by a bath [i.e., a bain-marie] can be of different degrees, as the bath can be lukewarm or boiling hot, which correspond to the first and second degrees of heat; with ashes, or sand, one can make a heat from the first up to the third degree; with a naked fire one can produce all four degrees.[40]

The distinction of four degrees of heat was first formulated in the context of the alchemical tradition. In a classic alchemical textbook, namely the influential *Lexicon alchemiae* (1612) – defined by James A. Ruffner as 'a key to the meaning of various arcane terms used by Paracelsus and his followers' – the alchemist Martin Ruland the Younger codified the four degrees as follows: 'The first degree is very slow, such as the warmth of a small fire (*ignaviusculi*) [...]. The second degree is more intense, so much that it strikes the hand more evidently, although it does not affect the organ with much strength. [...] The third degree causes injuries if touched [...]. The fourth is the highest and most destructive degree'.[41] In Ruland's text, the definition of the degrees of heat was based on the bodily sensations of hot and cold, which were then abstracted and generalised: 'Those [the degrees of fire] were defined [by the ancient philosophers] not only through the senses, but both through the effect [of fire] on things and the judgment of the [philosophers'] senses, especially sight and touch'.[42] The first degree, for instance, 'is described by touch as that upon which it is always possible to hold your finger, and which, despite the variations of intensity, should

---

39 De Sgobbis Antonio, *Nuovo et universale theatro farmaceutico* (Venice, nella Stamparia Iuliana: 1667) 57b.

40 Ibidem. On this point, see also Wilson, *A Compleat Course of Chymistry* 3: 'The several heats requir'd in chymical operations [...] have their first, second, third, and fourth degrees of fire'.

41 Ruland Martin, *Lexicon alchemiae sive dictionarium alchemisticum* (Frankfurt, Z. Palthenius: 1612) 261–262. See Ruffner J.A., "Agricola and Community: Cognition and Response to the Concept of Coal", in Osler M.J. – Farber P.L. (eds.), *Religion, Science, and Worldview. Essays in Honor of Richard S. Westfall* (Cambridge: 1985) 321.

42 Ruland, *Lexicon alchemiae* 261.

WITHOUT A THERMOMETER: THE TECHNICAL KNOWLEDGE OF HEAT   103

never cause any strong or violent impression'.[43] The distinctions formulated in Ruland's *Lexicon* were repeated by several other authors throughout the early modern age. In his 1657 *Lexicon chymicum*, William Johnson included the distinction of four degrees of heat, 'such as the ancient philosophers described them'.[44] He then reported *verbatim* Ruland's words about the role of the senses and of understanding in discovering such degrees. Interestingly, Johnson also added an example to clarify in which sense a fire – specifically the bain-marie – could have 'all four [degrees]': '1. The first is such that water produces very little heat. 2. The second is when [heat] is already felt more sharply. 3. The third is burning. 4. The fourth corrupts with the most intense heat, that which one can observe more evidently in the case of oil'.[45] Ruland's and Johnson's definition of the degrees of heat were structurally subjective, insofar as they delegated the measure to the individual practitioner's bodily skills. Already in 1611, in his *Institutiones medicinae*, the physician and alchemist Daniel Sennert pointed out that the distinction of fire degrees established by alchemists depended dramatically on the contingencies of practice: '[...] the time and space of any degree cannot be defined [abstractly], but much depends on the expertise of the maker (*in peritia artificis*) and on the knowledge of the matter to be treated'.[46] Sennert's advice to deal with this inconvenience was never to make the fire too strong, since 'if something is once corrupted by the violence of fire, its integrity cannot be restored'.[47]

Besides the determination of heat intensity based on the senses, other strategies emerged that were aimed to reduce the subjective element of the alchemical method. The evolution of dictionary entries on the degrees of fire over the seventeenth century is a testimony to this fact. From reading the 1661 edition of Thomas Blount's *Glossographia* – a popular English dictionary – one learns that 'in physick [degree] signifies a proportion of heat, or cold, moysture or driness in the nature of simples; and there are four such proportions or degrees'.[48] The method to distinguish between these four degrees is exclusively based on the practitioner's body: 'The first degree is so small, that it can scarce be

---

43   Ibidem.
44   Johnson William, *Lexicon chymicum, cum obscuriorum verborum, et rerum hermeticarum* (London, W. Nealand: 1657) 108.
45   Ibidem, 109.
46   Sennert Daniel, *Institutionum medicinae libri v* (Wittenberg, W. Meisner for Z. Schürer: 1611) 1145a.
47   Ibidem.
48   Blount Thomas, *Glossographia: or a Dictionary Interpreting all such Hard Words of Whatsoever Language, now Used in Our Refined English Tongue* (London, T. Newcomb for G. Sawbridge: 1661) unnumbered page [*Degree*].

perceived. The second, that which manifestly may be perceived without hurting the sense. The third, that which somewhat offends the sense. The fourth, which so much offends, that it may destroy the body'.[49] In the anonymous *Glossographia Anglicana Nova* of 1707, which relies heavily on Blount's work as well as on other sources, one finds an entry dedicated to the *Degrees of Fire in Chymistry* – deeply inspired by the corresponding entry of John Harris's 1704 *Lexicon Technicum* – where the distinction between the four degrees is formulated much differently than in the 1661 *Glossographia*:

> The first [degree] is made by two or three coals, and is the most gentle of all; the second is made with four or five coals, or only just to warm the vessel, but so that you may endure your hand upon it for some time. The third degree, is when there is heat to make a pot boil that contains 5 or 6 quarts of water. The fourth is as great a heat as can be made in the furnace.[50]

Although the reference to the bodily dimension did not disappear, it was coupled with other criteria aiming to objectify the scale of heat intensity. On the one hand, there was a reference to the quantity of coal used to light the fire; on the other, the author mentioned some basic technical operations – such as to bring water to the boiling point – that also included quantifications ('5 or 6 quarts of water'). The same pattern can be found in other coeval dictionaries, such as the famous *Cyclopaedia* by Ephraim Chambers (originally published in 1728), whose entry *Degree* (in the fifth edition) featured a section on chemical fire:

> Chemists distinguish four degrees of fire, or heat: The first, is two or three coals. The second, that of four or five coals, or rather so much as is

---

49 Ibidem.
50 *Glossographia Anglicana Nova: or, a Dictionary, Interpreting such Hard Words of Whatever Language, as Are at Present Used in the English Tongue* (London, D. Brown et al.: 1707) unnumbered page [*Degrees of Fire in Chymistry*]. On the role of the *Glossographia Anglicana Nova* in the diffusion of scientific knowledge in the eighteenth century see Layton D., "Diction and Dictionaries in the Diffusion of Scientific Knowledge: An Aspect of the History of the Popularization of Science in Great Britain", *The British Journal for the History of Science* 2.3 (1965) 224: 'The dictionary, like most of its predecessors, was written with the object of "instructing the Ignorant", but it was characterized by a special interest in science, and was an advance on previous works including scientific terms, in that it made an attempt to familiarize contemporary scientific knowledge in a cheap and handy form'.

WITHOUT A THERMOMETER: THE TECHNICAL KNOWLEDGE OF HEAT   105

sufficient to warm a vessel sensibly; yet so, as that the hand may be held on it a considerable time. The third degree, is when there is a fire capable of boiling a vessel of five or six pints of water. The fourth, is when there is fire enough for a furnace.[51]

The marginalization of bodily knowledge in the definition of the degrees of heat that one finds in late-seventeenth-century dictionary entries can *a fortiori* be observed in the scientific literature (courses, treatises, etc.) of the same period, namely in the works on which the dictionary entries were based. In the above-cited *Cours de chymie*, Lémery did not only describe the first two degrees of heat by the number of coals ('to make a fire of first degree one needs two or three small warm coals [...]. For a fire of second degree three or four coals are required [...]'), but also expressed the fourth degree in terms of the fuel used to produce it: 'For the fourth degree, one has to use coal and wood, in order to obtain a most violent fire'.[52] The definition of heat degrees through fuel was sometimes coupled with the criterion of the colour of fire, as shown in the distinction proposed by Wilson in the abovementioned *Compleat Course of Chymistry*. Whereas the first two degrees were explained by referring to the quantity of coal used to run the fire – 'the first degree, is a handful of small cole, or three of four charcole, of the thickness of a man's finger, well kindled. The second degree, is six or seven such charcole kindled' – the description of the last two included colour: 'The third degree, is such a one as will make the fire-place of the furnace, of a worm [warm] red. The fourth degree, is such as will cause the fire-place of the furnace to be of a white heat, or the most extream fire you can make'.[53]

Other authors suggested methods to distinguish the four degrees of heat based on technical operations. The seventeenth-century chemist Estienne de Clave, for instance, in his *Cours de chimie* (1646), mentioned the rapidity in the execution of the operation of distillation as a relevant criterion to establish a hierarchy of fire intensities: 'The four degrees of fire are distinguished [...] based on distillation: if at each contact [of the flame] (*à chaque attouchement*) a drop falls into the receptacle, we call it fourth degree; if it falls after two [contacts], it is the third degree; if it falls at the twentieth, it is the second degree;

---

51   Chambers Ephraim, *Cyclopaedia: or, an Universal Dictionary of Arts and Sciences*, vol. 1 (London, D. Midwinter et al.: 1741) unnumbered page [*Degree, in chemistry*].

52   Lémery, *Cours de chymie* 24–25. The third degree is always defined through the reference to the boiling point of water.

53   Wilson, *A Compleat Course of Chymistry* 3–4.

if it falls only after thirty or forty [contacts], then it is called first degree'.[54] A more articulated discussion was provided by Johann Moritz Hoffmann in the *Acta laboratorii chemici altdorfini* (1719). Hoffmann granted that the four degrees of heat could be distinguished only by means of the senses, particularly the resistance of the hand to the fire: '[...] the first degree is characterised by the mild and moderate heat which never harms the hand touching it; the second by the sharper [heat] which the hand can endure with difficulty; the third is more intense and causes injures; the fourth appears as entirely destructive'.[55] The sensible distinction, however, as Hoffmann stressed, was rarely brought up in practice (*rarius in usum vocatur*) and more common was for practitioners to refer to the ventilation of the furnace to assess the intensity of heat: 'Rather, the artisans (*artifices*) actually examine the degrees based mostly on the very government of fire, such that if only one register or fan (*registrum sive ventilabrum*) is open, the fire is of first degree; if two [are open], it is of second degree; if three, of third degree; if four, they call it a fire of fourth degree'.[56]

Throughout the eighteenth century, the thermometer was progressively introduced into practice to enhance the accuracy of the extant methods to measure heat intensity. Until the end of the century, however, the use of a thermometer never entirely ruled out the four-degree scale, nor the reliance on bodily and technical knowledge. One notable example is provided by Boerhaave's chemical textbook, the *Elementa chemiae* (1732; translated into English by the Epsom physician Timothy Dallowe in 1735). Boerhaave stressed that chemical practitioners should be able to distinguish between the different intensities of fire in order to 'direct and keep up [...] a fire'.[57] While the 'ancient chemists' did not elaborate a reliable method to measure the intensity of heat, the modern ones could rely on 'those beautiful thermometers of the ingenious Fahrenheit'.[58] Boerhaave identified six degrees and defined them in terms of degree intervals on the Fahrenheit thermometric scale. The first degree begins at 'number 1 in Fahrenheit's thermometer, and ends at the degree 80', while the

---

54   De Clave Estienne, *Le cours de chimie* (Paris, O. de Varennes: 1646) 18–19. On De Clave's chemistry course see Joly B., *Descartes et la chimie* (Paris: 2011) 29–31.

55   Hoffmann Johann Moritz, *Acta laboratorii chemici altdorfini, chemiae fundamenta, operationes praecipuas et tentamina curiosa, ratione et experientia suffulta, complectentia* (Nuremberg – Altdorf, apud haeredes J.D. Tauberi: 1719) 18.

56   Ibidem. The same distinction can be found in previous texts, such as Werner Rolfinck's *Chimia in artis formam redacta* (Jena, S. Krebs: 1661) 108.

57   Boerhaave Herman, *Elements of Chemistry*, trans. T. Dallowe, vol. 1 (London, J. Pemberton et al.: 1735) 241.

58   Ibidem.

WITHOUT A THERMOMETER: THE TECHNICAL KNOWLEDGE OF HEAT   107

second 'is supposed to begin at the 40th degree' until 'about 94'.[59] The third and fourth are also described in quantitative terms, respectively as the degree that 'begins at the degree 94, and reaches as far as 212, in which water generally begins to boil' and as that which 'may be reckoned from the degree 212 to 600'.[60] The definitions of the fifth and sixth degrees, however, are not exclusively thermometric but significantly rely on qualitative descriptions. The fifth degree starts 'at the degree 600, and reach[es] as far as that which is capable of melting iron', while the sixth 'comprehends the whole compass of the dioptrical and catoprical fire [...] which hardly any body is able to resist'.[61] As a matter of fact, even when the heat degrees were quantified through the thermometer, Boerhaave listed the technical operations that one could perform with a given degree, so as to integrate the quantitative appraisal with the qualitative one. Thus, the third degree is said to perform:

> The distillation of the distilled oils, and medicinal waters of vegetables. The sanguineous serous juices of animals coagulate in boiling water into a mass that will bear to be cut asunder; whilst all their solids are destroyed by it, and reduced to a thick, tenacious liquid. And hence it is absolutely destructive to all animals.[62]

Likewise, the fourth degree was defined as the one 'within which limits, all oils, saline *lixivia*, mercury, and oil of vitriol, recede from the fire, are carried upwards, and by this means distilled. In this too, lead and tin are put in fusion, and may be mixed together'.[63] The use of technical characterizations served two purposes. The first was of a pedagogic nature: in his chemical lectures, Boerhaave must have felt the need to help students understand the heat degrees with more tangible examples than the abstract thermometric measures. The second concerns the accuracy of the degrees' descriptions. The thermometric definitions encompassed a large span of Fahrenheit degrees (from 94 to 212, or from 212 to 600), and sometimes overlapped (as in the case of the first and second degrees). The qualitative descriptions, referring to specific technical operations, allowed Boerhaave to characterise heat degrees in operational terms, providing better-defined contours to the broader thermometric

---

59   Ibidem, 241–242.
60   Ibidem.
61   Ibidem.
62   Ibidem.
63   Ibidem.

indications. In this sense, paradoxically enough, one might say that the qualitative determinations were more accurate than quantitative ones.

Further on in the eighteenth century, another famous chemistry course presented an approach to the study of the degrees of fire analogous to Boerhaave's *Elementa*, namely the course held by Guillaume-François Rouelle at the Paris botanical garden (the Jardin du Roi) between 1742 and 1768.[64] In a seminal essay on eighteenth-century chemistry and sensuous technology (already mentioned above), Lissa Roberts has argued that a core feature of Rouelle's teaching consisted in training 'aspiring chemists [...] to monitor the progress of chemical processes and make chemical determinations by sensibly examining qualitative characteristics'.[65] The section of the course on heat degrees, however, presents the reader with a more complex picture. In a transcription datable around 1757, preserved at the Bibliothèque Interuniversitaire de Santé of Paris, one can read that Rouelle 'determines the different degrees of fire through the accidents of the operations themselves, except for the first which he measures with the thermometer'.[66] The first degree 'starts when ice melts and ends at the middle degree of boiling water', namely at 50 °C, and is the only one that Rouelle determined with a thermometer (probably a Réaumur's thermometer).[67] The second was until the boiling point of water, while the third and fourth were defined based on diverse technical operations and sensible qualities.

> The third degree goes from boiling water to the point where the bars holding the retort start to redden, the oils, tallows, salts, quicksilver, condensed acids boil [...], tin and lead melt [...]. The fourth degree starts when the third ends, and extends to the point where everything is burnt;

---

64 On Rouelle's teaching activities at the Jardin see Lehman C., "Innovation in Chemistry Courses in France in the Mid-Eighteenth Century: Experiments and Affinities", *Ambix* 57.1 (2010) 3–26. On chemical training in eighteenth-century France see Bensaude-Vincent B. – Lehman C., "Public Lectures of Chemistry in Eighteenth-Century France", in Principe L. (ed.), *New Narratives in Eighteenth-Century Chemistry* (Dordrecht: 2007) 77–96; Lehman C., "Les multiples facettes des cours de chimie en France au milieu du XVIIIe siècle", *Histoire de l'éducation* 130 (2011) 31–56.

65 Roberts, "The Death of the Sensuous Chemist" 506.

66 Rouelle Guillaume-François, *Cours de chymie*, vol. 1, Bibliothèque Interuniversitaire de Santé of Paris (BIS), Ms 5021, fol. 45r. See also Lehman C., "Le cours de chimie de Guillaume-François Rouelle" (2004) online: rhe.ish-lyon.cnrs.fr/cours_magistral/expose _rouelle/expose_rouelle_complet.php (accessed on 30 August 2022).

67 BIS, Ms 5021, fol. 45r.

at this degree several metals are destroyed, only gold, silver, copper and iron resist to it.[68]

Rouelle's definitions of the heat degrees – except for the first, determined with a thermometer – could only make sense for an audience acquainted with the operations and the chemical properties of the substances cited, or for students meant to acquire this knowledge by experience.

While the natural-philosophical courses of Boerhaave and Rouelle instantiated a partial detachment from the purely qualitative distinction of heat degrees, they nevertheless incorporated qualitative determinations – namely references to technical operations and sensible characteristics – which were far from the quantification ideal associated with modern science. Interestingly, such determinations were not perceived as less accurate than quantitative ones, as they seemed to specify the wide ranges of thermometric degrees by the reference to common facts of experience, well-known to practitioners working with fire.

The contemporary reader will perhaps be surprised to observe that practical knowledge and qualitative distinctions still played a central role in the artisanal sphere and natural philosophy alike, in an age when quantification was increasingly considered to be central to scientific activity, and thermometry was the object of an increasing number of expert inquiries. To provide an explanation for the reasons behind the persistence of "technical knowledge", one should consider the epistemological arguments elaborated by several eighteenth-century authors to defend the pertinence of this mode of knowledge, over against the advancement of the modern science of heat.

3 Artisanal Epistemology and Quasi-quantification

In the entry *Feu* of the sixth volume (1756) of the *Encyclopédie*, Venel provided a detailed discussion of the degrees of fire. Venel criticised the seventeenth- and early-eighteenth-century technical vocabulary of heat, based on the 'type of heated substance' and on the 'different mediums between the bodies and the fire', considered as insufficient, vague, and useless. He observed, however, that modern chemists had reconsidered and 'rectified' the four-degree partition, by reducing it to 'a small number of fixed expressions, established on the rational

---

68   Ibidem.

knowledge of the effects of fire, and largely sufficient for practice'.[69] In the version codified by Venel, the first degree of heat corresponds to the interval from 'the melting point of water', namely o °C, until the degree 'which gives us a sensation of warmth'; the first degree is also called the 'cold' degree.[70] The second degree goes from the 'heat perceivable by our body' until the heat 'almost sufficient' to boil water. With this degree of fire, as Venel stressed, one can perform 'digestions, infusions [...] desiccations of plants and of animal substances, evaporations, distillations, and all pharmaceutical preparations executed with a bain-marie'.[71] The third degree was the only invariable one, as it corresponded to the degree of boiling water, namely 100 °C. Again, Venel characterised it through the reference to operations that could be executed with such heat: 'We perform at this degree all decoctions of vegetal and animal substances, the distillation of plants with water, the firing of plasters which contain lead oxide (*chaux de plomb*) which one does not want to burn'.[72] The invariable nature of the third degree made it 'extremely convenient in practice', particularly as far as 'pharmaceutical and economic uses' were concerned. The fourth and last degree of heat, on the contrary, was very broad and ill-defined. It encompassed 'all the latitude from boiling water until the extreme violence of fire'.[73]

In the discussion of the fourth degree, Venel presented his epistemological views, which shed light on the resistance of the 'technical knowledge' paradigm in the early modern science of heat. Such views are best understood as a codification of the vernacular (or artisanal) epistemology paradigm, discussed at the beginning of this chapter.[74] Venel stressed that the broad extension of the last degree of heat could scare the unexperienced chemistry student, especially if used to the methods of 'experimental physics' (viz. the physics practiced in academic laboratories), as they would believe it necessary to produce 'exact measures [by using] thermometers and pyrometers well calibrated, and well tested'.[75] However, once having acquired some experience in the field, any student would realise that such exact measures are useless; the relevant degree

---

69 Venel Gabriel François, "Feu", in Denis Diderot – Jean Le Rond d'Alembert (eds.), *Encyclopédie, ou Dictionnaire raisonné des sciences, des arts et des métiers*, vol. 6 (Paris, Briasson et al.: 1756) 610b.

70 Ibidem.

71 Ibidem, 611a.

72 Ibidem. Marti L., "Nouvelles approaches en sciences humaines et sociales des pratiques et representations des poids et mesures", *Histoire & mesure* 35.1 (2020) 3–14.

73 Ibidem, 611a/b.

74 See in particular *supra*, footnote 7.

75 Venel, "Feu" 611b.

WITHOUT A THERMOMETER: THE TECHNICAL KNOWLEDGE OF HEAT

of heat to be used for a certain operation can be known more immediately through 'exercise, impression at first glance (*le coup d'œil*), or the workman's instinct (*l'instinct d'ouvrier*)'.[76] Here, Venel referred to the embodied nature of artisanal knowledge, characterised as immediate and implicit, which was far more efficient to work with fire than any measurement protocol aiming at accurate determinations. Transforming matter through the action of heat was a matter of bodily gestures and technical skills; to fire practitioners, the demand for precision was nothing but a burden.

One could object that besides the inexperienced students, precision instruments could be necessary to execute delicate and novel operations. Venel replied that *in abstracto* the objection holds, but *de facto* 'the cases in which it would be necessary to resort to such expedients are so rare, if not purely speculative, and therefore they do not constitute the foundation of the art: *rara non sunt artis*'.[77] To master the arts of fire, one should not dwell on trivial exceptions, but rather learn the regular and routine series of gestures that are, at least in the majority of cases, the most effective ones. In sum, Venel considered it useless to prescribe the systematic use of a thermometer to determine heat degrees, as demanded by scared beginners and pedantic academics alike, if not to satisfy a curiosity that had no grounds in practice.

The epistemological views expressed in Venel's entry were the synthesis of ideas widespread among late-seventeenth and early-eighteenth-century authors. William Salmon, an independent physician, apothecary, who also 'cast horoscopes, and professed alchemy', stressed in his much-reprinted *Pharmacopœia Londinensis* (1678) that the infinite degrees of fire in between the four traditional ones 'are not so easily declared'.[78] They could only be found 'by practice and experience', as every 'ingenious artist' knows.[79] A few decades later, in the preface to the 1751 edition of the *Chymische Untersuchungen Welche fürnehmlich von der Lithogeognosia*, the pioneer of pyrochemistry Johann Heinrich Pott stressed that measurements of heat degrees with a thermometer were most often 'superfluous subtleties (*überflüssige Subtilitaeten*)'.[80] The only requirements for effective fire management were 'to know one own's

---

76   Ibidem.

77   Ibidem.

78   Salmon William, *Pharmacopoeia Londinensis. Or, the New London Dispensatory* (London, J. Dawks: 1702) 832. For Salmon's biography see Lee S. (ed.), *Dictionary of National Biography*, vol. 50 [Russen-Scobell] (London, Smith, Elder and co.: 1897) 209b.

79   Salmon, *Pharmacopoeia Londinensis* 832.

80   Pott Johann Heinrich, *Chymische Untersuchungen Welche fürnehmlich von der Lithogeognosia* (Berlin – Potsdam, C.F. Boss: 1751) 3.

furnaces, and to keep track of time'.[81] The reference to time as a crucial element for the good organization of practice was typical of activities to which economy – understood as the concern of making the most of the available resources – was central, as the examples of cooking and lighting show.[82]

Elsewhere in the *Encyclopédie*, namely in the entry *Eaux distillées* (not signed) of the fifth volume, while describing the procedure to prepare distilled water, the author referred to Boerhaave's distillation method, pointing out that 'he is obliged to measure the heat degree that he uses, which is very uncomfortable in practice'.[83] An alternative method – that did not rely on a 'naked fire' (as Boerhaave did), but on a sort of bain-marie – was considered easier as it avoided exact measurement: the practitioner was thus not 'obliged to manage their fire with painstaking attention'.[84] The quantity of liquid to be distilled was also not weighed accurately, but rather 'determined by an observation relayed from one artisan to another (*transmise d'artiste à artiste*), and preserved in pharmacopoeias'.[85]

As we see, several early modern authors, while discussing the issue of fire management (in chemistry, craftsmanship, etc.), considered the knowledge embedded in bodily gestures and technical operations as preferable to that generated by measuring tools, particularly the thermometer. This practical knowledge was preferred insofar as it was considered more efficient than rigid measuring protocols, which were seen by many as too speculative and abstract vis-à-vis the concrete and straightforward nature of the "technical approach". The vagueness of heat degrees defined by reference to practice – which one might locate at the origin of the eventual triumph of thermometry – concealed in fact an implicit knowledge that was widely shared across communities of practitioners. The implicit knowledge of heat, also conveyed by the emergence of a technical vocabulary, was relatively stabilised for the community of reference, and could therefore have a normative character.[86] Whilst

---

81   Ibidem.

82   On economy (or "thrift") see Werrett S., *Thrifty Science: Making the Most of Materials in the History of Experiment* (Chicago – London: 2019). On time economy as a principle of organisation of early modern practices see Storni M., "Time Thrift and Economic Science in the Eighteenth Century", *Science in Context* (forthcoming).

83   [Anonymous], "Eaux distillées", in Denis Diderot – Jean Le Rond d'Alembert (eds.), *Encyclopédie, ou Dictionnaire raisonné des sciences, des arts et des métiers*, vol. 5 (Paris, Briasson et al.: 1755) 196b.

84   Ibidem.

85   Ibidem.

86   This goes against what Harald Witthöft has stated on the ancient approach to measurement, regarded as purely qualitative and relative, over against the modern one, abstract and normative. See Witthöft H., "Ökonomie, Währung und Zahl – Wirtschaftgeschichte

WITHOUT A THERMOMETER: THE TECHNICAL KNOWLEDGE OF HEAT

based on a minor form of abstraction, the four-degree scale ultimately represented a form of measurement in its own right, which one might describe as "quasi-quantification". If actual quantification requires the use of specifically designed measuring tools, the reference to a rigidly determined reference scale, and – in Thomas S. Kuhn's words – the systematic production of 'actual numbers', the quasi-quantification typical of non-thermometric approaches was characterised by a codification of standards, the formulation of *sometimes* rigorous distinctions between different degrees of heat, and the reliance on technical operations reproducible by any practitioner.[87]

In fine, it is important to notice that the quasi-quantification of heat degrees did not precede (chronologically) nor "anticipate" (conceptually) the advent – and systematic use – of the thermometer, but coexisted with it, and often presented itself as a viable alternative to it. This chapter had the ambition to show that the study of the technical knowledge of heat – the development of a technical vocabulary as well as the non-thermometric strategies to measure heat intensities – is a key to reassess the history of heat and underscore its complexity and plurality. This is in fact a crucial step to undermine linear narratives *à la* Maxwell, that are still conditioning our present understanding of the question.

Bibliography

Primary Sources

Dictionnaire universel français et latin, vulgairement appelé Dictionnaire de Trévoux, tome 4 [F=JAM] (Paris, Compagnie des Libraires Associés: 1771).

Glossographia Anglicana Nova: or, a Dictionary, Interpreting such Hard Words of Whatever Language, as Are at Present Used in the English Tongue (London, D. Brown et al.: 1707).

[Anonymous], "Eaux distillées", in Diderot Denis – d'Alembert Jean Le Rond (eds.), *Encyclopédie, ou Dictionnaire raisonné des sciences, des arts et des métiers*, vol. 5 (Paris, Briasson et al.: 1755) 196a–198b.

---

und historische Metrologie. Ein Literatur- und Forschungsbericht 1980 bis 2007", *Vierteljahrschrift für Sozial- und Wirtschaftgeschichte* 95.1 (2008) 27. For a critical perspective see Chambon G. – Marti L., "Nouvelles approaches en sciences humaines et sociales des pratiques et representations des poids et mesures", *Histoire & mesure* 35.1 (2020) 3–14.

87 Kuhn T.S., *The Essential Tension: Selected Studies in Scientific Tradition and Change* (Chicago – London: 1977) 180. Lorraine Daston has also stressed that actual quantification 'is necessarily a sieve: if it did not filter out local knowledge such as individual skill and experience [...], it would lose its portability' (Daston L., "The Moral Economy of Science", *Osiris* 10 (1995) 9).

Arnaud E.R., *Introduction a la chymie, ou à la vraye physique* (Lyon, chez Claude Proste: 1650).

Blount Thomas, *Glossographia: or a Dictionary Interpreting all such Hard Words of Whatsoever Language, now Used in Our Refined English Tongue* (London, T. Newcomb for G. Sawbridge: 1661).

Boerhaave Herman, *Elements of Chemistry*, trans. T. Dallowe, vol. 1 (London, J. Pemberton et al.: 1735).

Chambers Ephraim, *Cyclopaedia: or, an Universal Dictionary of Arts and Sciences*, vol. 1 (London, D. Midwinter et al.: 1741).

De Clave Estienne, *Le cours de chimie* (Paris, O. de Varennes: 1646).

De Sgobbis Antonio, *Nuovo et universale theatro farmaceutico* (Venice, nella Stamparia Iuliana: 1667).

Du Monstier Arthur, "Différences qui se rencontrent entre les chymies de Le Febvre et de Glaser", in Le Febvre Nicaise, *Cours de chymie, pour servir d'introduction à cette science*, tome 5 (Paris, Rollin: 1751) 296–408.

Furetière Antoine, *Dictionnaire universel, contenant tous les mots français tant vieux que modernes*, tome 2 (The Hague, Arnout and Reinier Leers: 1690).

Glaser Christophe, *Traité de la chymie, enseignant par une brève et facile méthode toutes ses plus nécessaires préparations* (Paris, chez l'Auteur: 1667).

Hoffmann Johann Moritz, *Acta laboratorii chemici altdorfini, chemiae fundamenta, operations praecipuas et tentamina curiosa, ratione et experientia suffulta, complectentia* (Nuremberg – Altdorf, apud haeredes J.D. Tauberi: 1719).

James Robert, *A Medicinal Dictionary, including Physic, Surgery, Anatomy, Chymistry and Botany, in All Their Branches Relative to Medicine*, volume 2 (London, T. Osborne: 1745).

Johnson William, *Lexicon chymicum, cum obscuriorum verborum, et rerum hermeticarum* (London, W. Nealand: 1657).

Le Febvre Nicaise, *Traicté de la chimie*, tome 1 (Paris, Thomas Iolly: 1660).

Lémery Nicolas, *Cours de chymie, contenant la manière de faire les opérations qui sont en usage dans la médecine par une méthode facile, avec des raisonnements sur chaque opération* (Paris, chez l'Auteur: 1675).

Lewis William, *A Course of Practical Chemistry: In Which Are Contained All the Operations Described in Wilson's Complete Course of Chemistry* (London, J. Nourse: 1746).

Marin François, *Les Dons de Comus, ou les délices de la table*, 3 tomes (Paris, Prault: 1739).

Menon Joseph, *Les soupers de la Cour, ou l'art de travailler toutes sortes d'alimens pour servir les meilleures tables, suivant les quatre saisons*, tome 1 (Paris, L. Cellot, 1778).

Pivati Gianfrancesco, *Nuovo dizionario scientifico e curioso sacro-profano*, tome 4 [F] (Venezia, B. Milocco, 1747).

Pott Johann Heinrich, *Chymische Untersuchungen Welche fürnehmlich von der Lithogeognosia* (Berlin – Potsdam, C.F. Boss: 1751).

Rolfinck Werner, *Chimia in artis formam redacta* (Jena, S. Krebs: 1661).

Rouelle Guillaume-François, *Cours de chymie*, vol. 1, Bibliothèque Interuniversitaire de Santé of Paris (BIS), Ms 5021, online: www.biusante.parisdescartes.fr/histmed/medica/cote?ms 05021_23.

Ruland Martin, *Lexicon alchemiae sive dictionarium alchemisticum* (Frankfurt, Z. Palthenius: 1612).

Salmon William, *Pharmacopoeia Londinensis. Or, the New London Dispensatory* (London, J. Dawks: 1702).

Sennert Daniel, *Institutionum medicinae libri V* (Wittenberg, W. Meisner for Z. Schürer: 1611).

Shaw Peter, *Chemical Lectures, Publickly Read at London, in the Years 1731 and 1732; and since at Scarborough, in 1733; for the Improvement of Arts, Trades, and Natural Philosophy* (London: J. Shuckburgh: 1734).

Venel Gabriel François, "Feu", in Diderot Denis – d'Alembert Jean Le Rond (eds.), *Encyclopédie, ou Dictionnaire raisonné des sciences, des arts et des métiers*, vol. 6 (Paris, Briasson et al.: 1756) 609a–612b.

Wilson George, *A Compleat Course of Chymistry, containing near Three Hundred Operations* (London, at the Author's House and by Walter Kettilby: 1699).

Secondary Sources

Bensaude-Vincent B. – Lehman C., "Public Lectures of Chemistry in Eighteenth-Century France", in Principe L. (ed.), *New Narratives in Eighteenth-Century Chemistry* (Dordrecht: 2007) 77–96.

Bernasconi G., "Cuisine et cultures du feu: de l'âtre au 'feu enveloppé' (XVIIIe–début du XIXe siècle)", *Food and History* 20.2 (2022) 109–128.

Chambon G. – Marti L., "Nouvelles approaches en sciences humaines et sociales des pratiques et representations des poids et mesures", *Histoire & mesure* 35.1 (2020) 3–14.

Chang H., *Inventing Temperature: Measurement and Scientific Progress* (Oxford – New York: 2004).

Daston L., "The Moral Economy of Science", *Osiris* 10 (1995) 2–24.

Kuhn T.S., *The Essential Tension: Selected Studies in Scientific Tradition and Change* (Chicago – London: 1977).

Layton D., "Diction and Dictionaries in the Diffusion of Scientific Knowledge: An Aspect of the History of the Popularization of Science in Great Britain", *The British Journal for the History of Science* 2.3 (1965) 221–234.

Lee S. (ed.), *Dictionary of National Biography*, vol. 50 [Russen-Scobell] (London: 1897).

Lehman C., "Le cours de chimie de Guillaume-François Rouelle" (2004) online: rhe.ish-lyon.cnrs.fr/cours_magistral/expose_rouelle/expose_rouelle_complet.php (accessed on 29 August 2022).

Lehman C., "Innovation in Chemistry Courses in France in the Mid-Eighteenth Century: Experiments and Affinities", *Ambix* 57.1 (2010) 3–26.

Lehman C., "Les multiples facettes des cours de chimie en France au milieu du XVIIIe siècle", *Histoire de l'éducation* 130 (2011) 31–56.

Maxwell J.C., *Theory of Heat* (London: 1871).

Powers J.C., "Measuring Fire: Herman Boerhaave and the Introduction of Thermometry into Chemistry", *Osiris* 29.1 (2014) 158–177.

Roberts L., "The Death of the Sensuous Chemist: The 'New' Chemistry and the Transformation of Sensuous Technology", *Studies in History and Philosophy of Science Part A* 26.4 (1995) 503–529.

Smith P.H., *The Body of the Artisan: Art and Experience in the Scientific Revolution* (Chicago – London: 2004).

Smith P.H., "Epistemology, Artisanal", in Sgarbi M. (ed.), *Encyclopedia of Renaissance Philosophy* (Cham: 2018) 1–9, online: https://doi.org/10.1007/978-3-319-02848-4_1182-1.

Werrett S., *Thrifty Science: Making the Most of Materials in the History of Experiment* (Chicago – London: 2019).

Witthöft H., "Ökonomie, Währung und Zahl – Wirtschaftgeschichte und historische Metrologie. Ein Literatur- und Forschungsbericht 1980 bis 2007", *Viereljahrschrift für Sozial- und Wirtschaftgeschichte* 95.1 (2008) 25–40.

CHAPTER 4

Working with Fire: Remedy Making from the Shop to the Garden in the Apothecaries' Guild (Eighteenth Century, Paris)

Bérengère Pinaud

Abstract

In this chapter, I will present the multiple aspects of fire management in the apothecaries' daily lives in eighteenth-century Paris. The apothecary emerges as a singular type of artisan and as a double-faceted figure through his multiple uses of fire, since he worked both on his own and as part of a larger guild at the same time. By "traveling" between the apothecary's shop and the communal garden, I argue that fire becomes a scientific, technical, economical, teaching, and social object simultaneously. Using post-mortem inventories and the *Traité de pharmacie* written by the apothecary-chemist Guillaume-François Rouelle (1703–1770), I will first look at the place of remedy making, whose function fluctuates between the laboratory and the kitchen, where fire is a practical instrument as well as an instrument for the pursuit of knowledge. Then, drawing on administrative ledge books (finances and deliberations), I will focus on fire as a "shared material" by the corporation in the context of the newly built chemistry laboratory in the communal garden located in rue de l'Arbalète in 1700. At the logistical level, fire becomes a technology at the heart of an artisanal network as it requires artisans (masons, carpenters ...) to set up furnaces, ovens, and even a portable furnace. At the corporative level, fire management entails collective questions beyond remedy making such as finances, supplies, and utensil damage and fixing, and it gathers a socio-professional group around a teaching activity.

Keywords

apothecaries – experiments – laboratory – chemistry – domestic spaces – guild – investment

© KONINKLIJKE BRILL BV, LEIDEN, 2025 | DOI:10.1163/9789004521766_006

In the early modern period, an apothecary's daily practice consisted of various operations requiring fire. In order to produce remedies or other chemical substances, one of the main agents of transformation was heat: one might consider the infusion of simples in boiling water, the heating of preparations overnight on smouldering ashes, or the distillation of spirits. Since the Middle Ages more generally, fire was ubiquitous in the apothecary's practice – amongst other operations, it was necessary to perform decoction, infusion, distillation, sublimation, maceration, pulverization, dissolution, conservation in sugar or honey, and cooking comfits – and practitioners needed to be trained to manage it appropriately in all its degrees and intensities, from lukewarm baths to naked fire.[1]

Pharmacy, which referred to the knowledge of the therapeutic properties of natural substances, was only one part of the apothecary's profession, that included the confection of remedies and the selling of medical and non-medical goods such as confectionaries and jams, oils and vinegar, chemical substances, roots, simples, candles or fruits. As such, in eighteenth-century Paris, the practice of pharmacy varied amongst apothecaries depending on the training they had received and who they worked for. Guild apothecaries had the status of *maîtres apothicaires* because they had passed examinations after their apprenticeship and had made a financial contribution to the guild for membership.[2] Examinations were held in public before an assembly of physicians from the Parisian Faculty of Medicine and warden apothecaries from the Parisian guild. Candidates had to show aptitude in following the three fundamental steps to prepare remedies: choosing the plants, preparing them, and mixing the elements together.[3] In Paris, the master-apothecaries belonged to the grocers' guild (*épiciers*). Both grocers and apothecaries shared a communal office established on rue des Lombards as well as certain financial resources. However, each profession was responsible for its own sub-guild, which also came with independent funds and management. For instance, the apothecaries' guild owned a garden located on rue de l'Arbalète in Paris, the Jardin des apothicaires. Towards the end of the eighteenth century, the guild of the grocers-apothecaries was progressively splitting, and when the Parisian

---

1 Aulesa C.V., "Defining 'Apothecary' in the Medieval Crown of Aragon", in Sabaté F. (ed.) *Medieval Urban Identity. Health, Economy and Regulation* (Newcastle: 2015).
2 For details of the apothecaries' training, see Brockliss L. – Colin J., *The Medical World of Early Modern France*, (Oxford – New York: 1997).
3 Bibliothèque interuniversitaire de Pharmacie (BIUP), Paris: Registre 9, pièce 68, fol. 1 (1748).

WORKING WITH FIRE: REMEDY MAKING FROM THE SHOP TO THE GARDEN 119

guilds were dissolved in 1777, the Jardin des apothicaires became the College of Pharmacy.[4]

Not all the apothecaries belonged to the guild: those working outside of it included the "royal apothecaries" (working for the royal court), "privileged apothecaries" (who obtained a privilege issued by the king to sell one specific remedy or recipe), "simple apothecaries" (running a shop not associated with the guild), "nun apothecaries" (practicing in a convent and furnished by the apothecaries' guild), "hospital apothecaries" (for example working at the Hôpital Général, Hôtel-Dieu), or "military apothecaries" (for example at the Hôpital militaire).[5] Although these apothecaries' profession was not fundamentally different from that of the apothecaries of the guild, their practice either took place in the context of a single institution or they had to limit their sales to one patented medicine. The guild apothecaries, on the contrary, had publicly shown their abilities and financially contributed to the guild throughout their lives; this gave them the right to control the town's remedy market.

This paper focuses on the *maîtres apothicaires* with their unique status conferred by the guild. They constituted a unified group from a technical, scientific, and financial perspective. As artisans, working on raw materials with fire and associated techniques and tools, remained the mainstay of their daily practice well into the eighteenth century. The analysis of the apothecaries' everyday use and understanding of fire allows us to grasp the complexity of their practices. I will argue that the apothecaries' fire can be referred to as 'artisanal fire' because it was shaped by the very artisanal nature of their daily activities. These took place in the house, the laboratory, and the guild's communal quarters. In each location, fire was alternatively a material to work with to prepare the medicines, the object of enquiry during practice to produce a certain understanding of matter, or a force damaging to materials over time which required a specific setting to function. Moreover, the study of the apothecaries' practice makes it possible to observe their connections with other practitioners (masons, glassmakers, potters) within the city.

To begin with, I analyse probate inventories to investigate the location of fire in the apothecaries' houses (section 1). Despite varying wealth, the inside

---

4 Planchon F.G., *Le Jardin des Apothicaires de Paris* (Paris: 1895); Simon J., *Chemistry, Pharmacy and Revolution in France, 1777–1809* (London: 2016).

5 Lunel A., *La maison médicale du roi : XVIe–XVIIIe siècle* (Seyssel: 2008); Rivest J., *Secret Remedies and the Rise of Pharmaceutical Monopolies in France During the First Global Age* (Ph.D. dissertation, Johns Hopkins University: 2016); Strocchia S.T., "The Nun Apothecaries of Renaissance Florence: Marketing Medicines in the Convent", *Renaissance Studies* 25.5 (2011) 627–647.

of the houses testifies to the presence of fire in several rooms, for example the kitchen and even the attic.

However, post-mortem inventories present limits: they are neither exhaustive nor representative of a whole social group, and most of all, they are not informative on actual practices. For this reason, I investigate class notes on daily practice to study the way activities were carried on as the apothecaries prepared remedies (section 2). Beyond functions, these sources are informative on the uses of specific instruments and fire itself. They also demonstrate that, over the course of their practice, the apothecaries drew a certain understanding of fire from the observation of its actions on substances and utensils. This understanding was based on the artisanal and medical nature of their operations and along the lines of eighteenth-century chemical explanations of natural phenomena.

Lastly, I focus on the administrative ledgers that apothecaries kept in order to record the expenses and the meeting minutes of their guild (section 3). Throughout the century, the purchase of certain materials and devices reveals the logistical dimension of fire management. Apothecaries heavily relied on other artisans' expertise with fire to furnish and maintain the newly built laboratory, for example glassmakers, masons, vessel makers. A complementary question at stake was to keep the equipment in a good state because fire was a force that was damaging to materials over time. But most of all, the management of fire at the Jardin des apothicaires was also at play on an administrative level, and directed at making this institution a new centre of the apothecaries' practice.

1   Domestic Fire: Rooms, the Household, and Materials

Probate inventories give a broad view of what apothecaries' houses contained and, to a certain extent, how the business was run. These documents were established after a member's death in order to clarify the legacy. They aimed at describing everything owned by the deceased, including the wife's or the husband's goods. As a consequence, wives' inventories are equally informative because of the joint ownership that marital status gave. Inventories always started with the cellar, then the kitchen, following with the adjacent rooms and upper floors. The shop constituted its own entity and was inventoried at the end before the personal papers with the help of nearby apothecaries to estimate the prices. A typical apothecary household was composed of the *maître apothicaire*, his wife, and their children. Beyond the family core, the

WORKING WITH FIRE: REMEDY MAKING FROM THE SHOP TO THE GARDEN   121

apothecaries could take an apprentice and/or a maid. In most cases, they were all present during the inventorying.

Historians have often restricted their research on pharmacy practice to the shop: its setting, what it contained, and how it looked from the outside.[6] Some studies drawing from probate inventories have taken the house as a whole in order to stress the urban, social, and political dimension of the apothecaries' practice.[7] A selection of Parisian apothecaries' post-mortem inventories, spanning from 1693 to 1770, are informative on the logistics around fire by mentioning the location of certain objects, rooms, and people involved in the making of remedies on a daily basis.[8] Even though inventories present a multitude of apothecaries' profiles, poor and rich, as well as house configurations, they stress that the pharmacy trade required fire all the time.[9]

Cellars stand out as a storage place for fire-feeding material. In cases where the apothecaries' houses contained one, it stored wood and charcoal next to a few wine or vinegar bottles.[10] Cellars were places of storage connected to the above level that consisted of, but were not limited to, the shop and a kitchen. By feeding with heating materials the activities taking place upstairs, they were part of an operational chain. Kitchens were the second room to be looked at. Notaries would list all the objects and pieces of furniture, starting with the fireplace and then around it. The apothecaries' kitchens, just like in any house, contained the usual objects to keep the fire kindled: andirons, a shovel, tongs, a pair of pincers, including pins and bellows. Kitchens included drip trays, pots,

---

6   Wallis P., "Consumption, Retailing, and Medicine in Early-Modern London", *The Economic History Review* 61.1 (2008) 26–53; Fors H., "Medicine and the Making of a City: Spaces of Pharmacy and Scholarly Medicine in Seventeenth-Century Stockholm", *Isis* 107.3 (2016) 473–494; Douwes A., *Visiting Pharmacies: An Exploratory Study of Apothecary Shops as Public Spaces in Amsterdam, c. 1600–1850* (Master's dissertation, University of Utrecht: 2020).

7   De Vivo F., "Pharmacies as centres of communication in early modern Venice", *Renaissance Studies* 21.4 (2007) 505–521; Shaw J. – Welch E., *Making and Marketing Medicine in Renaissance Florence*, Clio Medica 89 (Amsterdam – New York: 2015).

8   Archives Nationales (AN), Paris: MC/ET/I/197 (Gilette Pijard, 1693), MC/ET/CXIII/211 (François Rousseau, 1705), MC/ET/II/368 (Françoise Renard, 1711), MC/ET/I/267 (Jacqueline Rohais, 1716), MC/ET/XII/386 (Joseph Second, 1724), MC/ET/XXIX/376 (Geneviève Baudrier, 1724), MC/ET/V/463 (Marie Françoise Cardon, 1751), MC/ET/XVII/801 (Adrien Machon), MC/ET/XXVII/258 (Jérôme Pébernard, 1751), MC/ET/XCI/1084 (Guillaume-François Rouelle, 1770).

9   On disparity of wealth, see: Brockliss – Colin, *The Medical World* 480–549; Lafont O., "Une courte lettre révélatrice des préoccupations des apothicaires du XVIIIe siècle", *Revue d'Histoire de la Pharmacie* 85.313 (1997) 35–40.

10   AN: CXIII/211, II/368, I/267, XII/386, XXIX/376, XCI/1084.

cauldrons, pans, saucepans, necessary for both cooking and remedy making activities. They often contained an iron, indicating that the room was never restricted to these two activities but was a living space where people carried out other heat-related house maintenance activities as well. In some cases kitchens contained a water dispenser – one happening to be located in a room adjacent to it.[11] Beyond the lengthy lists of miscellaneous utensils, certain objects entailed specific practices and logistics that were part of remedy making on a daily basis. For example, the apprentices or the apothecaries could use resting beds and mattresses in kitchens or in rooms nearby, while keeping an eye on the fire.[12] The kitchen was thus part and parcel of the apothecaries' activities. It held a connecting role in the house and covered the activities of remedy making, heating, cooking, and house maintenance at once.

However, the medical activities around fire were not restricted to these two rooms and could extend to the attic. For instance, the apothecary Joseph Second had an iron furnace and a few pans available there as well as a rest bed. Patrice Bret and Catherine Lanoë have demonstrated that housework spread over all the floors in artisans' house during the early modern period.[13] In some situations, wealthier apothecaries owned a laboratory. Guillaume-François Rouelle's was located in his courtyard and additional fire-related activities took place there – possibly teaching, experiments, and chemical operations.[14] In one case, a notary found a foot warmer (*chauferette de comptoir*) in the kitchen, a sort of box covered in iron that was filled with heat to keep one's feet warm.[15] It is unsure whether the object was restricted to the kitchen or taken back and forth between the kitchen and the shop. During the cold months of winter it was probably used by people busy with stationary tasks.[16]

As the inventories highlight, the shop was central in remedy making but limited to certain activities, such as the fast preparation of dry remedies, trade, and account keeping. Inventories would typically list the utensils and the medical substances contained in the shops (the last room examined), giving indications of the type of trade. For example, Jérôme Pébernard's items included a

---

11 AN: II/368, XII/386, XXIX/376, XCI/1084.
12 AN: CXIII/211, II/368, I/267, XXIX/376, V/463, XXVII/258.
13 AN: XII/386. Bret P. – Lanoë C., "La formation d'un espace de travail entre sciences et arts et métiers", *Actes des congrès nationaux des sociétés historiques et scientifiques* 127.7 (2006) 139–154; Bret – Lanoë, "Laboratoires et ateliers" 149–155.
14 AN: XCI/1084.
15 AN: II/368.
16 Roche D., "Le froid et le chaud", *Histoire des choses banales* (Paris: 1997) 134–139.

WORKING WITH FIRE: REMEDY MAKING FROM THE SHOP TO THE GARDEN   123

variety of cooking pots and casseroles, and he correlatively sold many sorts of jams and confectionaries.[17]

In recent years, historians have shown that households and housework management were central in early modern knowledge production.[18] As in any domestic space of the period, all the members of the apothecaries' household also had a role in the pharmacy trade and house maintenance. Following the apothecary François Rousseau's death, the apprentice and the housemaid guided the notaries through the house, confirming their key role in the chain of work. There was a hierarchy of practical knowledge (or know-hows) that apprentices were to learn during their training and, regarding the activities requiring fire, they probably carried out a variety of tasks based on their experience.[19] Some preparations needed constant care over long hours. It was often the apprentice's job to watch the fire during the night; and to that end, apprentices were conveniently lodged close to the back room.[20] Montagne's room, who was Rousseau's apprentice, was located under a staircase and over a passage leading to the shop.[21] Regarding wives and widows, it remains unclear what tasks they would carry out, though widows could only keep the shop running on the condition that they hired an apprentice. Furthermore, through the listing of ordinary life objects and the mention of people, inventories tend to show that apothecaries' households were small-scale businesses. According to Philip Rieder, apothecaries were also 'entrepreneurs' because they set up a shop and invested financial and social capital to keep it running over generations: they kept systematic account of their transactions, had family take over the business (sons, sons-in-law) to avoid bankruptcy, and continually hired apprentices as both a complementary workforce, and to keep the craft alive.[22] The apothecaries' houses were dedicated to the remedy trade and everyone contributed to a chain of operation that was active day and night; and these operations required fire all the time.

Inventories are valuable sources because they keep lists of objects and furniture that were either too common, damaged, or rudimentary to be worth preserving through time. Yet they remain silent about the people's actual practices

---

17   AN: XXVII/258.
18   Leong E., *Recipes and Everyday Knowledge: Medicine, Science, and the Household in Early Modern England* (Chicago: 2018); Werrett S., *Thrifty Science: Making the Most of Materials in the History of Experiment* (Chicago: 2019).
19   Klein U., "Apothecary-Chemists in Eighteenth-Century Germany", in Principe L.M. (ed.) *New Narratives in Eighteenth-Century Chemistry* (Dordrecht: 2007) 113–114.
20   Rieder, *Materia medica* 234.
21   AN: CXIII/211.
22   Rieder, "Figure de l'apothicaire" 209.

when it came to fire management. Even pharmacopoeias, which became major references amongst the apothecaries between the sixteenth and the eighteenth centuries, remained an ideal picture of the profession. They were meant to guide the practice and did highlight how techniques of fire management were central in making remedies. Nonetheless, they hardly provide any indication of whether and how these techniques were actually employed in practice. Notes from pharmacy and chemistry classes are sources closer to the practice and help in documenting the actual operations that were carried out in laboratories, shops, attics or kitchens.

2   Writing Fire: Recording Daily Practice

Although the number of chemistry courses increased during the second half of the eighteenth century, only a small number of class notes kept by pharmacy students who attended these classes have survived. Recently, J. Perkins has contended that class notes are a valuable type of source to understand the chemical theories and practices of the century.[23] For apothecaries, learning to use fire precisely started with practice. Many of them attended chemistry courses that were taught by apothecaries themselves. Recorded using first-person discourse, some of the extant class notes were thus very close to the apothecaries' direct experience of fire.

Amongst them, Guillaume-François Rouelle's (1703–1777) production stands out. Rouelle was chemistry demonstrator at the Jardin du Roi during the 1750s; he probably used his teaching experience to start his own private courses at his house located in place Maubert.[24] The few extant versions of his *Cours de chimie* are nowadays scattered in different archives in France and the United Kingdom.[25] To date, private and chemistry courses have been well researched; however, Rouelle's pharmacy classes and his *Traité de pharmacie* have received

---

23   Perkins J., "Chemistry Courses and the Construction of Chemistry, 1750–1830", *Ambix* 57 (2010) 1.

24   Rappaport R., "G.-F. Rouelle: An Eighteenth-Century Chemist and Teacher", *Chymia* 6 (1960): 68–101.

25   BIUP: Ms 16 "Notes sur la chimie"; Ms 17 "Chimie de Rouelle"; Ms 19–20 "Leçons de chimie"; Ms 5021 "Cours de chimie". Wellcome Library (London): MSS.4281–4282 "Cours de Chimie". On Diderot's notes: Jacques J., "Le 'Cours de chimie' de G.-F. Rouelle recueilli par Diderot", *Revue d'histoire des sciences* 38.1 (1985) 43–53. About Rouelle's chemistry: Secrétan C., *Un aspect de la chimie prélavoisienne: (le cours de G.-F. Rouelle)* (Lausanne: 1943); Viel C., "L'enseignement de la chimie et de la matière médicale aux apothicaires aux XVIIe et XVIIIe siècles", *Revue d'Histoire de la Pharmacie* 87.321 (1999) 63–76.

far less attention.[26] The manuscript is an interface between the practical use of chemical knowledge of fire, the daily pharmacy practice, and the use of several objects identified in inventories. It shows that preparing simples required the frequent use of fire; conversely, fire was observed and examined in order to understand its nature and its action on matter using the analytical framework provided by chemistry theories. These practices were ultimately turned into a body of knowledge passed down to students in order to be invested again in remedy preparation, and so on.

In the eighteenth century, apothecaries increasingly employed chemical methods and theories concurrently with the well-established Hippocratic and Galenic conceptions already at the basis of their medical practice.[27] They used chemical processes not only for most of their daily operations, but also as a way to explain natural phenomena, including operations requiring heat or fire. The complete title of Lemery's first famous book, *Cours de chymie contenant la manière de faire les opérations qui sont en usage dans la médecine*, testifies to how chemistry was incorporated as a central technique in the preparation of remedies.[28] Along with water, air, and earth, fire was classified both as an element and as a natural instrument; it was a principle of heat used to change the forms of matter.[29] Rouelle dedicated over more than twenty pages to this topic and a section on the degrees of fire in his *Cours de chimie*.[30] He established the degrees by the 'accidents of the operations' – what physically happened to matter in contact with heat – except for the first degree that he measured with a thermometer. The first degree, for example, 'begins where ice ended' and finished at the 'boiling water average degree'. It was in this 'latitude of heat that nature operates on the vegetal kingdom', because it is where plants grew and produced fruits, and 'this degree is also used to prepare flower essences … and tinctures'.[31] This underlines how much the apothecaries' use of chemistry was

---

26   Brockliss – Jones, *The Medical World*; Lehman, "Les multiples facettes"; Bensaude-Vincent B. – Lehman C. "Public Lectures of Chemistry in Mid-Eighteenth-Century France", in Principe L. (ed.) *New Narratives in Eighteenth-Century Chemistry* (Dordrecht: 2017) 77–96; Perkins J., "Chemistry Courses, the Parisian Chemical World and the Chemical Revolution, 1770–1790", *Ambix* 57.1 (2010): 27–47.

27   Perkins, "Chemistry Courses and the Construction" *Ambix* 57 (2010) 1.

28   Lémery Nicolas, *Cours de chymie contenant la manière de faire les opérations qui sont en usage dans la médecine par une méthode facile* (Paris, chez l'Auteur: 1675).

29   Lehman, "Mid-Eighteenth-century Chemistry in France as Seen Through Student Notes from the Courses of Gabriel-François Venel and Guillaume-François Rouelle", *Ambix* 56.2 (2009): 163–189.

30   BIUP: Ms 19 "Cours de chimie", fols. 11–23.

31   BIUP: Ibidem, f°23–24. See also Roberts, "The Death of the Sensuous Chemist: The 'New' Chemistry and the Transformation of Sensuous Technology", *Studies in History and*

first and foremost based upon the creation and the observation of substances' reactions with each other in order to produce remedies.[32]

Just as his *Cours de chimie*, several versions of Rouelle's *Traité de pharmacie* remain. In the format of class notes, they were titled differently and written by various people around the 1750s, mostly medical students and natural philosophers; only a few of them remain.[33] They contain minor variations possibly due to the note-taking process itself, or to bits of information missed during copying. One of them is an in-quarto book and contains more than 650 pages divided into two sections – the chemistry classes (400 pages) then the pharmacy classes (250 pages).[34] Similar to famous pharmacopoeias, such as Charas's *Pharmacopée royale* and Lemery's *Pharmacopée universelle*, the manuscript starts with preliminary observations about the pharmacy art stages (picking the right plants, drying/preparing them, mixing them) and the typical preparations (infusions, decoctions, medicinal wines, powders ...).[35] Then it rigorously follows the traditional order of recipes adopted by the Parisian *Codex Medicamentarius*, a book issued by the Faculty of Medicine and containing the list of medical recipes officially recognised by the city.[36] The manuscript is a collection of Rouelle's comments – approval or critical – on every recipe of the codex by referring to various European pharmacopoeias (codices of Lyon, Bath, Brussels), other specialists' practice (herbalists, physicians, surgeons), his own observations, and even hearsay. Thus, the notes could be used during practice as a twofold reference guide criss-crossing the Parisian codex with Rouelle's complementary advice. They remained unpublished, which could be explained by the spontaneous style of spoken lessons and their content sometimes very critical towards certain practitioners or books such as the codex.

Philosophy of Science 26 (1995) 503–529.

32   Simon J., "Analysis and the hierarchy of nature in eighteenth-century chemistry", *British Journal for the History of Science* 35.1 (2002) 1–16; Idem, "Pharmacy and Chemistry in the Eighteenth Century: What Lessons for the History of Science?", *Osiris* 29.1 (2014) 283–297.

33   BIUP: Ms 18 "Précis du cours de pharmacie"; Ms 21 "Traité de pharmacie"; Ms 262 "Fournier. Cours de pharmacie" (1748).

34   BIUP: Ms 262 (1748). See also: Lemay P. – Janot M.-M., "Le cours de pharmacie de Rouelle", *Revue d'Histoire de la Pharmacie* 45.152 (1957) 17–21.

35   Charas Moyse, *Pharmacopée royale galénique et chimique* (Paris: Laurent D'Houry, 1681); Lemery Nicolas, *Pharmacopée universelle contenant toutes les compositions de pharmacie qui sont en usage dans la Médecine* (Paris, Laurent D'Houry: 1697).

36   The most recent version of the codex preceeding Rouelle's classes: Martinenq Jean-Baptiste-Thomas, *Codex Medicamentarius, seu pharmacopoea parisiensis, ex mandato Facultate medicinae parisiensis* (Paris, Guillelmum Cavelier: 1748).

WORKING WITH FIRE: REMEDY MAKING FROM THE SHOP TO THE GARDEN   127

The third section of the *Traité de pharmacie* – in the preliminary part about the fundamentals of pharmacy – is about the techniques of drying and highlights that the apothecaries used multiple forms of fire depending on the nature of the things to dry: leaves, flowers, roots, barks, fruits, animal parts. Rouelle introduced the topic by stating that 'the preservation of plants consists almost only in drying them'.[37] Rouelle started this section with the roots, explaining that in most cases these were dried under the sun, except the mucilaginous ones which needed the 'étuve'.[38] According to the *Dictionnaire de Trévoux*, an eighteenth-century 'étuve' was an enclosed container that was 'heated up in order to get the substance inside sweating', and under which a portable stove was placed. This technique, close to modern ovens, was also used by refineries, milliners, and confectioners.[39]

Then Rouelle moved on to the flowers. Certain flowers had such a delicate texture that one had to place them between two pieces of paper to keep their colour. In this case, sun was not strong enough, therefore they had to be placed next to a fire or on top of an oven.[40] Fruits were traditionally dried in 'étuves', but could not be preserved for more than a year or worms would settle. There Rouelle called upon the contemporary naturalist René-Antoine Ferchault de Réaumur's technique (1683–1757) that prevented this by resorting to a heat used in refining. As for animal substances, the 'étuve' was required though not sufficient to prevent worms, and Rouelle preserved them in vinegar. The fact that Rouelle referred to Réaumur demonstrates that knowledge and practices circulated also between the apothecaries and natural philosophers about the study of natural materials, the degrees of heat, and the conditions for life.[41]

Drying by using fire and diverse heating processes ensured that the materia medica was preserved for longer periods. The preservation of natural substances was a hot topic in seventeenth-century natural philosophy especially for apothecaries and physicians who were invested with these questions due to their medical occupation.[42] However, it was inseparable from the preservation of aspect, colour, and taste of the substances. For example, Rouelle explained that flowers needed special care. Certain species such as violets could not keep

---

37   BIUP: Ms 21, "Traité de pharmacie" 55.

38   Ibidem, 56.

39   "Étuve", *Dictionnaire de Trévoux* (Paris, Compagnie des libraires associés: 1771) 922.

40   BIUP: Ms 21, "Traité de pharmacie" 57–61.

41   Bertucci P., *Artisanal Enlightenment: Science and the Mechanical Arts in Old Regime France* (New Haven – London: 2017), 52–79.

42   Cook H.J., "Time's Bodies. Crafting the Preparation and Preservation of Naturalia", in Smith P.H. – Findlen P. *Merchants and Marvels. Commerce, Science, and Art in Early Modern Europe* (London – New York: 2002), 223–247.

their colours because of a fermentative movement of their water. In this case, a 'very vivid heat' ('chaleur très vive') just like that of the 'étuve' was necessary. They had to be placed next to an oven, though even after this they could not be preserved for more than three or four months.[43] Rouelle added that certain people wrongly dried the plants under the shade instead of the sun in order to preserve the aromatic smell. He corrected this by wrapping the plants in fabric that would be hung in the attic, where the heat was greater.[44] These different techniques stress the variety of objects and spaces for work in the apothecaries' houses: kitchen, attic, outside, perhaps by a window. They also indicate that the apothecaries worked with relatively low temperatures, based on the resistance the medical substances could undergo: never going higher than oven heat or fires used in distillation.

Once the theoretical grounds of the remedy making were expounded, Rouelle commented on every single recipe drawn from the Parisian codex, which stipulated what recipes were allowed and their steps. In theory the Codex laid out a stable and unique method, but in practice the techniques of cooking were fluid from one apothecary to another. In some cases, the *Traité de pharmacie* not only shows Rouelle's personal practice, but also the knowledge shared between the apothecaries and other practitioners. The section on the syrups condenses a good number of thoughts on fire techniques because the first step was to infuse or boil the plants. For example, to prepare the violet syrup, Rouelle left the flowers to infuse overnight on burning ashes or on a 'mild fire' ('feu léger') and filtered the substance in the morning; by comparison, he observed that in dispensaries flowers were infused then boiled.[45] On poppy syrup, Rouelle acknowledged Zuvelfer's method of boiling the plant for a short time instead of five hours (as stated in the Parisian codex); according to him it gave an unbearable flavour and decomposed the poppy.[46] Often, Rouelle simply taught his own way using first-person discourse: about the apple syrup, he concentrated the sap in a hot water bath, following the way of the Ancients who used the remedy to make the action of certain plants milder.[47] Now and then, Rouelle strictly condemned certain practices, especially of merchants or herbalists, but most of the time he compared them to see the pros and cons of different techniques. Apple, poppy, and violet syrup recipes are but a few examples showing that fire techniques were not confined to printed official

---

43   BIUP: Ms 21, "Traité de pharmacie" 59–60.

44   Ibidem, 58–29.

45   Ibidem, 210.

46   BIUP: Ms 21, "Traité de pharmacie" 216.

47   Ibidem, 225.

WORKING WITH FIRE: REMEDY MAKING FROM THE SHOP TO THE GARDEN 129

books but drew from a shared knowledge that Rouelle had his students write down during private classes.

Rouelle's observations from his daily practice coupled with his knowledge of chemical processes provided him with multiple opportunities to claim that he had a better understanding of substances and matter than other practitioners. He did so in his discussion of Altherno Fernilli syrup regarding confectioners, drawing a line between two types of analysis.[48] Both apothecaries and confectioners noticed a 'disadvantage' of cooking jams in copper pans. Yet Rouelle disagreed with the confectioners' explanation – according to them, scrubbing off the jams when they were cold caused the copper deposit in the jams, and their evidence was that the pans were shining when the jams were boiling hot. Instead, Rouelle argued that the copper deposit in jams only happened because acids of boiling preparations had an effect on metals: if jams flew easily, it was because heat made them more liquid; and copper maslin pans shone only because they were eroded. His explanation involved different relations between various elements: acids, copper, and the action of fire. His didactic argument aimed at correcting a certain explanation of a phenomenon according to his chemical knowledge. The fact that Rouelle tried to explain what should stand as the right chemical analysis of fire actually did not mean that the apothecaries necessarily had better expertise in chemical fire in practice; in fact, other practitioners such as distillers were revealed to employ a wider range of technical processes with alembics.[49]

Lastly, on a written level, fire also took a symbolic form. Contrary to other extant manuscripts of Rouelle's pharmacy classes, the student who recorded the *Traité de pharmacie* chose to write the elements with their chemical symbols. He frequently used these symbols, for example in the following passage on the infusion of violets:

> On met le double d'eau [triangle inversé] parce que l'infusion de ces fleurs dans un vaisseau de grès ou de terre [triangle inversé barré] vernissé qu'on puisse boucher exactement avec des vessies et du papier en les laissant en infusion pendt[*sic*] la nuit sur les cendres [E] ou à un feu léger [triangle] [...] et le sirop est fait.[50]

---

48   Ibidem, 221–22.
49   Spary E.C., "Distilling Learning", *Eating the Enlightenment: Food and the Sciences in Paris* (Chicago: 2012) 146–194.
50   BIUP: Ms 21, "Traité de pharmacie" 210.

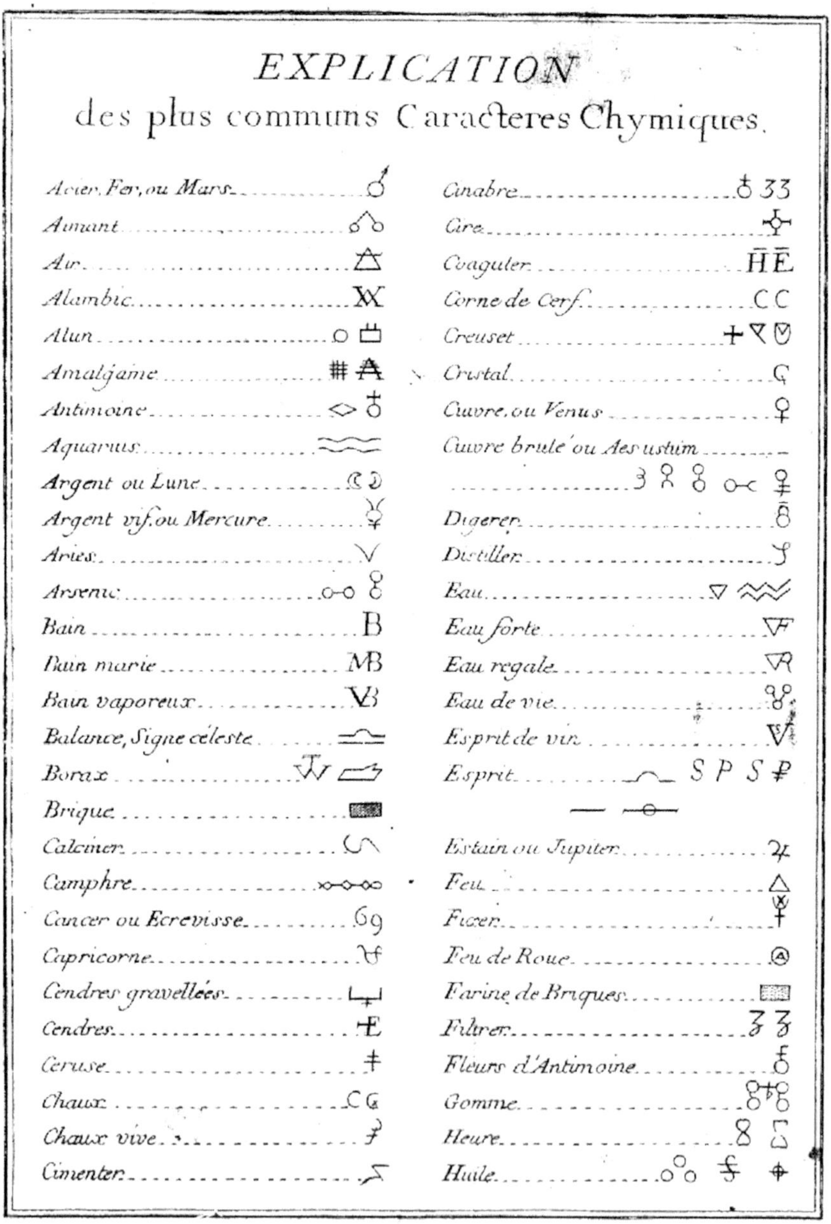

FIGURE 4.1 Nicolas Lémery, *Cours de chimie* (Paris, Théodore-Hyacinthe Baron: 1757), 790–1
COPYRIGHT: BIBLIOTHÈQUE NATIONALE DE FRANCE

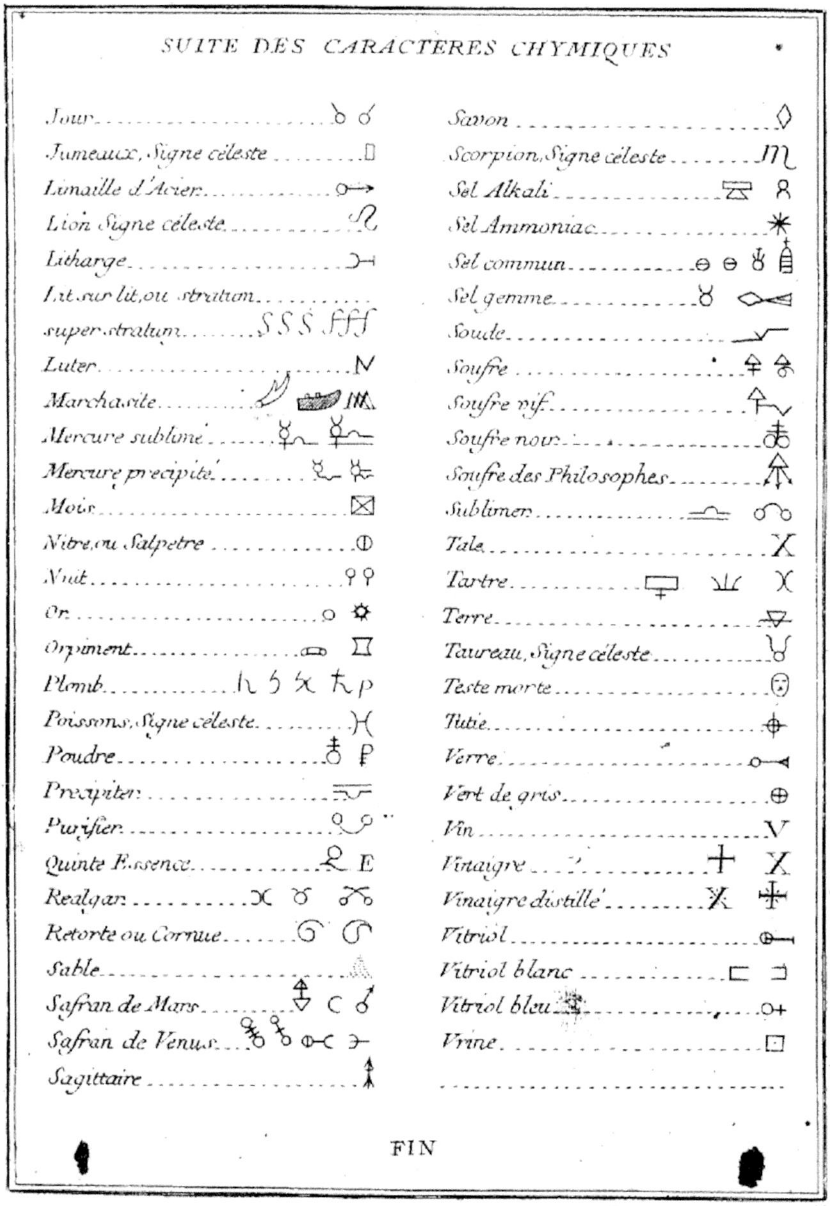

WORKING WITH FIRE: REMEDY MAKING FROM THE SHOP TO THE GARDEN   131

FIGURE 4.2   Nicolas Lémery, *Cours de chimie* (Paris, Théodore-Hyacinthe Baron: 1757), 790–1
COPYRIGHT: BIBLIOTHÈQUE NATIONALE DE FRANCE

One puts in twice as much water [downward triangle] because the infusion of these flowers in a stoneware or varnished terracotta [crossed out downward triangle] vessel that one can seal with bladders and paper leaving them infusing ovr[sic] night on ashes [E] or at low fire [upward triangle] [...] and the syrup is done.

These symbols were common and codified in eighteenth-century chemistry. Lemery's *Cours de chimie* displayed a long list of them, from chemical substances and principles to actions and techniques [Figs. 4.1–4.2]. The above quote used the symbols for fire and tin, but many more actions and elements were declined into written symbols to refer to chemical practices, heating techniques, and understanding of matter, such as 'to boil', 'hot water bath', alembic', 'steam bath', 'to distil', 'burnt copper'. It is unsure why the other notebooks did not use them, perhaps it was no more than a personal preference in the note-taking process.

The initiative in writing down the practice in the *Traité de pharmacie* brings to the front the handicraft, the techniques, and the knowledge of fire that the apothecaries employed in their everyday work. The apothecaries' fire helped to shape their laboratories, kitchens, and houses into hybrid and versatile places, at the crossroads of craft, artisanship, and theoretical knowledge.[51] Additionally, the manuscript demonstrates that the apothecaries' knowledge and techniques of fire were both collective and built through individual experience, which was peculiar to artisanal spaces.

3 Shared Fire: Preparing Fire for Teaching

At the beginning of the century, apothecaries collaborated to establish a laboratory in their communal garden, the Jardin des apothicaires, located at rue de l'Arbalète in Paris [Fig. 4.3]. The garden was purchased in 1624 by the apothecaries' guild, which gradually set up there throughout the eighteenth century. It was situated on the left bank of the river Seine, close to the countryside and not far from the famous Jardin du roi. Historians have paid great attention to chemistry courses given by apothecaries throughout the century, insisting on the theoretical content and the conflicts with the physicians of the Parisian faculty of medicine – the doctors being traditionally entrusted with the teaching

---

51 Bret P. – Lanoë C., "Laboratoires et ateliers, des espaces de travail entre sciences et arts et métiers, XVIᵉ–XVIIIᵉ siècle", in Hilaire-Pérez L. – Simon F. – Thébaud-Sorger M. (eds.), *L'Europe des sciences et des techniques* (Rennes: 2016) 149–155.

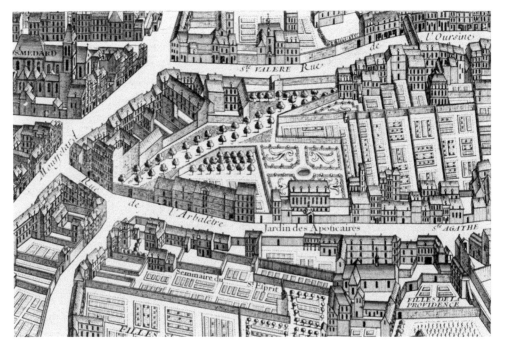

FIGURE 4.3   Michel-Étienne Turgot (patron) and Louis Brétez (designer), "Jardin des apothicaires", *Plan de Paris*, (1734–1739), plate 7
COPYRIGHT: WIKIMEDIA COMMONS

of botany, materia medica, and chemistry.[52] Yet the material considerations ensuring the existence of the laboratory and the project of teaching chemistry has received less attention. Considering this, I argue that fire became a shared resource amongst numerous practitioners, from the setting up and the artisans' expertise to the maintenance of utensils in contact with fire.

Regularly filled, the financial ledgers and the minutes are a key source concerning the administrative and daily life of the garden – for example the hiring of gardeners, purchases, refurbishment works, summer examinations. In the same way, they recorded the expenses made in 1700 to build the laboratory and afterwards to fix tools or to do renovation works, as well as the other practitioners (masons, glassmakers, potters) who were paid to build material. As such, these ledgers highlight the existence of a collaborative knowledge between the apothecaries and other practitioners around fire subject to economics and logistics.[53]

---

52   Perkins, "Chemistry Courses" 27–47; Lehman, "Les multiples facettes" 31–56.
53   BIUP: Registre 37 "Délibérations (1677–1735)"; Registre 38 "Délibérations (1735–1776)"; Registre 153 "Livre pour les comptes du jardin" (1694–1774).

Claude Biet was the warden of the financial resources during the year 1700.[54] The expense chapter unfolds an enumeration of instruments, furniture, and buildings over four pages. Even though the purchases are tracked with no order, they can be distributed into four categories. First, many objects were bought to meet the requirements of the chemistry laboratory and the classes, for instance the twenty-five wicker chairs. A second category of objects was in relation to pharmacy, such as mortars of different sizes and materials (iron, white marble), one of them being old (and seemingly used). They might have been for the classes or to have a set of utensils already onsite when apothecaries prepared public remedies (e.g., theriac). A more logistical third category includes sponges, scissors, wicker and iron rings. Lastly, the fourth category of objects directly relates to fire. It is the most extensive and includes: one large chimney, several furnaces, two pots (*marmites*) of different sizes, one portable furnace, some terracotta furnaces, twigs, one pan, one shovel, one iron spoon, six iron spatulas, two cartloads of sand for the alembic (*athanor*), some sandstone vessels, four terrines, and one varnished terracotta alembic.

Though still about fire-related utensils, this list is quite different from what can be found in the inventories. The objects were more numerous or made with other materials. The items show the logistics of maintaining fire and the chain of operations during chemistry classes. For instance, the provision of sand, absent from houses, was the necessary condition for heating up the alembic. In the houses, the apothecaries owned alembics, but the one in the communal garden worked with large amounts of sand (the sand fire technique). Equally, the portable furnace might have been easy to carry, perhaps outside the laboratory. Some items were bought in quantity (two, four, six) and could have served to demonstrate processes using vessels of different sizes, or simply in case they were broken. Building the laboratory also granted the *maîtres apothicaires* access to carry out operations they could not do at home, either because they did not have the necessary sets of tools, furniture, and buildings, or because they lacked the time. This was explicitly stated during a meeting held in May 1700, when the warden apothecaries announced that 'any colleague will be allowed to work there, publicly or in private for all the chemical and even galenic operations'; the wardens also added that 'the preparation that will happen to be made during the said course shall be distributed to the colleague apothecaries ... whose time or trade did not allow to perform chemistry'.[55] The diversity of fire-related objects stresses how difficult it is to track down every single object that the apothecaries had access to. However,

---

54 BIUP: Registre 153 "Livre pour les comptes du jardin" 19–24.

55 BIUP: Registre 37, 3 May 1700, fols. 57v–58v.

WORKING WITH FIRE: REMEDY MAKING FROM THE SHOP TO THE GARDEN   135

setting up a new source of fire in the garden by building a laboratory added another two new functions of fire in the apothecaries' practice: the laboratory was becoming a place of teaching and of production on a wider scale.

Beyond mere objects, the financial ledgers can be read to trace the origins of the material used to build the laboratory. Claude Biet reported the name of practitioners called on for specific orders (vessel or glass furnishers, furnace builders). Regarding the objects and utensils, the apothecaries called on five different artisans: sir Jamois (mason) for 'une grande cheminée dans notre laboratoire' ('a large fire place in our laboratory') and 'la construction des fourneaux' ('the construction of the furnaces'); sir Lepicard (terracotta potter, *potier de terre*) for 'plusieurs fourneaux et autres ouvrages fournis pour notre laboratoire' ('several furnaces and other works for our laboratory') and a 'fourneau portatif' ('portable furnace'); sir Didier and sir Legrand (earthenware potters, *faïenciers*) for 'plusieurs vaisseaux de verre' ('several glass vessels'); and sir Cautier (profession not indicated) for 'plusieurs ustensiles ou vaisseaux' ('several utensils or vessels'). On their own, the financial ledgers are a source that can bring to the fore the names of artisans specialising in making objects resistant to fire who might not have left any trace otherwise. What these practitioners did and knew is rather absent from the history of instrument makers, and yet they created new objects and new functions based on an external demand paired with their own expertise.[56]

More than lists of names, the warden apothecary Biet sometimes indicated in what context the apothecaries interacted with these practitioners. For example, he specified the provenance and condition of purchase. On one occasion, some stoneware vessels were directly purchased 'à la halle' ('at the market'). Now and then, orders were split between various artisans for similar utensils. As we saw above, the construction of furnaces required different expertise, probably because they were made from different materials or in different formats. In one case, the construction was time-consuming and required one or more apothecaries around to guide the artisans in the building of a tailor-made structure: for 'la grande cheminée et les fourneaux' ('large chimney and the furnaces'), 'la présence était nécessaire depuis le matin jusqu'au soir' ('presence was needed from morning to evening').

It is clear from these records that knowledge of fire existed through networks of specialised artisans and was never owned by or specific to one type of artisan. Fire was a material that needed collaboration to be shaped into a certain use or management. The building of the laboratory in the Jardin des

---

56   Stewart L., "Science, Instruments, and Guilds in Early-Modern Britain", *Early Science and Medicine* 10.3 (2005): 392–410.

apothicaires drew similarities with the trading zones identified by Pamela Long across the period between the fifteenth and the seventeenth centuries.[57] These spaces of work were characterised by the connection of several artisans and practitioners exchanging their expertise for a common project. For the construction of the laboratory in the Jardin, masons, potters, glassmakers, and the apothecaries contributed to tailor objects or arrangements following specific instructions. This type of situation was probably more frequent than we might think. In Charas's *Pharmacopée royale*, the sections on furnaces and mortars explained the ideal measurements and the way to produce a quality mortar (from the type of earth to the recipe itself).[58] Charas's book suggests that the apothecaries' knowledge went beyond the materia medica and how to prepare it; they were also acquainted with the type of material their utensils and constructions such as furnaces were made of; this was because in fact it guaranteed the quality of what they produced.

Another ledger, the guild's meeting minutes, indicates that the maintenance around fire was again a central topic around the decade 1760. During the years 1763–1764, important works took place in the garden to build an upper floor and a natural history cabinet.[59] Even more expenses were incurred for the laboratory again: the apothecaries paid a coppersmith for an alembic and a mason for some works.[60] In resonance with the expenses recorded in the financial ledgers, one meeting held on 23 August 1762 shows that the apothecaries had been complaining about the lack of quality utensils to carry out operations and lessons properly, then agreeing on the absolute need to buy more materials for their laboratory:

> Mrs. les gardes ont dit [...] que le laboratoire étant absolument dépourvu d'ustensiles de la profession, ils étaient contraints de faire de grosses dépenses pour le meubler chaque année, que depuis quatre ans il en avait beaucoup coûté à chacun des associés, que les bassines et alambics avaient toujours par le transport beaucoup souffert, que d'ailleurs ils croient convenable qu'il y eût dans ce laboratoire des alambics de différentes grandeurs qui apparteinssent à la compagnie dont on ferait inventaire, que les évaporations des différents sels étaient très dispendieuses tant par le nombre de vaisseaux de verre, de grès et de terre

---

57 Long P., *Artisans/Practitioners and the Rise of the New Sciences, 1400–1600* (Corvallis: 2011).

58 Charas, *Pharmacopée royale* 89.

59 BIUP: X,6.

60 BIUP: Registre 153 "Livre pour les comptes du jardin", page 125: Sieur De Lesne *chaudronnier*, Sieur Veudelet *maçon*.

WORKING WITH FIRE: REMEDY MAKING FROM THE SHOP TO THE GARDEN 137

qui s'y cassent que par la quantité de charbon qu'on est obligé de consommer pour perfectionner ces opérations soit au bain marie soit au feu de sable, que sur ces considérations ils prient messieurs les anciens d'autoriser les gardes à faire faire aux dépends de la compagnie les alambics propres au travail qui s'exécute pendant le cours de chimie et une bassine d'argent pour les évaporations de la continance d'environ vingt-quatre pintes, promettant que le tout sera porté sur l'inventaire de la compagnie, jamais prêté à aucun membre et très soigneusement ménagé.[61]

Messrs. the Wardens said [...] that the laboratory being absolutely lacking utensils of the craft, they were forced to make large expenditures to furnish it each year. That for four years, it had cost a lot to each associate, that bowls and alembics had always suffered from transportation, that for that matter they believe it appropriate that there would be in that laboratory alembics of various sizes that would belong to the guild, the inventory of which one would make. That the evaporations of the various salts were expensive as much for the number of glass, stoneware, and earthenware vessels breaking there, as for the quantity of coal that one needs to consume in order to perfect these works either in a hot water bath or sand fire. That upon this account, they beg the senior gentlemen to allow the wardens to have alembics suitable for the work performed during the Chemistry class made at the expense of the guild and one silver bowl for the evaporations containing about twenty-four pints, promising that everything will be put on the guild's inventory, never lent to any member and very carefully spared.

Utensils were damaged over time. Although it was common practice, there is no explicit mention of second-hand use nor priority for fixing things.[62] This could be due to the nature of the record itself, seeking to appear in desperate need of new materials because of deterioration. External causes included transportation while the technical causes referred to operations requiring various heat-based techniques (sand fire, hot water bath, coal); as a result, the fellows could not carry out their work normally. In this situation, the apothecaries' management of fire was tied to maintenance over long periods. Because of fire, utensils were cracked or needed replacement. However, these observations were part of a financial argument, which was to get more material for the laboratory. And the reason why the apothecaries were requiring new material

---

61   BIUP: Registre 38, 23 August 1762, fols. 26–27.
62   Werrett, "The Power of Lasting" and "The Broken World", *Thrifty science* 83–108, 109–128.

was their attempt to open the chemistry class again in spite of the prohibition by the Faculty of Medicine of Paris. A 1760-minute confirms that around this period one topic at hand was to renovate the main building and make the amphitheatre bigger, which had become too small for the crowd of students.[63] The apothecaries dramatically complained about the damage of fire to justify the renovation and purchase of new material.

Finally, the passage results in the specific decision that everything must belong to the guild and never lent, placing fire management at the administrative level. On one hand, the apothecaries' claim was about financially managing the consumption, the supply, the logistics, the distribution, and the sharing of fragile utensils, On the other hand, it was about locating it in one unique place, namely the communal garden. This was making the guild both the keeper of the tools and an established place of teaching. This was allowed by the constant financial contributions made by apothecaries and apprentices every year. Over the century, the investment was such that it went from 513 livres in 1700 to 20,764 livres in 1774.[64] Buying alembics and silver bowls aimed at making the Jardin des apothicaires a reserve for the apothecaries. The laboratory was designed to run courses and to enable the apothecaries to produce batches of remedies for busy fellows. This contributed at positioning the garden as an administrative and scientific gathering point for the guild. Very practically, fire management was thus dealt with during the guild's meetings and through their finances so that the communal business and the garden's resources kept running smoothly. As a result, the garden grew as part of a corporate system integrated in an urban space where apothecaries were frequently invested both as artisans and as political agents.[65]

4 Conclusion

The three types of sources I have examined – probate inventories, the *Traité de pharmacie*, and guild's ledgers – demonstrate that fire was not just a primary material. It was a material taking very different forms and showing professional associations around it, contributing to a social, material, scientific, and economic history of fire. By studying these sources both for their content and their context of production, it is possible to bring out multiple aspects of the

---

63   BIUP: Registre 38, 18 September 1760, fol. 21.
64   BIUP: Registre 153 "Livre pour les comptes du jardin", fol. 24 (1700), fol. 149 (1774).
65   In Italy: De Vivo, "Pharmacies as Centres" 505–521; Shaw – Welch, *Making and Marketing Medicine*. In France: Marraud M., "Permanences et déplacements corporatifs dans la ville", *Histoire & mesure* 25.1 (2010) 3–46.

WORKING WITH FIRE: REMEDY MAKING FROM THE SHOP TO THE GARDEN 139

practice and the knowledge of fire in artisanal circles. More specifically, they highlight the breadth of the apothecaries' techniques of fire management. In the house, the family members shared and carried out the tasks of the business in multiple rooms depending on the type of heat required. The apothecaries' considerations on what objects to use, how, and where, found a balance between a shared artisanal knowledge and individual practice. Handwritten class notes stressed that the apothecaries embodied one chemical conception of fire that was produced by hands-on observation during daily practice. Finally, the apothecaries' fire also bore a corporate dimension tied to questions of supply, logistics, and maintenance during the construction of their communal laboratory and afterwards. The ledgers showed the apothecaries' sociability in play when it came to setting up sources of fire in their communal garden. The construction of the laboratory could only be achieved in virtue of other artisans' expertise in fire (masons, glassmakers, potters).

In this essay, I argued that the apothecaries' fire was multi-faceted on one the hand, and both practical and theoretical on the other hand. This specificity is to be found in the versatility of their practice. Philip Rieder argued that the apothecaries were 'artisans, practitioners and entrepreneurs' all at once, highlighting their hybrid profile. Indeed, the apothecaries constituted a group of practitioners interested in various topics of natural knowledge and defined by their identity as craftsmen. During their practice, they required fire at every step, and even more so when they incorporated chemical methods in recipes. Fire spanned from being a heating tool, a source of knowledge on matter, a written symbol, the core of teaching projects, and subject to economic and social considerations. While the apothecaries used fire in a diversity of places – the shop, the kitchen, the attic, the laboratory, and the communal garden – it was constantly directed at the production of remedies and at the transmission of knowledge, both working in tandem.

Bibliography

Primary Sources
Archives
Archives Nationales (Paris)
    MC/ET/I/197 (Gilette Pijard, 1693),
    MC/ET/CXIII/211 (François Rousseau, 1705),
    MC/ET/II/368 (Françoise Renard, 1711),
    MC/ET/I/267 (Jacqueline Rohais, 1716),
    MC/ET/XII/386 (Joseph Second, 1724),
    MC/ET/XXIX/376 (Geneviève Baudrier, 1724),

MC/ET/V/463 (Marie Françoise Cardon, 1751),

MC/ET/XVII/801 (Adrien Machon, 1751),

MC/ET/XXVII/258 (Jérôme Pébernard, 1751),

MC/ET/XCI/1084 (Guillaume-François Rouelle, 1770).

Bibliothèque interuniversitaire de Santé-Pharmacie (Paris)

Ms 16. "Notes sur la chimie, d'après Mr Rouelle, maître-apotiquaire, à Paris, de l'Académie royalle des sciences et démonstrateur royal du Jardin du roy. 1754." (1754–1755).

Ms 17. "Chimie de Rouelle" (1759).

Ms 18. "Précis du cours de pharmacie, donné par M. Rouelle, l'aîné, académicien, apoticaire, chimiste et démonstrateur de chimie au Jardin royal à Paris" (1780).

Ms 19–20. "Leçons de chimie, par Rouelle" (no date).

Ms 21. "Traité de pharmacie. Cours de Rouelle" (no date).

Ms 262. "Fournier. Cours de pharmacie par Monseigneur Rouelle, apothicaire, & de l'Académie royale des sciences. Paris, 1748" (1748).

Ms 5021. G.-F. Rouelle, "Cours de chymie" (1757–1780).

Registre 9, pièce 68. "Arrêt décrétant que les 'gagnants-maîtrises de l'Hôpital Général' ne pourront être admis dans le corps des apothicaires qu'après avoir subi les examens de rigueur" (1748).

Registre 14, pièce 67. "Réponse des épiciers-apothicaires au Mémoire de la Faculté de médecine concernant la Botanique (1771)".

Registre 37. "Délibérations de la Communauté des apothicaires-épiciers, du 28 mai 1677 au 21 février 1735".

Registre 38. "Délibérations de la Communauté des apothicaires-épiciers, du mardi 3 janvier 1736 au 14 juin 1776".

Registre 153. "Livre pour les comptes du jardin" (1694–1774).

Box X, pièce 6. "Travaux de maçonnerie (par Dernelle), de charpente (par Brulée), de couverture (par Esnault) (1763–1764)".

Wellcome Library (London)

MSS.4281–4282. "Cours de Chimie. Rédigé d'après les Leçons de Mr. Rouelle et transcrit en 1762" (1762). Volume 1: Cours de Chimie. 1761. Volume 2: Cours de Chymie rédigé d'après les Leçons de Mr. Rouelle, et transcrit en 1762. Commencé le 2 février de cette présente année sur les cahiers de Mr. Vincent de Billi, venus de M. Huet par M. David qui les tenoit de M. Diderot.

Printed Works (before 1800)

Charas Moyse, *Pharmacopée royale, galénique et chimique*, (Paris, Laurent D'Houry: 1681).

Dictionnaire de Trévoux, (Paris, Compagnie des libraires associés: 1771).

Lemery Nicolas, *Cours de chymie contenant la manière de faire les opérations qui sont en usage dans la médecine par une méthode facile. Avec des raisonnements sur chaque opération; pour l'instruction de ceux qui veulent s'appliquer à cette science*, (Paris, Nicolas Lemery: 1675).

Lémery Nicolas, *Pharmacopée universelle, contenant toutes les compositions de pharmacie qui sont en usage dans la Médecine: tant en France que par toute l'Europe ; leur vertus, leurs doses, les maniéres d'opérer les plus simples & les meilleures : avec un lexicon pharmaceutique, plusieurs remarques, et des raisonnemens sur chaque opération : par Nicolas Lémery*, (Paris, Laurent D'Houry: 1697).

Martinenq Jean-Baptiste-Thomas, *Codex Medicamentarius, seu pharmacopoea parisiensis, ex mandato Facultate medicinae parisiensis*, (Paris, Guillelmum Cavelier: 1748).

Turgot Michel-Étienne. *Plan de Paris / Dessiné et gravé, sous les ordres de Messire Michel Etienne Turgot [...] Levé et dessiné par Louis Bretez, gravé par Claude Lucas, et écrit par Aubin* (1734–1739).

Secondary Sources

Alexandre-Bidon D., *Dans l'atelier de l'apothicaire : histoire et archéologie des pots de pharmacie, XIIIe–XVIe siècle* (Paris: 2013).

Aulesa C.V., "Defining 'Apothecary' in the Medieval Crown of Aragon", in Sabaté F. (ed.) *Medieval Urban Identity. Health, Economy and Regulation* (Newcastle: 2015).

Bensaude-Vincent B. – Blondel C. (ed.), *Science and spectacle in the European Enlightenment* (Aldershot: 2008).

Bertucci P., *Artisanal Enlightenment: Science and the Mechanical Arts in Old Regime France* (New Haven – London: 2017).

Bret P. – Lanoë C., "La formation d'un espace de travail entre sciences et arts et métiers", *Actes des congrès nationaux des sociétés historiques et scientifiques* 127 (2006) 139–154.

Bret P. – Lanoë C., "Laboratoires et ateliers, des espaces de travail entre sciences et arts et métiers, XVIe–XVIIIe siècle", in Hilaire-Pérez L. – Simon F. – Thébaud-Sorger M. (eds.), *L'Europe des sciences et des techniques. Un dialogue des savoirs, XVe–XVIIIe siècle* (Rennes: 2016) 149–155.

Brockliss L. – Colin J., *The Medical World of Early Modern France* (Oxford – New York: 1997).

Bustarret C., "Usages des supports d'écriture au XVIIIe siècle : une esquisse codicologique", *Genesis. Manuscrits – Recherche – Invention* 34 (2012) 37–65.

Butterworth E., "Apothecaries Cornets: Books as Waste Paper in the Renaissance", *MLN* 133 (2018) 891–913.

Cavallo S. – Gentilcore D., "Introduction: Spaces, objects and identities in early modern Italian medicine", *Renaissance Studies* 21 (2007) 473–479.

Cook H., *Matters of Exchange: Commerce, Medicine, and Science in the Dutch Golden Age* (New Haven: 2007).

Daston L. – Park K., *Wonders and the order of Nature 1150–1750* (New York: 1995).

De Vivo F., "Pharmacies as centres of communication in early modern Venice", *Renaissance Studies* 21 (2007) 505–521.

Douwes A., *Visiting Pharmacies: An Exploratory Study of Apothecary Shops as Public Spaces in Amsterdam, c. 1600–1850* (Master's dissertation, University of Utrecht: 2020).

Hilaire-Pérez L. – Simon F. – Thébaud-Sorger M., *L'Europe des sciences et des techniques, XVe–XVIIIe siècle : un dialogue des savoirs* (Rennes: 2016).

Findlen P. – Smith P.H., *Merchants and Marvels. Commerce, Science, and Art in Early Modern Europe* (London – New York: 2002).

Fors H., "Medicine and the Making of a City: Spaces of Pharmacy and Scholarly Medicine in Seventeenth-Century Stockholm", *Isis* 107 (2016) 473–494.

Gentilcore D., "'For the Protection of Those Who Have Both Shop and Home in this City': Relations Between Italian Charlatans and Apothecaries", *Pharmacy in History* 45 (2003) 108–122.

Gentilcore D., *Food and Health in Early Modern Europe. Diet, Medicine and Society, 1450–1800* (London – New Delhi – New York – Sidney: 2016).

Golinski Jan, *Science as Public Culture: Chemistry and Enlightenment in Britain, 1760–1820* (Cambridge: 1992).

Hannaway O., *The Chemists and the Word: The Didactic Origins of Chemistry* (Baltimore – London: 1975).

Jacques J., "Le 'Cours de chimie' de G.-F. Rouclle recueilli par Diderot", *Revue d'histoire des sciences* 38 (1985) 43–53.

Marie Janot M.-M. – Lemay P., "Le cours de pharmacie de Rouelle", *Revue d'Histoire de la Pharmacie* 45 (1957) 17–21.

Jaussaud P., *Pharmaciens au Muséum : chimistes et naturalistes*, Archives (Paris: 2019).

Klein U. – Lefèvre W., *Materials in Eighteenth-Century Science: A Historical Ontology*, Transformations: Studies in the History of Science and Technology (Cambridge, USA: 2007).

Klein U. – Spary E. (ed.), *Materials and Expertise in Early Modern Europe: Between Market and Laboratory* (Chicago – London: 2010).

Kovacs A., "Au-delà de la diététique", *Histoire, médecine, santé*, 17 (2021) 9–21.

Lafont O., "Une courte lettre révélatrice des préoccupations des apothicaires du XVIIIe siècle", *Revue d'Histoire de la Pharmacie* 85.313 (1997) 35–40.

Lafont O., "Nicolas Lémery, providence des bibliophiles", *Revue d'Histoire de la Pharmacie* 96.363 (2009) 267–276.

Laughran M.A., "Medicating with or without 'Scruples': The 'Professionalization' of the Apothecary in Sixteenth Century Venice", *Pharmacy in History* 45 (2003) 95–107.

Lehman, "Mid-Eighteenth-century Chemistry in France as Seen Through Student Notes from the Courses of Gabriel-François Venel and Guillaume-François Rouelle", *Ambix* 56.2 (2009) 163–189.

Lehman C., "Les multiples facettes des cours de chimie en France au milieu du XVIIIe siècle", *Histoire de l'éducation* 130 (2011) 31–56.

Leong E., "Making Medicines in the Early Modern Household", *Bulletin of the History of Medicine* 82 (2008) 145–168.

Leong E., *Recipes and Everyday Knowledge: Medicine, Science, and the Household in Early Modern England* (Chicago: 2018).

Long P., *Artisans/Practitioners and the Rise of the New Sciences, 1400–1600* (Corvallis: 2011).

Lunel A., *La maison médicale du roi : XVIe–XVIIIe siècles* (Seyssel: 2008).

Marraud M., "Permanences et déplacements corporatifs dans la ville", *Histoire & mesure* 25 (2010) 3–46.

Marraud M., *Le pouvoir marchand : corps et corporatisme à Paris sous l'Ancien régime* (Ceyzerieu: 2021).

Perkins J., "Chemistry Courses and the Construction of Chemistry, 1750–1830", *Ambix* 57 (2010) 1–1.

Perkins J., "Chemistry Courses, The Parisian Chemical World and The Chemical Revolution, 1770–1790", *Ambix* 57 (2010) 27–47.

Planchon F.G., *Le Jardin Des Apothicaires de Paris* (Paris: 1895).

Principe L.M. (ed.), *New Narratives in Eighteenth-Century Chemistry*, Archimedes (Dordrecht: 2007).

Principe L.M., "Rêves d'or. La surprenante longévité de l'alchimie au cœur de la chimie", *L'actualité chimique* 424 (2017): 68–71.

Pugliano V., "Pharmacy, Testing, and the Language of Truth in Renaissance Italy", *Bulletin of the History of Medicine* 91 (2017) 233–273.

Rappaport R., "G.-F. Rouelle: An Eighteenth-Century Chemist and Teacher", *Chymia* 6 (1960): 68–101.

Rieder P. – Zanetti F., "Le remède et ses usages historiques (1650–1820)", *Histoire, médecine et santé* 2 (2012) 9–19.

Rieder P. – Zanetti F. (ed.), *Materia medica: savoirs et usages des médicaments aux époques médiévales et modernes* (Geneva: 2018).

Rivest J., *Secret Remedies and the Rise of Pharmaceutical Monopolies in France During the First Global Age* (Ph.D. dissertation, Johns Hopkins University: 2016).

Roberts L., "The Death of the Sensuous Chemist: The 'New' Chemistry and the Transformation of Sensuous Technology", *Studies in History and Philosophy of Science* 26 (1995) 503–529.

Roberts L. – Werrett S. (ed.), *Compound Histories: Materials, Governance and Production, 1760–1840* (Leiden – Boston: 2017).

Roche D., *Histoire des choses banales : naissance de la consommation dans les sociétés traditionnelles, XVIIe–XIXe siècle* (Paris: 1997).

Secrétan C., "Un aspect de la chimie prélavoisienne: (le cours de G.-F. Rouelle)", *Mémoires de la société vaudoise des Sciences naturelles* 7 (1943) 219–444.

Shaw J. – Welch E., *Making and Marketing Medicine in Renaissance Florence*, Clio Medica 89 (Amsterdam – New York: 2015).

Simmons A., "Trade, Knowledge, and Networks: the Activities of the Society of Apothecaries and its Members in London, c.1670–c.1800", *British Journal for the History of Science* 52 (2019) 273–296.

Simon J., "Analysis and the Hierarchy of Nature in Eighteenth-Century Chemistry", *British Journal for the History of Science* 35 (2002) 1–16.

Simon J., "Pharmacy and Chemistry in the Eighteenth Century: What Lessons for the History of Science?", *Osiris* 29.1 (2014) 283–297.

Simon J., *Chemistry, Pharmacy and Revolution in France, 1777–1809* (London: 2016).

Smith P.H., *The Body of the Artisan: Art and Experience in the Scientific Revolution* (Chicago – London: 2004).

Spary E.C., *Eating the Enlightenment: Food and the Sciences in Paris* (Chicago: 2012).

Stewart L., "Science, Instruments, and Guilds in Early-Modern Britain", *Early Science and Medicine* 10.3 (2005): 392–410.

Strocchia S., "The Nun Apothecaries of Renaissance Florence", *Renaissance Studies* 25 (2011) 627–647.

Viel C., "L'enseignement de la chimie et de la matière médicale aux apothicaires aux XVIIᵉ et XVIIIᵉ siècles", *Revue d'Histoire de la Pharmacie* 87 (1999) 63–76.

Wallis P., "Consumption, Retailing, and Medicine in Early-Modern London", *The Economic History Review* 61 (2008) 26–53.

Warolin C., *Le cadre de vie professionnel et familial des apothicaires au XVIIᵉ siècle* (Ph.D. dissertation, Sorbonne Paris IV: 1994).

Werrett S., *Thrifty Science: Making the Most of Materials in the History of Experiment* (Chicago – London: 2019).

PART 2

Early Modern Fire Technologies

∵

CHAPTER 5

The Kitchen Fire (Eighteenth and Early Nineteenth Century)

Gianenrico Bernasconi

Abstract

The technology of fire management, which is an essential aspect of food transformation, is still largely understudied. An analysis of eighteenth-century cooking books, dictionaries, and discussions of fuel, showcases the knowledge of heat degrees, the possibility of choice among sources of fuel, the technical gestures and manipulations used to control fire – a task which appears to have been of primary importance. The eighteenth-century cooking practice is marked by the transition from the fire of the hearth of the fireplace to the fire enclosed in the metallic structure of the stove and the oven. The study of the projects of "cooking machines" submitted to the Paris Academy of Sciences displays some of the key steps leading to a "domestication" of fire – a process in which the innovation of heating systems and steam engines had an equally relevant place. All these efforts aimed in fact at finding a better way of using fire and economizing fuel, as a solution to the wood shortage that endured throughout the eighteenth-century. The projects for a new technology of fire also reveal the need of saving time and rationalizing food preparation, which is connected to the emergence of a bourgeois idea of cooking at the turn of the nineteenth century.

Keywords

fire – fireplace – cooking – technology – artisanal knowledge

Several historians have examined the domestic uses of fire in the early modern period, but they have generally privileged heating and lighting over cooking practices.[1] The processing of food required specific skills to reduce fuel con-

---

1 Jandot O., *Les délices du feu. L'homme, le chaud et le froid à l'époque moderne* (Ceyzérieu: 2017); Castelluccio S., *L'éclairage, le chauffage et l'eau aux XVIIᵉ et XVIIIᵉ siècles* (Montreuil: 2016). On the kitchen, see Williot J.-P. (ed.), *Du feu originel aux nouvelles cuissons. Pratiques,*

sumption, to guarantee the continuity of cooking procedures, and to respect the different passages of food preparation while preserving the qualities of the various ingredients, according to the precepts of the art of taste and of dietetic literature. The techniques of fire management were based on a material culture and on a specific know-how, of which only traces have survived. Such techniques are therefore an extremely complex object of inquiry – that which explains, along with the supposed immobility of these techniques, the shortfall in the scholarly literature. Between the second half of the eighteenth century and the beginning of the nineteenth century, the fire used in the kitchen underwent an important transformation: from the open space of the hearth, it was enclosed in the metal frames of ovens and stoves, which captured the flame and conducted heat.[2] Although it was not an invention of the Enlightenment, as proved by the ovens and furnaces published in *Holzsparkunst* (1618) by the German Franz Kessler (1580–around 1650),[3] the enclosure of fire is one of the major issues in the history of technology in the eighteenth century, given the energy crisis associated with wood shortages, which demanded a more efficient use of heat, and stimulated by the experiments to control the power of fire.[4]

In the kitchen, the novelty of ovens and stoves did not only lie in a better use of fuel. Enclosing fire in a metal container also made it possible to give autonomy to the cooking process, which heralded the transition from a 'technique of gestures' to a 'technique of the machine', a transition that made it possible to economize time and work, and even to experiment with new forms of food preparation in which culinary art played a marginal role. Of course, this phenomenon took hold slowly and encountered resistance. The fire in the hearth still fascinated people, the chimney was still very much present in homes of the first decades of the nineteenth century, and projects for ovens and furnaces took time to develop into a real industry. However, as Sigfried Giedion has pointed out, the enclosure of the kitchen fire opened up a new page in the

---

techniques, rôles sociaux (Bruxelles: 2015); Montanari M., "Cuocere, non cuocere, cucinare. Il fuoco nelle pratiche e nelle ideologie alimentari dell'alto Medioevo", in *Il fuoco nell'alto Medioevo* (Spoleto: 2013) 711–730.

2 Brewer P.J., *From Fireplace to Cookstove. Technology and the Domestic Ideal in America* (Syracuse: 2000).

3 This text had been translated into French: Keslar François, *Epargne-bois, c'est-à-dire nouvelle & par-ci devant non-commune ni mise en lumiere, invention de certains & divers fourneaux artificiels* (Oppenheim, Jean-Théodore de Bry: 1619).

4 Kitsikopoulos H., *Innovation and Technological Diffusion. An Economic History of Early Steam Engines* (London: 2019); Gras A., *Le choix du feu. Aux origines de la crise climatique* (Paris: 2007).

THE KITCHEN FIRE (EIGHTEENTH AND EARLY NINETEENTH CENTURY) 149

history of domestic technology, and in the process of transforming fire into heat or work.[5]

This article aims to document this passage, exploring at first the techniques of fire in the hearth with the aim of restoring their richness, which can be seen in the language of the recipes, in the material culture and in the vernacular knowledge of the different fuels. I will also highlight the problems associated with this technical culture, which required a great investment of time and work. In the second part of the article, the issues of fire enclosure will be studied through the analysis of projects of ovens and furnaces presented to the Paris Academy of Sciences during the eighteenth century, and of the reports written by academicians charged with examining such projects, which are outstanding documents for investigating the history of technical knowledge about fire and cooking in that period. Examples from the early nineteenth century will also be used to complete the analysis. The chapter will focus on devices, often portable, that contained fire in metal frames.

The aim of this study is to show that domestic techniques, and particularly those concerning food preparation, were not only the convergence point of knowledge and practices relating to fire management, but more generally represented catalyzers of innovation, well beyond the simple research for solutions to afford an economy in fuel.

1 Fire in the Hearth: the Techniques of Fire Management

According to the German physician and chemist Johann Heinrich Pott (1692–1777), the 'kitchen fire' is a 'coarse fire', an 'impure fire'; one of its main properties is that it 'spreads out on all sides', which distinguishes it from the rotating movement of the 'pure fire'.[6] Despite its coarse character, the kitchen fire has its place in scholarly classifications. Jean Henri Samuel Formey (1711–1797) in his *Encyclopédie portative, ou science universelle, à la portée de toute le monde* (1758), published a list of fires, in which the kitchen fire, which 'takes its substance from wood, which is used for boiling, cooking, and roasting', is the only 'artisanal' fire to be mentioned alongside the 'solar fire', which produces 'light and heat on the earth', 'natural fire', 'which is found in men, in

---

5 Giedion S., *Mechanization Takes Command: a Contribution to Anonymous History* (New York: 1969) 527.
6 Pott Johann Heinrich, "Supplément de quelques remarques qui touchent la dissertation du feu d'un auteur moderne fort célèbre", in *Lithogéognosie ou examen chimique des pierres et des terres* (Paris, chez Jean-Thomas Herissant: 1753) 405.

animals, and in plants', or again the 'fire of war and justice', the Greek fire, the 'fire of joy, or fireworks', the 'underground fires, of which are the mountains vomiting fire' and those 'aerial, of which are the ardens, the falling stars'.[7] Even the chemist Gabriel-François Venel (1723–1775) attributed to cooking a central place in the practical knowledge of fire. In the article *Feu (chimie)* published in 1756 in the *Encyclopédie*, Venel provided a synthesis of the knowledge and techniques of fire, which he defined as one of the 'materials or principles of the composition of bodies' and as the principle of heat.[8] While the article describes in detail 'the art of chemical fire', it does not omit to recall that 'the true books of this science are the laboratories of the chemists, the various factories where mines, metals, salts, stones, earths, etc. are worked by means of fire; the shops of all the workmen who practice chemical arts, as dyers, enamellers, distillers, etc., and the office and kitchen can furnish many useful lessons on this point'.[9]

However ordinary it might seem, the kitchen fire thus attracted the attention of eighteenth-century scientists, though they never described this fire very precisely. The techniques associated with the management of the cooking fire in the hearth belong in fact to those everyday practices that it is extremely difficult to document. In order to understand these techniques, it is useful to follow the approach of the so-called *Historische Praxeologie*, which combines the analysis of material documents with discursive traces to reconstruct the history of practices.[10] From this methodological perspective, cooking literature offers an interesting starting point for the inquiry. If recently scholarly work attests to the interest in recipe books for understanding the modes of codification and transmission of practical knowledge in the domestic space, the role of this literature as a source for a history of cooking techniques should be further explored.[11]

A survey of seventeenth- and eighteenth-century cooking literature, especially from France, makes it possible to identify the evolution of patterns in

---

7   Formey Jean Henri Samuel, *Encyclopédie portative ou science universelle, à la portée de tout le monde* (Berlin, chez tous les librairies: 1758) 62–63.

8   Venel Gabriel François, "Feu", in Diderot Denis – Le Rond d'Alembert Jean, *Encyclopédie, ou Dictionnaire raisonné des sciences, des arts et des métiers*, vol. 6 (Paris, chez Briasson et al.: 1756) 609–612.

9   Ibidem, 612.

10   Haasis L., Rieske C. (eds.), *Historische Praxeologie. Dimensionen vergangenen Handelns* (Paderborn: 2015); voir aussi Bernasconi G., "Pour une archéologie des pratiques. Mesure du temps, corps et prestation (XVIIIᵉ–XXᵉ siècle)", *Socio-anthropologie* 40 (2019) 247–262.

11   Leong E., *Recipes and Everyday Knowledge. Medicine, Science, and the Household in Early Modern England* (Chicago: 2018); Wall W., *Recipes for Thought. Knowledge and Taste in the Early Modern English Kitchen* (Philadelphia: 2016); Di Meo M., Pennell S. (eds.), *Reading and Writing Recipe Books, c.1500–1800* (Manchester: 2013).

THE KITCHEN FIRE (EIGHTEENTH AND EARLY NINETEENTH CENTURY)   151

the vocabulary of fire, which showcases an increasing complexity over time.[12] While in the seventeenth century the terms 'fire (*feu*)'[13] and 'small fire (*petit feu*)' were the most used, the eighteenth century enriched the vocabulary of fire with other terms, such as 'moderate fire (*feu modéré*)', 'very small fire (*très petit feu*)', 'clear fire (*feu clair*)', 'lively fire (*feu vif*)', 'slow fire (*feu lent*)', 'hot ashes (*cendres chaudes*)', and 'soft fire (*feu doux*)'.[14] Dictionaries also attest to this vocabulary. In 1727, Antoine Furetière's *Dictionnaire universel* defines 'small fire (*petit feu*)' as a term designating the 'cooking of certain things, which are prepared with a fire that is not violent and that acts only slowly', while a 'good fire (*bon feu*)' is, on the contrary, 'a large and ardent fire that acts strongly'.[15] The *Dictionnaire de l'Académie* of 1798 mentions 'hell fire (*feu d'enfer*)', which is used to describe a 'very hard' fire employed for grilling.[16] The use of these terms in cooking books follows a combinatorial logic – 'put it on hot ashes on a very low fire',[17] 'cook gently over hot ashes on a low flame'[18] – which suggests a double function of these qualifications. They designate both the materiality of fire – such as in the case of 'hot ashes', 'lively fire', and 'clear fire' – and a degree of heat – such as in 'low fire', 'very low fire', 'moderate fire', and 'high fire'. Different cooking operations are associated with a particular degree of heat: to 'stew (*mitonner*)' means 'to cook or heat something over a low fire in some

---

12   For the seventeenth century, these fire qualifications are taken from La Varenne François Pierre, *Le cusinier françois* (Paris, chez Pierre David: 1651), et de Lune Pierre, *Le cuisinier* (Paris, chez Pierre David: 1656) and for the eighteenth century from [Marin François], *Suite des Dons de Comus, ou l'art de la cuisine, réduite en pratique* (Paris, chez Pissot, Didot, Brunet fils: 1742). For an analysis of the vocabulary of fire, see also the article by Marco Storni, "Without a Thermometer: The Technical Knowledge of Heat in the Early Modern Age", in this volume.

13   For terms that were difficult to translate, the original French version has been added in brackets.

14   Alexandre-Bidon D., *Une archéologie du goût: céramique et consommation* (*Moyen Âge-Temps modernes*) (Paris: 2005) 219. This vocabulary of heat gradation is also found in other practices associated with the use of fire, such as chemistry, which in the early modern age shared the material culture and the practising spaces with cooking: see Werrett S., *Thrifty Science. Making the Most of Materials in the History of Experiment* (Chicago – London: 2019), 42 and 50–55; Guerrini A., "The Ghastly Kitchen", *History of Science* 54.1 (2016) 71–97; Cooper A., "Homes and Households", in: Park K., Daston L. (ed.), *Cambridge History of Science*, vol. 3: *Early Modern Science* (Cambridge: 2006) 224–237.

15   Furetière Antoine, *Dictionnaire Universel contenant tous les mots françois*, vol. 2 ([La Haye: 1727] Hildesheim – New York: 1972), "feu".

16   *Dictionnaire de l'Académie françoise … éd. revue, corrigée et augmentée par l'Académie elle-même*, vol. 1 (Paris, chez J.J. Smits: 1798) 580.

17   [Marin], *Suite des Dons de Comus*, vol. 2, 80.

18   Ibidem, vol. 2, 404.

sauce or liquor',[19] to 'sweat (*suer*)' means 'to let out moisture',[20] while 'to simmer (*mijoter*)' means 'to cook slowly and gently',[21] which requires a tempered heat, often achieved through the use of hot ashes. If the recipes prescribe a specific degree of heat, they also provide information about its variation during cooking. The recipe for 'plain rice (*riz au naturel*)' in Pierre de Lune's *Le cuisinier* (1653) is an example of this variation:

> After having hulled your rice well, wash it with hot water, and make it dry in the oven or in front of the fire, and put it with boiling milk, and let it cook on a low fire, when it is swollen crush it with a wooden spoon, put back boiling milk, season it with salt, and put sugar in it, and keep it on hot ashes.[22]

To change the intensity of fire, one can use specific substances, such as fat, which makes the fire hotter, or salt, which puts it out.[23] Sometimes, the attention is not focused on the degree of fire but on the constancy of the heat, as in the recipe provided in the *Suite du Dons de Comus* to make candied cherries. These must be cooked 'over a good, even fire', a fire that requires a special care: 'make sure that your fire is not neglected'.[24] As the cooking often lasts several hours, the fire requires continuous control from the kitchen staff in order to maintain a constant degree of heat.[25] This control includes gestures such as blowing, lighting, fanning, de-flaming, extinguishing, maintaining, covering the fire, for which specific tools exist, such as andirons, shovels, pincers, tweezers or even bellows.[26] The fire is lit with a wick or a match, the embers are handled, moved, covered, in order to favor or hinder combustion, or adjust and distribute the heat. The ember holder (*porte-braise*), 'a large iron shovel or pan with a perforated lid', is used to move embers,[27] while the 'ember clamp'

---

19   Furetière, *Dictionnaire universel*, "mitonner".

20   Diderot – d'Alembert, *Encyclopédie ou Dictionnaire raisonné*, vol. 15, "suer" 625.

21   *Dictionnaire de l'Académie française*, vol. 2, "mijoter" 105.

22   De Lune Pierre, *Le cuisinier* (Paris, chez Pierre David: 1653) 333.

23   Tieron, "Lettre à M. ..., Administrateur de l'Hôpital de ...", *Journal de Paris*, 7 September 1783, 1131.

24   [Marin], *Suite du Dons de Comus*, vol. 2, 10.

25   Ibidem, vol. 2, 54 and 70.

26   For a description of these gestures and tools, see Pierre Richelet, *Dictionnaire françois, contenant généralement tous les mots tant vieux que nouveaux* (Chez Jean Elzevir, Amsterdam: 1706) "feu" 353, *Dictionnaire de l'Académie françoise*, vol. 1 (Paris: 1777) "feu" 482, See also Roche D., *Le peuple de Paris: essai sur la culture populaire au XVIIIᵉ siècle* (Paris: 1998) 185.

27   Inventaire générale des monuments et des richesses artistiques de la France, *Objets civils domestiques: vocabulaire* (Paris: 1984) 456.

THE KITCHEN FIRE (EIGHTEENTH AND EARLY NINETEENTH CENTURY)   153

(*resserre-braises*), a 'U-shaped or arched metal blade',[28] is used to hold them in order to applicate the *cale-pot*, a 'semi-circular metal or pottery instrument [...] allowing to wedge a container in the embers of the hearth'.[29] Sometimes heated tools are used to complete the cooking process, a technique often used in the *Cuisinier françois* de La Varenne, especially to prepare eggs: to 'fry egg whites, and put them on your soup, redden the shovel on the fire, and pass it over them to complete the cooking'.[30] Other pieces of cookware can also be used for a specific purposes, such as the '*tourtière*', 'which heats up from above and below',[31] or the '*pommier*', 'a semi-circular container' which, when placed near the hearth, allows an indirect use of the heat.[32] There are also various devices for controlling the cooking process by varying the distance from the fire, such as the rack, which was replaced in the eighteenth century by the tripod, or the grilling fork, whose long iron rod allowed the meat to be exposed to the flame from all sides.[33]

Fuel, which is rarely mentioned in cookery books, offers another path for the study of the knowledge and skills needed to manage fire. In the literature, we find a specification of the types of wood that are best for cooking food, since the quality of fuel, as the *Almanach des gourmands* reminds us at the beginning of the 19th century, 'is important for a good kitchen'.[34] The *Dictionnaire universel de commerce* notes that among the 'driftwood (*bois flotté*)' namely that which 'is transported on trains and (...) on the rivers', there is one variety, the '*bois de traverse*', 'pure beech, without bark, which is destined for bakers and pâtissiers, who use it to heat their ovens'.[35] In *Les délices de la campagne*, the following advice is given regarding oven heating: 'In order that the heat penetrates the walls, the splinters of large dry wood, particularly beech, are much better than

---

28   Ibidem, 470.

29   Ibidem, 464.

30   Varenne, *Le cusinier françois* 161.

31   Diderot – d'Alembert, *Recueil de planches, sur les sciences, les arts libéraux, et les arts méchaniques* (Paris, chez Briasson et al.: 1763), "Confiseur", planche II.

32   Quellier F., "La cuisson des fruits en France à l'époque moderne", in Williot (ed.), *Du feu originel* 120.

33   Pardhaillé-Galabrun A., *La naissance de l'intime. 3000 foyers parisiens, XVIIe–XVIIIe siècles* (Paris: 1988) 288–289.

34   *Almanach des gourmands: servant de guide dans les moyens de faire excellente chère ... par un vieil amateur*, 1812, 324.

35   Savary De Bruslon Jacques, *Dictionnaire universel de commerce, contenant tout ce qui concerne le commerce qui se fait dans les quatre parties du monde ... Ouvrage posthume de Sr Jacques Savary Des Bruslon*, vol. 1 (Amsterdam, chez les Jansons: 1726) 380.

the faggot'.[36] Henri-Louis Duhamel du Monceau (1700–1782), author of important works on wood and the forest in the 18th century, confirms that the '*bois de traverse*' is preferred 'by cooks, pastry cooks, bakers and brewers', because it 'burns well, makes a nice flame and produces little smoke'.[37] Similarly, hornbeam is also appreciated for cooking as it 'makes a good, clear fire that lasts and gives good heat, good charcoal and good ashes'.[38] However, the choice of wood was not only based on quality criteria, but above all on economic requirements, as Steven Kaplan has shown for the bakers of the eighteenth century.[39] One can therefore imagine that many of them used the '*bourrée*', namely 'the smallest and worst wood, which catches fire quickly, but which lasts little: it is used to heat the oven'.[40] Sometimes it is the size of the logs that is indicated, as Jourdan Le Cointe does in his *Cuisine de santé* (1790): 'The wood used for roasting must be dry and of medium size, about the same size as the upper arm; it must flare up and give a clear fire, which penetrates the meat by degrees without scorching it or roasting its surface quickly before the interior is cooked for the same reason'.[41] Even the arrangement of the logs in the fireplace can be the subject of particular attention. The agronomist Alexis Costa de Beauregard (1726–1797) remarked that 'throughout Italy, where wood is generally scarce [...]. Instead of holding the fire against the hearth, which causes almost half of this initial heat to be lost, it is the end of the logs that is there. And the fire is directed along the length of the kitchen in such a way that almost all the heat is used up'.[42]

The availability of wood therefore depends on the income of the individual or on their place of residence. It is a good that is not stored, especially in cities like Paris, where only 10% of the 3,000 inventories studied by Daniel Roche mention a wood reserve.[43] Although – as recalled by Venel in the *Instruction sur l'usage de la houille* (1775) – 'wood is, in all countries [...] the common,

---

36   De Bonnefons Nicolas, *Les délices de la campagne, suitte du Jardinier françois* (Amsteldan [sic], chez Raphael Smith: 1655) 5.

37   Duhamel de Monceau Henri-Louis, *Du transport, de la conservation et de la force des bois* (Paris, chez L.F. Delatour: 1767) 29.

38   Liger Louis, *La nouvelle maison rustique, ou Economie générale de tous les biens de campagne*, vol. 1 (Paris, chez la veuve Savoye: 1768) 830.

39   Kaplan S.L., *Le meilleur pain du monde: les boulangers de Paris au XVIIIᵉ siècle* (Paris: 1996) 99, and Kaplan S.L., "The times of bread in eighteenth-century Paris", *Food and Foodways: Explorations in the History and Culture of Human Nourishment* 6.3–4 (1996) 292.

40   Diderot – d'Alembert, *Encyclopédie ou Dictionnaire raisonné*, vol. 6, "fagot" 371.

41   Le Cointe Jourdan, *La cuisine de santé* (Paris, chez Briand: 1790) 15.

42   Costa de Beauregard Alexis, *Essai sur l'amélioration de l'agriculture dans les pays montueux et en particulier dans la Savoye* (Chambéry, de l'imprimerie de M.F. Gorrin: 1774) 250.

43   Roche D., *Histoire des choses banales. Naissance de la consommation, XVIIᵉ–XIXᵉ siècle* (Paris: 1997) 146.

THE KITCHEN FIRE (EIGHTEENTH AND EARLY NINETEENTH CENTURY)   155

ordinary and exclusive aliment of fire',[44] there were regions where other fuels
were used, such as in England where, from the 16th century onwards, the use
of coal and peat for cooking was widespread, especially in poor households,
because of the high price of wood for domestic use.[45] In the eighteenth century,
in response to the shortage of wood, the use of these fuels was also proposed
in France, but its introduction was not very successful.[46] In 1772, the States of
the Province of Languedoc drew up a body of instructions on the use of coal.
It was in this context that Venel published the abovementioned *Instruction sur
l'usage de la houille* where he devoted an entire chapter to cooking. In order
to prove the advantages of coal, Venel compared coal fire with wood fire in
cooking, using the example of roasting. Venel praised the constancy of the coal
fire, which 'once it has reached its blazing state, perseveres in the same state,
without the addition of any new material, for several hours', which allows carrying out cooking operations 'without any particular attention on the part of
the roaster, and consequently without him being able to fail by his negligence'.
The wood fire, on the other hand, has the disadvantage of 'becoming uneven,
because it consumes its aliment very quickly; that to maintain it, it is necessary
to put some [wood] back into the hearth at all times; that this operation varies
the activity of the fire, and finally that the negligence of the roaster on this
point, can cause this fire to work by impetus, or alternations of slowing down
and excess, which can only be detrimental to the operation'.[47]

However, the characteristics of coal are not always suited to the requirements of cooking, as Johann Friedrich von Pfeiffer (1718–1787) stressed with
regard to its use in baking: 'those who would like to use purified earthen charcoal for objects that require prompt heat and a lively and flaming fire, such as,
for example, a baker's oven, would be deceived in their expectations, because
purified earth coal and peat do not give, as doesn't charcoal, a blazing fire, but
only a fixed fire, that is to say, in a state of blazing or ardent ignition'.[48]

---

44   Venel Gabriel François, *Instructions sur l'usage de la houille* (Avignon, chez Gabriel
      Regnault: 1775) II.
45   Pennell S., *The Birth of the English Kitchen, 1600–1850* (London – New York: 2016) 61–69;
      Zylberberg D., "Fuel Prices, regional diets and cooking habits in the English industrial
      revolution (1750–1830)", *Past & Present* 229 1 (2015) 91–122.
46   Abad R., "L'Ancien Régime à la recherche d'une transition énergétique?", in Bouvier Y.
      Laborie L. (eds.), *L'Europe en transitions. Energie, mobilité, communication, XVIIIe–XXIe
      siècles* (Paris: 2016) 56 and ff.
47   Venel, *Instructions sur l'usage de la houille* 329–330.
48   Pfeiffer Johann Friedrich, *Histoire du charbon de terre et de la tourbe* (Paris, chez P. de
      Lormel: 1787) 182. This work was published in German in 1775 under the title *Geschichte
      der Steinkohlen und des Torfs*.

Wood remained the main domestic fuel in France until the mid-nineteenth century, certainly for reasons of coal supply, but also because of a mistrust of coal, as Jean-François Clément Morand (1726–1784) remarked when he spoke of 'the distance that one should observe when one approaches the meat to be roasted' with coal, since 'it seems quite reasonable to think that by not observing a certain rule, which is easily acquired through practice, the bituminous emanations, by penetrating into the meat, could soften it, and even alter its taste'.[49]

The archives and cookbooks only provide a fragmentary knowledge of the techniques of fire control in the hearth, which emerge from the diversity of the denominations, the relationship between the degree of heat and the method of cooking, the richness of the material culture, and the knowledge of fuels. These traces also reveal the importance of the work required to manage this irregular fire, which must be constantly fed and controlled.

2 Enclosing Fire: Kitchen Stoves and Ovens

On 7 September 1783, the *Journal de Paris* published a letter from Tieron, bursar of the Collège Royal de Louis-le-Grand, which carries an exceptional description of daily work in a kitchen in the second half of the eighteenth century:

> The kitchen was like a hell in the afternoon of the days when meat was eaten. After having used part of the morning to dry the wood near a track, and having split it into pieces, it was arranged in the whole length of the chimney. Six courses of spits, one on top of the other, which had to run together, being continually stopped by the hooks, forced the cooks to work hard. [...] [T]he necessity of a clear and burning fire made them continually throw away the fats from the pans, and spread salt by the handful. From an operation as laborious as it was expensive, the result was that every day the meat was burnt, and the others half-cooked: in short, the discontent was universal and very well founded.[50]

In the rest of his letter, Tieron explains the benefits of the introduction of the 'new stoves invented and built by Sieur Roché de Chavagnac', which 'not only

---

49 Morand Jean-François-Clément, *L'art d'exploiter les mines de charbon de terre*, in *Description des arts et métiers, faites et approuvées par messieurs de l'Academie des sciences*, vol. 2 (Paris, chez Saillant, Nyon et Desaint: 1776) 3.

50 Tieron, "Lettre à M. ..., Administrateur de l'Hôpital de ..." 1131.

THE KITCHEN FIRE (EIGHTEENTH AND EARLY NINETEENTH CENTURY)   157

occupy the least possible space and protect the service from all fears of fire', but also make it possible to save 'half on the consumption of wood' and to carry out the service 'without fatigue and with the greatest accuracy'.[51] The criticism of fire in the hearth, which emerges from this description of the kitchen of the Collège Louis-Le-Grand, regularly recurs in projects for kitchen ovens of the eighteenth century. In January 1761, while presenting his *Nouveau Foyer de cuisine* in the *Journal oeconomique*, Thomas-Ignace Vanière, author of a *Cours de latinité* published in two volumes between 1759 and 1773, remarked that 'the kitchen chimney is, without question, one of the most interesting parts of our houses, because of the considerable expense involved, the disastrous accidents that can and do occur in it, and its intimate and daily connection with health, since it is our own substance that is prepared in it'. The kitchen fire is however poorly controlled: 'The inconveniences resulting from this vicious form are so considerable, that it is difficult to understand, when we examine them closely, how so many centuries have passed without any thought being given to remedying them'.[52] The complaint about the poor management of the kitchen fire is still found at the beginning of the nineteenth century, a time when projects for ovens and stoves multiplied. In *L'art d'économiser le bois de chauffage et tous les autres combustibles* (1827), published by César Gardeton (1786–1831), one finds a French translation of an anonymous English article whose author complains that the fire in the hearth is 'too open', 'consumes much more wood than the chamber stoves' and is easily dispersed. Although the pot is on the fire, and although the fire, by placing the wood well, envelops the pot, so to speak, and the flame rises around it, the air nevertheless, which comes freely from all sides, seizes the greatest force of the fire, makes it rise and drives it out through the chimney.[53]

From the eighteenth century onwards, several projects for ovens and stoves aiming to correct these defects were proposed. Although a technique of enclosed fire had been used in baking and pastry-making since antiquity, and in cooking since the Renaissance following the introduction of the oven called *potager*, the proliferation of metal stoves met a need to save fuel that remained urgent throughout the age of the Enlightenment. However, the study of the projects of such machines presented to the Academy of Sciences shows that the use of fire went beyond the sole issue of fuel consumption, and became a

---

51   Ibidem.
52   *Nouveaux fourneaux économiques et portatifs*, tiré à part de la *Gazette de Santé* du dimanche 1ᵉʳ octobre 1780 (Paris, chez la Veuve Ballard & Fils: 1780).
53   Gardeton César, *L'art d'économiser le bois de chauffage et tous les autres combustibles* (Paris, chez L. Cordier et al.: 1827) 50–51.

field of experimentation with new forms of food preparation, which did not require the presence of a cook.

3 Multifunctionality and Fuel Economy

In the eighteenth century, the spread of the term '*économie*' signals the emergence of a technological discourse to which scholars, inventors, entrepreneurs and even consumers contributed. The term 'economic' refers to a consumer ethic in opposition to the ostentatious behavior of the aristocratic elite, but also announces a process of innovation in the rational use of resources, which affects both the productive and the domestic spheres.[54] The scarcity of wood, attributed to the development of fire manufacture (glassmaking, forge, etc.), the increase in domestic consumption and even the poor forest management made the saving of fuel through economic machines one of the major technical issues of the second half of the eighteenth and the first half of the nineteenth century.[55] This is the context of the new ovens and furnaces, which testify to the spread of a better knowledge of combustion and especially of the circulation of heat. Well before Antoine-Laurent Lavoisier's (1743–1794) research on combustion and fuels[56] or that of Benjamin Thompson, Earl of Rumford (1753–1814) on heat and kitchen stoves,[57] the publication of *La mécanique du feu* in 1713 by Nicolas Gauger (ca. 1680–1730), marked – as Olivier Jandot has stressed – a 'fundamental milestone in the history of the conquest of heat'.[58] Gauger's work, a central text in the historical process of 'materialisation of aerial fluids', which Marie Thébaud-Sorger has examined in depth,[59] inspired the work of scientists such as Benjamin Franklin (1706–1790) and had a great influence on the emerging discipline of 'caminology' in the nineteenth

---

54 For a discussion of the notion of "economy" see Storni M., "Time Thrift and Economic Science in the Eighteenth Century", *Science in Context* (forthcoming); Werrett, *Thrifty Science*; Spary E., *Feeding France. New Sciences of Food, 1760–1815* (Cambridge: 2014).

55 Abad, "L'Ancien Régime à la recherche d'une transition énergétique?".

56 Lavoisier Antoine, "Expériences sur l'effet comparé de différents combustibles", *Histoire de l'Académie royale des sciences, avec les mémoires de mathématique & de physique. Année 1781* (Paris: 1781) 379–394.

57 Thompson comte de Rumford Benjamin, *Mémoires sur la chaleur* (Paris, chez Firmin Didot: 1804–1805).

58 Jandot, *Les délices du feu* 221.

59 Thebaud-Sorger M., "Capturing the Invisible: Heat, Steam and Gases in France and Great Britain, 1750–1800", in Roberts L., Werrett S. (eds.), *Material, Governance and Production, 1760–1840* (Leiden: 2017) 85–105; Miller D.P., "A New Perspective on the Natural Philosophy of Steams and Its Relation to the Steam Engine", *Technology and Culture* 61.4 (2020) 1135.

century. Although Jandot has shown the importance of *La mécanique du feu* for the improvement of heating systems during the second half of the eighteenth century, it is still to be noted that Gauger's research also influenced the projects for the enclosure of fire in ovens and kitchen stoves. Indeed, the circulation of heat conditioned by Gauger through the use of metal frames, and a mastery of ventilation, allows the same fire to be used for the preparation of different foods. This transfer of knowledge from heating to cooking is also the activity of inventors who dealt with various problems concerning domestic techniques.[60]

In the reports of the commissioners charged by the Paris Academy of Sciences with the expertise of these cooking machines, the importance of the devices designed to govern the circulation of heat is highlighted. On 1 July 1739, the commissioners of the Academy examined a 'machine [...] to cook many different dishes at the same time', invented by François Fresneau, 'former surgeon of the king's ships and surgeon major of the hospitals'.[61] This stove, made of a hard plate and intended for ships, allowed 'considerable savings in the consumption of wood or coal'. It occupied little space on ships, did not weigh much, and could thus be moved easily. It was multifunctional, since Fresneau integrated 'a pot for soup, [...] a saucepan placed on a sand bath, another saucepan that fits on a water bath or steam bath, and a sheet metal tower where a lantern with four horizontal divisions can be placed, on which pies can be put to make four pieces of pastry at once'.[62] The cooking of several dishes at the same time was made possible by the arrangement of the metal parts of the oven, 'the flame is intercepted by an inclined piece of sheet metal', the 'fire', understood here as heat, 'circulates in the tower to cook the meats, without the meats themselves being damaged by the immediate action of the flame'. Furthermore, by means of baths – 'sand bath', 'bain-marie' or 'steam bath' – Fresneau managed to moderate the heat 'by an intermediary between the fire and the material'.[63] This technique underscores the proximity of cooking and chemistry, in which, as the famous chemist Nicolas Leméry (1645–1715) stressed, baths are used so that 'the vessel is gently heated'.[64] On 26 October 1742, the letters patent were published, by which the King granted Fresneau an exclusive privilege for thirty years, registered by Parliament on 11 February 1743, to 'build, sell and sell on credit [...] a stove invented by him

---

60   Gallo E., "La cuisson chez les inventeurs pionnier d'appareils de chauffage (1780–1830)", in Williot (ed.), *Du feu originel aux nouvelles cuissons* 215–226.

61   Archives de l'Académie des sciences, Procès-verbal de la séance du 1 juillet 1739.

62   Ibidem.

63   Diderot – d'Alembert, *Encyclopédie, ou Dictionnaire raisonné*, vol. 2, "Bain" 21.

64   Lémery Nicolas, *Cours de chymie contenant la manière de faire les opérations qui sont en usage dans la médecine, par une méthode facile* (Paris, chez Delespine: 1701) 55–56.

to carry out the different operations of cooking with a single fire'.[65] About ten years later, in 1752, Vanière, whose criticisms of fire in the hearth have been cited above, submitted for the Academy's approval 'a new kitchen stove', for which he obtained an exclusive privilege registered at the Parliament of Paris on 4 September 1760.[66] Vanière's stove was presented as a box two and a half feet long and two feet wide.[67] It was made of a 'wooden frame filled with panels, lined internally with tin. Its bottom, which hosts the fire inside, is made of a strong metal sheet'. With his invention, Vanière could economize 'the use of fire as much as possible, since he makes it serve four different sides, and he uses reflected heat'. Its advantage is the control of ventilation, as it 'can be governed very easily as in all the other furnaces by giving more or less passage to the air, either from below or from above'. Like Fresneau's stove, Vanière's stove is characterized by its multi-functionality, as it is capable of cooking different dishes at the same time, through a new management of fire. This resemblance was acknowledged by the Academy's commissioners, who concluded, however, that the two stoves differed in the type of fuel used. In fact, unlike Fresneau, Vanière used coal, which yielded a 'fire that is always proportionate [...] to the number and volume of the pieces that are roasted in it'.[68] Over the eighteenth century, other oven projects were presented to the Academy of Sciences, as well as economical metal stoves whose function was to cook several dishes at the same time, while reducing fuel consumption. This is the case of Nivert's 'portable stove (*cuisine portative*)' of 1782,[69] a tin or leather box which 'contains three large stoves which can be used on three sides at once, in addition to the fact that their lower part still serves to heat the upper part of the oven'. Another example is that of Pierre Tranoy's 'new, economical and portable kitchen stove (*nouveau foyer de cuisine, économique et portatif*)', which consisted of a piece of wooden furniture the interior of which is 'lined with English tinplate', and has the advantage 'that it is suitable for making roasts at the same time or separately, and for making them perfect on all occasions [...], it can also be used to make boiled dishes, stews, fish, chops, pies and biscuits'.[70] Tranoy underlines the economic character of his fireplace: 'with two sols of coal one can do what another could not do with twenty sols of wood'. Enclosed in metal frames, fire is transformed into heat, which the oven captures and directs to different

---

65   Archives de l'Académie des sciences, Procès-verbal de la séance du 20 février 1743.
66   Archives de l'Académie des sciences, Procès-verbal de la séance du 23 décembre 1752.
67   Ibidem.
68   Ibidem.
69   Archives de l'Académie des sciences, Procès-verbal de la séance du 4 septembre 1782.
70   Archives Nationales de France, O1294, Tranoy, Nouveau foyer de cuisine économique et portatif.

THE KITCHEN FIRE (EIGHTEENTH AND EARLY NINETEENTH CENTURY)

vessels, thus multiplying the effects of this machine, making it possible to cook several dishes at once and save fuel.

4 Saving Time: the Autonomy of Processes

In the eighteenth century, the measurement of time represents a broader phenomenon other than the sole question of the awareness and internalisation of time on which historiography has focused.[71] Antide Janvier (1751–1835), in his famous *Etrennes chronométriques pour l'an 1811*, drew attention to the fact that time measurement is not only 'the rule of society', which 'gives the first signal for prayer, opens the gates of cities, convenes assemblies, and starts work', but it also has 'uses in the positive sciences or in experiments'.[72] Through timepieces, processes are quantified by measuring their duration. Marco Storni's recent research has shown that the economy of time is one of the criteria that emerged during the eighteenth century as relevant to the mastery of machines in order to measure their performance by quantifying the duration of a process or its acceleration.[73] With kitchen ovens and stoves, saving time takes a different form. It is not only a parameter useful in quantifying machine performance, but a use value that might attract consumers, who were promised discharge from the time required for food preparation, thanks to the autonomy of these machines. The technology of the enclosed fire relieved cooks from the burden of governing the fire, and thus reduced the role of individual skill in cooking practice. The multifunctionality of cooking several dishes at once and even the choice of fuel also made it possible to save working time. A fire fueled by wood is in fact very demanding in terms of attention, insofar as it requires the cook to be constantly on the lookout: logs must be added and the flame adapted to the cooking method. Gabriel François Venel, in praising the use of charcoal in cooking, observed that the regularity and duration of its combustion freed the *rôtisseur* from the constraint of fire control.[74] These advantages of economical ovens and stoves heralded the emergence of a new bourgeois kitchen during the eighteenth century. In 1746, the publication of *La Cuisinière bourgeoise*, the

---

71 Verhoeven G., Blondé B., "Against the Clock. Time Awareness in Early Modern Antwerp, 1585–1789", *Continuity and Change* 28.2 (2013) 213–244; Voth H.-J., *Time and Work in England (1750–1830)* (Oxford: 2000); Dohrn-van Rossum G., *History of the Hour. Clocks and Modern Temporal Orders*, (Chicago – London: 1996).

72 Janvier Antide, *Etrennes chronométriques pour l'an 1811* (Paris, chez Courcier et al.: 1810) 122.

73 Storni, "Time Thrift and Economic Science".

74 Venel, *Instructions sur l'usage de la houille* 329–330.

most successful gastronomic book of the Enlightenment, inaugurated a simpler and more rational cooking practice.[75] It was not only a question of adapting recipes to the budget of the bourgeoisie, but also of taking into account a kitchen staff that was much smaller than that of the aristocratic hotels.

The advertisements, which present the public with economical ovens and stoves, make it possible to grasp the socio-technical dimension associated with the passage to an enclosed fire. In the *Journal oeconomique* of 1761, Vanière describes the use of his kitchen stove, which allows the 'least skillful' cook, 'after one or two experiments', 'never to miss the cooking point'. The cook's schedule is thus freed from the old constraints thanks to this machine which 'no longer requires the presence of the cook, as in our kitchens, where he is obliged almost always to have the shovel or the tongs in his hand when the spit is turning, and to share, in spite of having them, the heat of his fire with the meats which are roasting in there'.[76] In the rest of the article, Vanière explains the spirit of the work he is carrying out both on a literary level in his *Cours de latinité* and in his activity as an inventor: 'the dearest object of my ambition is to serve my country, and even to be, if I can, useful to the whole universe, by spreading facilities, conveniences, pleasures, and the saving of money and time, in intellectual and mechanical operations'.[77] In 1780, in an advertisement published in the *Gazette de santé*, Nivert used similar arguments to present his portable stove: 'We can see how useful and convenient such a stove can be for preparing food, the cooking of which requires no maintenance of fire, no care, no expense, no attention, and which can be left either in a fireplace or in a courtyard, and always with the certainty that they will be cooked to the appropriate point'.[78]

These projects for 'economical and portable stoves', whose objective, summarized in Henri-Gabriel Duchesne's (1739–1822) *Dictionnaire de l'industrie*, was to 'find an easy and convenient way of preparing food without difficulty, care or embarrassment',[79] thus confirm the emergence of a bourgeois cuisine,

---

75   Girard A., "Le triomphe de la 'la cuisinière bourgeoise'. Livres culinaires, cuisine et société en France aux XVIIᵉ et XVIIIᵉ siècle", *Revue d'histoire moderne et contemporaine* 24.4 (1977) 497–523, see also Forino I., *La cucina. Storia culturale di un luogo domestico* (Torino: 2019) 71.

76   Vanière, "Nouveau foyer de cuisine, oeconomique et portatif", *Journal oeconomique*, 1761, 30.

77   Ibidem, 31.

78   *Nouveaux fourneaux économiques et portatifs*, tiré à part de la *Gazette de Santé* du dimanche 1ᵉʳ Octobre 1780 (Paris, chez la Veuve Ballard & Fils: 1780).

79   Duchesne Henri-Gabriel, *Dictionnaire de l'industrie, ou Collection raisonnée des procédés utiles dans les sciences et dans les arts*, vol. 3 (Paris, chez Poigné et al: 1800, 3rd ed.) 96.

THE KITCHEN FIRE (EIGHTEENTH AND EARLY NINETEENTH CENTURY)   163

which required less work and therefore less staff than the aristocratic one. In particular, the use of coal in stoves was, according to Jean-François-Clément Morand,

> A great advantage for day laborers, or for poorly off individuals, without servants, or who have only one. This fire is not subject to variation or disturbance [...]; it relieves the former from having to look after the fire while their food is cooking: for the latter, their servants are not distracted from the other occupations of the household.[80]

At the beginning of the nineteenth century, the question of time saving continued to play an important role in the public presentation of kitchen stoves, whose economy was part of a more general project of social reform. In 1821, Charles Harel (1772–1853), manufacturer of economical stoves, for which he had improved a project of his friend, the famous chemist and philanthropist Antoine-Alexis Cadet de Vaux (1743–1828),[81] but also an admirer of the philosopher and reformer Charles Fourier (1772–1837) and a member of the Fourierist movement,[82] published a *Description des divers appareils propres à économiser le temps et le combustible*. In this text Harel attributed to the machines, which use steam for cooking, a social function that is realized through the saving of time.[83] Harel explains the 'manner of conducting the oven', which can cook a pot with only two loads of coal provided that the circulation of air is well controlled. Having respected these instructions 'one can then leave one's pot for 4 to 5 hours; this advantage is invaluable to many people, and it is the one that made me sell the most ovens'.[84] Like Morand, Harel evokes the organization of the household 'of people of mediocre fortune', to whom his appliances are addressed:

---

80   Morand Jean-François-Clément, "Mémoires sur les feux de houille, ou charbon de terre", in Morand Jean-François-Clément, *L'art d'exploiter les mines de charbon de terre*, partie 2, section 4, Description des arts et métiers, faites et approuvées par messieurs de l'Academie des sciences (Paris, chez Saillant, Nyon et Desaint: 1776) 4.

81   Cadet-de-Vaux Antoine-Alexis, *Fourneau-Potager économique ... suivi d'Observations sur l'application de cet appareil à tous les besoins du ménage et sur un poêle-fourneau par M. Harel* (Paris, chez Collas et al.: 1807).

82   Desmars B., "Harel Charles (Louis)", http://www.charlesfourier.fr/ (accessed on 2 November 2022).

83   Harel is also the author of *Ménage sociétaire son bien-être en diminuant sa dépense* (1839), a work in which he lays the foundations for residential utopias.

84   Harel Charles, *Description des divers appareils propres à économiser le temps et le combustible* (Paris, chez l'auteur: 1821) 14.

In this class, one often has only a servant; one sends them on errands, to do the housework, to take the children for walks; here one has nothing to fear from their absence, since it is proven by experience that food exposed to steam cannot burn, and that one can leave the stove for four to five hours. If this system is convenient for people who have only a maid, it is even more so for people who have none, or who only have a housekeeper.[85]

The enclosure of fire and the change of fuel gave autonomy to the machines and therefore to food processing. The fire no longer had to be constantly controlled, the heat became more regular, which freed the time of those in charge of the kitchen.

5 Artificial Spaces: Steam Cooking

In the *Encyclopédie*, Jacques-François de Villiers (1727–1790) defines stoves as 'utensils intended to contain the fire, and to apply this element as an instrument to the substances that one wishes to change by its action'.[86] By enclosing fire, ovens and furnaces create an artificial space in which the substances are transformed by the effect of heat or by the intermediary of other substances, such as steam, which is adequately conveyed and regulated.[87]

Towards the end of the 1670s, Denis Papin (1647–1713), who was an assistant to Robert Boyle (1627–1691), invented the digester, a device for cooking solid substances at high pressure, which was used to soften bones, an animal part generally discarded in cooking, in order to obtain a gelatin used to prepare soups for the poor.[88] The digester was conceived by Papin as both a laboratory instrument to transform solids and to study new properties of heat, and as a kitchen machine that aimed to improve food practice through the introduction of a new economical cooking technique and new forms of recipe standardization.[89] The danger of a high-pressure cooking technique limited the use of the digester, which nevertheless continued to circulate throughout the eighteenth century both in laboratories and to be discussed by philanthropic

---

85   Ibidem, 29–30.
86   Diderot – d'Alembert, *Encyclopédie, ou Dictionnaire raisonné*, vol. 7, "fourneau" 233.
87   Miller, "A New Perspective on the Natural Philosophy of Steams" 1129–1148.
88   Storni M., "Denis Papin's Digester and Its Eighteenth-Century European Circulation", *The British Journal for the History of Science*, 54 (2021) 443–463; Spary, *Feeding France* 204–211.
89   Storni, "Denis Papin's digester" 447.

scholars who, from the last decades of the eighteenth century, multiplied their attempts to enrich the diet of the poor.

If the digester is a hermetically sealed space in which steam is used under pressure, during the eighteenth century there were many other projects of ovens and furnaces that introduced the use of steam in domestic cooking through simpler devices. In 1771, Etienne Mignot de Montigny (1714–1782) and Jean-Charles de Borda (1733–1799) proposed to the Academy to approve the oven of the Venetian doctor and inventor Bartholomew Dominiceti, active especially in England.[90] Known for his 'fumigatory baths' through which he treated patients with steam therapy, Dominiceti applied this technique to food preparation:

> A large copper pot is inserted into the mouth of the furnace, which serves only to contain water, which, when boiling, provides the steam that will heat the rest. In the cauldron is placed a pot which closes the opening exactly, and which leaves between the outer and inner walls of the cauldron a drain of two or three inches, where the steam can circulate, cooking the meat either in the water or in its own juice.[91]

In the conclusion of their report, the commissioners noted the similarities between Dominiceti's economic stove and Fresneau's stove approved in 1739, adding however that the former 'appears to us to be suitable for a number of uses to which Mr Fresneau's stove cannot be applied; moreover in Mr Fresneau's stove everything is heated over an open fire and in Mr Dominiceti's stove it is the steam that cooks the meats, so that they do not run the risk of burning'.

In 1780, Nivert's 'portable stove (*fourneau portatif*)' was announced to the public in an emphatic tone: 'we lack a treatise on chemical cooking [...] based on principles. In the meantime, here is an instrument capable of giving an idea of the action of water combined with fire on food'.[92] The dishes to be cooked in this furnace are placed in vessels of different materials (glass, porcelain or crystal), which are placed in the furnace where a lamp is lit in the hearth: 'The effect of heat which results from this, and whose degree is that of boiling water, is such that at the end of the ordinary time for cooking, the food is perfectly cooked'.[93] The cooking takes place through 'a sort of bath formed

---

90 Hilaire-Pérez L., *L'invention technique au siècle des Lumières* (Paris: 2000) 103–104.
91 Académie des sciences, pochette de séance de samedi 16 février 1771.
92 *Nouveaux fourneaux économiques et portatifs*, tiré à part de la *Gazette de Santé* du dimanche 1er Octobre 1780 (Paris, chez la Veuve Ballard & Fils: 1780).
93 Ibidem.

by the juice of the food, which being reduced to steam, penetrates, divides and cooks it'. The advantage is that 'the food retains all its flavour, without having to fear either a bad smell, or a burnt taste, or the effects of verdigris'. 'If this apparatus' – adds the author of the article – 'was hermetically sealed, it would be Papin's machine'.

The effect of steam in cooking food and its use in the kitchen were also researched by the pharmacist and agronomist Antoine Parmentier (1737–1813), who in his work on the potato described this phenomenon in detail: 'the water, having been reduced to steam, needs to be forced back to better penetrate the texture of each tuber, to penetrate and combine the constituent parts more perfectly, from which it results that they are cooked sooner and have more flavour'.[94] Parmentier completed his research by designing steam cooking devices, such as the American pot (*marmite américaine*), which he had the merit of 'naturalizing, in a way, [...] in France'. The purpose of this 'economic utensil' is 'to cook potatoes in such a way as to make them more palatable'. Thanks to the collaboration with the Franco-American writer Michel Guillaume Saint-Jean de Crèvecoeur (1735–1813), the American pot was improved following 'an idea of M. Parmentier', who

> Arranged the cauldron containing the water in such a way that it enters the furnace. The rim of the cauldron rests on the rim of the stove, a chimney leads through its lower opening into the hearth, and through its upper opening to the level of the stove, this arrangement spares the fuel, concentrates the heat and hastens the boiling.[95]

For steaming potatoes, Parmentier 'substituted a tin sieve for the horsehair sieve, fitted with two inwardly curved hands', which is 'pierced with holes, which increases the volume of steam from the water, especially as the machine is exactly closed'. The American pot is not only intended for cooking potatoes, but also 'for cooking vegetables and plants'.[96] At the beginning of the nineteenth century, Antoine Cadet-de-Vaux's stove, which was perfected by Harel, included the use of a diaphragm, 'a tinned sheet metal capsule, carried on three ties, with a hole in the rim', which allowed 'to hold anything that you want to steam', using the simmering stock of the pot-au-feu. In the works of Parmentier

---

94 Parmentier Antoine, *Recherches sur les végétaux nourissans* (Paris, De l'imprimerie royale: 1781) 128.

95 Ibidem.

96 "Description, usages et avantages de la Marmite Américaine", *Bibliothèque physico-économique, instructive et amusante* 1 (1788) 226–234, especially 228.

THE KITCHEN FIRE (EIGHTEENTH AND EARLY NINETEENTH CENTURY)   167

and Cadet-de-Vaux, steaming was seen as a new technique for preparing 'food-stuffs with their aroma, flavour and nutritional properties; substances whose principles are not resistant to the action of a long and strong boiling'.[97] The use of steam, often developed to improve the diet of the people, was applied to domestic uses, the results of which at the beginning of the nineteenth century were yet to be seen.

6   Conclusion

The enclosure of fire in ovens and kitchen stoves responded to the need for fuel economy, which was undoubtedly one of the major issues of domestic techniques during the eighteenth and early nineteenth century. However, by enclosing the fire, these appliances assumed an autonomy in the process of food transformation, which relieved the cook of the control of cooking, freeing up time for work, that which suited the emergence of a bourgeoisie that could not count on a very large domestic staff. Also, by enclosing fire, these ovens and stoves created artificial spaces in which substances were transformed by means of steam. In conclusion, it is important to return to the diffusion of these appliances and cooking techniques in early-nineteenth-century cooking practice. An initial examination of the archives shows the creation of man-ufactures, such as that of Charles Harel, who received a silver medal at the 1819 *Exposition des produits de l'industrie française* for his appliances produced in rue de l'Arbresec in Paris.[98] From the 1820s onwards, the use of ovens and stoves also appeared in domestic treatises and in cooking literature. In 1822, Madame Pariset's *Manuel de la maîtresse de maison*, composed in the form of letters, also witnesses the diffusion of these cooking technologies: 'I would have very little to say to you, my dear friend, about the furniture and utensils necessary for your kitchen. They are the same everywhere'.[99] In spite of this reluctance to give advice, the author does not hesitate to point out 'a piece of kitchen furniture of which I cannot say too much good about: I am referring to Harel's stove. It is as convenient as it is economical'; moreover, 'it cooks the dishes better, since it cooks them more quickly: it keeps them hot for a long

---

97   Cadet-de-Vaux, *Fourneau-potager économique* 24.
98   *Description des expositions des produits de l'industrie française, faites à Paris depuis leur origine jusqu'à celle de 1819* (Paris, Bachelier: 1824) 112–116.
99   Mme Pariset, *Manuel de la maîtresse de maison, ou lettres sur l'économie domestqiue* (Paris chez Audot: 1822) lettre VIII, 56.

time after they are cooked'.[100] Even cookery books recorded the appearance of ovens and stoves. *Le cuisinier parisien ou Manuel complet d'économie domestique*, published in 1828, gives the 'description and use of the economical ovens and utensils recently introduced into kitchens', among which Harel's stove is mentioned.[101] The use of the new machines and cooking technologies, however, did not immediately rule out the use of fire in the hearth; on the contrary, the two systems coexisted throughout the nineteenth century, as tradition and innovation in the kitchen remained deeply entangled.[102]

Bibliography

Primary Sources

Albert, *Le cuisinier parisien ou Manuel complet d'économie domestique* (Paris, chez Dufour & compagnie: 1828).

Almanach des gourmands: servant de guide dans les moyens de faire excellente chère … par un vieil amateur, 1803–1812.

Bonnefons Nicolas de, *Les délices de la campagne, suitte du Jardinier françois* (Amsteldan [sic], chez Raphael Smith: 1655).

Cadet-de-Vaux Antoine-Alexis, *Fourneau-Potager économique … suivi d'Observations sur l'application de cet Appareil à tous les besoins du ménage et sur un poêle-fourneau par M. Harel* (Paris, chez Collas et al.: 1807).

Costa de Beauregard Alexis, *Essai sur l'amélioration de l'agriculture dans les pays montueux et en particulier dans la Savoye* (Chambéry, de l'imprimerie de M.F. Gorrin: 1774).

De Lune Pierre, *Le cuisinier* (Paris, chez Pierre David: 1656).

"Description, usages et avantages de la Marmite Américaine", *Bibliothèque physico-économique, instructive et amusante* 1 (1788) 226–234.

Description des expositions des produits de l'industrie française, faites à Paris depuis leur origine jusqu'à celle de 1819 (Paris, Bachelier: 1824).

Dictionnaire de l'Académie Françoise, 2 vols. (Nismes, chet Gaude chez: 1777).

Dictionnaire de l'Académie Françoise, 2 vols. (Paris, chez J.J. Smits: 1798).

Diderot Denis – Le Rond d'Alembert Jean-Baptiste, *Encyclopédie, ou Dictionnaire raisonné des sciences, des arts et des métiers par une société de gens de lettres*, 35 vols. (Paris, chez Briasson et al.: 1751–1780).

---

100   Ibidem.

101   Albert, *Le cuisinier parisien ou Manuel complet d'économie domestique* (Paris, chez Dufour & compagnie: 1828).

102   Ritchie Garrison J., "Remaking the Kitchen, 1800–1850", Gaskell I., Carter S.A. (eds.), *The Oxford Handbook of History and Material Culture* (Oxford: 2020) 141.

THE KITCHEN FIRE (EIGHTEENTH AND EARLY NINETEENTH CENTURY)   169

Duchesne Henri-Gabriel, *Dictionnaire de l'industrie, ou Collection raisonnée des procédés utiles dans les sciences et dans les arts*, 6 vols. (Paris, chez Poigné et al: 1800, 3rd ed.).

Duhamel de Monceau Henri-Louis, *Du transport, de la conservation et de la force des bois* (Paris, chez L.F. Delatour: 1767).

Formey Jean Henri Samuel, *Encyclopédie portative ou science universelle, à la portée de tout le monde* (Berlin, chez tous les librairies: 1758).

Furetière Antoine, *Dictionnaire universel contenant tous les mots François*, 4 vols. (La Haye: 1727) (Hildesheim – New York: 1972).

Gardeton César, *L'art d'économiser le bois de chauffage et tous les autres combustibles* (Paris, chez L. Cordier et al.: 1827).

Harel Charles, *Description des divers appareils propres à économiser le temps et le combustible* (Paris, chez l'auteur: 1821).

Janvier Antide, *Etrennes chronométriques pour l'an 1811* (Paris, chez Courcier et al.: 1810).

Keslar François, *Épargne-bois, c'est-à-dire nouvelle & par-ci devant non-commune ni mise en lumiere, invention de certains & divers fourneaux artificiels* (Oppenheim, Jean-Théodore de Bry: 1619).

Lavoisier Antoine, "Expériences sur l'effet comparé de différents combustibles", *Histoire de l'Académie royale des sciences, avec les mémoires de mathématique & de physique. Année 1781* (Paris, de l'imprimerie royale: 1784) 377–390.

Le Cointe Jourdan, *La cuisine de santé* (Paris, chez Briand: 1790).

Lémery Nicolas, *Cours de chymie contenant la manière de faire les opérations qui sont en usage dans la médecine, par une méthode facile* (Paris, chez Delespine: 1701).

Liger Louis, *La nouvelle maison rustique, ou Économie générale de tous les biens de campagne* (Paris, chez la veuve Savoye: 1768).

Marin François, *Suite des Dons de Comus, ou l'art de la cuisine, réduite en pratique*, 3 vols. (Paris, chez Pissot, Didot, Brunet fils: 1742).

Morand Jean-François-Clément, "L'art d'exploiter les mines de charbon de terre", *Description des arts et métiers, faites et approuvées par messieurs de l'Académie des sciences*, 3 vols. (Paris, chez Saillant, Nyon et Desaint: 1768–1779).

Nouveaux fourneaux économiques et portatifs, tiré à part de la *Gazette de Santé* du dimanche 1er Octobre 1780 (Paris, chez la Veuve Ballard & Fils: 1780).

Mme Pariset, *Manuel de la maîtresse de maison, ou lettres sur l'économie domestique* (Paris chez Audot: 1822).

Parmentier Antoine, *Recherches sur les végétaux nourissans* (Paris, De l'imprimerie royale: 1781).

Pfeiffer Johann Friedrich, *Histoire du charbon de terre et de la tourbe* (Paris, chez P. de Lormel: 1787).

Pott Johann Heinrich, *Lithogéognosie ou examen chimique des pierres et des terres* (Paris, chez Jean-Thomas Herissant: 1753).

Richelet Pierre, *Dictionnaire françois, contenant généralement tous les mots tant vieux que nouveaux* (Chez Jean Elzevir, Amsterdam: 1706).

Savary De Bruslon Jacques, *Dictionnaire universel de commerce, contenant tout ce qui concerne le commerce qui se fait dans les quatre parties du monde ... Ouvrage posthume de Sr Jacques Savary Des Bruslon*, 4 vols. (Amsterdam, chez les Jansons: 1726–1732).

Rumford Thompson Benjamin comte de, *Mémoires sur la chaleur*, 2 vols. (Paris, chez Firmin Didot: 1804–1805).

Tieron, "Lettre à M. ..., Administrateur de l'Hôpital de ...", *Journal de Paris*, 7 septembre 1783.

Vanière, "Nouveau foyer de cuisine, oeconomique et portatif", *Journal oeconomique* (1761) 30.

Varenne François Pierre, *Le cusinier françois* (Paris, chez Pierre David: 1651).

Venel Gabriel François, *Instructions sur l'usage de la houille* (Avignon, chez Gabriel Regnault: 1775).

Secondary Sources

Abad R., "L'Ancien Régime à la recherche d'une transition énergétique?", in Bouvier Y. Laborie L. (eds.), *L'Europe en transitions. Energie, mobilité, communication, XVIIIe–XXIe siècles* (Paris: 2016) 23–84.

Alexandre-Bidon D., *Une archéologie du goût: céramique et consommation (Moyen Âge-Temps modernes)* (Paris: 2005).

Bernasconi G., "Pour une archéologie des pratiques. Mesure du temps, corps et prestation (XVIIIe–XXe siècle)", *Socio-anthropologie* 40 (2019) 247–262.

Brewer P.J., *From Fireplace to Cookstove. Technology and the Domestic Ideal in America* (Syracuse: 2000).

Cooper A., "Homes and Households", in: Park K., Daston L. (eds.), *Cambridge History of Science*, vol. 3: *Early Modern Science* (Cambridge: 2006) 224–237.

Di Meo M., Pennell S. (eds.), *Reading and Writing Recipe Books, c. 1500–1800* (Manchester: 2013).

Dohrn-van Rossum G., *History of the Hour. Clocks and Modern Temporal Orders*, (Chicago – London: 1996).

Forino I., *La cucina. Storia culturale di un luogo domestico* (Torino: 2019).

Giedion S., *Mechanization Takes Command: a Contribution to Anonymous History* (New York: 1969).

Girard A., "Le triomphe de la 'la cuisinière bourgeoise': livres culinaires, cuisine et société en France aux XVIIe et XVIIIe siècle", *Revue d'histoire moderne et contemporaine* 24 4 (1977) 497–523.

Gras A., *Le choix du feu. Aux origines de la crise climatique* (Paris: 2007).

Guerrini A., "The Ghastly Kitchen", *History of Science* 54.1 (2016) 71–97.

THE KITCHEN FIRE (EIGHTEENTH AND EARLY NINETEENTH CENTURY) 171

Haasis L., Rieske C. (eds.), *Historische Praxeologie. Dimensionen vergangenen Handelns* (Paderborn: 2015).

Hilaire-Pérez L., *L'invention technique au siècle des Lumières* (Paris: 2000).

Jandot O., *Les délices du feu. L'homme, le chaud et le froid à l'époque moderne* (Ceyzérieu: 2017).

Kaplan S.L., *Le meilleur pain du monde: les boulangers de Paris au XVIIIe siècle* (Paris: 1996).

Kitsikopoulos H., *Innovation and Technological Diffusion. An Economic History of Early Steam Engines* (London: 2019).

Leong E., *Recipes and Everyday Knowledge, Medicine, Science, and the Household in Early Modern England* (Chicago: 2018).

Miller D.P., "A New Perspective on the Natural Philosophy of Steams and Its Relation to the Steam Engine", *Technology and Culture* 61.4 (2020) 1129–1148.

Pardhaillé-Galabrun A., *La naissance de l'intime. 3000 foyers parisiens, XVIIe–XVIIIe siècles* (Paris: 1988).

Pennell S., *The Birth of the English Kitchen, 1600–1850* (London – New York: 2016).

Quellier F., "La cuisson des fruits en France à l'époque moderne", in Williot (eds.), *Du feu originel aux nouvelles cuissons. Pratiques, techniques, rôles sociaux* (Bruxelles: 2015) 113–127.

Ritchie Garrison J., "Remaking the Kitchen, 1800–1850", in Gaskell I. – Carter S.A. (eds.), *The Oxford Handbook of History and Material Culture* (Oxford: 2020).

Roche D., *Histoire des choses banales. Naissance de la consommation, XVIIe–XIXe siècle* (Paris: 1997).

Roche D., *Le peuple de Paris: essai sur la culture populaire au XVIIIe siècle* (Paris: 1998).

Storni M., "Denis Papin's Digester and Its Eighteenth-Century European Circulation", *The British Journal for the History of Science* 54.4 (2021) 443–463.

Storni M., "Time Thrift and Economic Science in the Eighteenth Century", *Science in Context* (forthcoming).

Thébaud-Sorger M., "Capturing the Invisible: Heat. Steam and Gases in France and Great Britain, 1750–1800", in Roberts L. – Werrett S. (eds.), *Material, Governance and Production, 1760–1840* (Leiden: 2017) 85–105.

Verhoeven G. – Blondé B., "Against the Clock. Time Awareness in Early Modern Antwerp, 1585–1789", *Continuity and Change* 28.2 (2013) 213–244.

Voth H.-J., *Time and Work in England (1750–1830)* (Oxford: 2000).

Wall W., *Recipes for Thought. Knowledge and Taste in the Early Modern English Kitchen* (Philadelphia: 2016).

Werrett S., *Thrifty Science. Making the Most of Materials in the History of Experiment* (Chicago – London: 2019).

Zylberberg D., "Fuel Prices, Regional Diets and Cooking Habits in the English Industrial Revolution (1750–1830)", *Past & Present* 229.1 (2015) 91–122.

CHAPTER 6

Fire Mechanics: Inventors and Promoters of Heating Systems (Eighteenth and Early Nineteenth Century)

Olivier Jandot

Abstract

Until the eighteenth century, the question of domestic heating, which was of utmost importance for everyday life, had rarely attracted the interest of artisans, inventors, and scientists. The first author to address specifically the question of heating systems was Nicolas Gauger, who published in 1713 the seminal work *La Mécanique du feu, ou l'art d'en augmenter les effets et d'en diminuer la dépense*. In this chapter, I will explore the main steps leading to the codification of a practical knowledge of domestic fire in the eighteenth century, as a key example to display the complexity of the path to technological invention. I will also present the tools and methods deployed by the actors involved in this technological research, whom I will call – in reference to the title of Gauger's book – "fire mechanics," namely specialists of domestic heating systems, trying to sketch a profile of this group of practitioners. Finally, I will show how the efforts of the "fire mechanics" contributed to the emergence, at the turn of the nineteenth century, of an actual market of heating machines, of which I will illustrate the main actors and their commercial strategies.

Keywords

fire – heating devices – home heating – comfort – fireplace – stove – invention

Until the beginning of the eighteenth century, the question of domestic heating, which was of the utmost importance for everyday life, had rarely attracted the interest of craftsmen, inventors, or scientists. From the end of the Middle Ages onwards, a fault line was established, which permanently and schematically divided Europe into two geographical areas using two different technical systems of domestic heating. From the banks of the Rhine to Moscow, the use of the closed fireplace dominated. The earthenware or tiled stove, because of

FIRE MECHANICS: INVENTORS AND PROMOTERS OF HEATING SYSTEMS 173

its mass, accumulates and diffuses heat, making it possible to fight effectively against the rigours of continental winters while obscuring the view of the fire. In Western Europe, the British Isles and North America, on the other hand, the open fireplace prevailed, the construction of which was the responsibility of masons and architects. The open fireplace is an artefact that is not so much a means of heating as a means of preventing the spread of fire and evacuating smoke. Its mediocrity and imperfections were well known, but the architects' thinking was mainly focused on the shapes of the fireplace, possibly on the question of smoke evacuation, but not on the question of improving the calorific power of fire and the diffusion of heat.[1]

The first author to take a specific interest in this question was Nicolas Gauger, author in 1713 of a seminal work: *La Mécanique du feu, ou l'art d'en augmenter les effets et d'en diminuer la dépense*.[2] The work was seminal in several respects. First, because it was the first to lay the foundations of a theoretical reflection on the physical phenomena at work in combustion and in the diffusion of heat from the domestic fireplace. Then, because it was the first to try to increase the efficiency of the domestic fire, i.e., to increase the heat released from the fireplace and to improve its diffusion without adding more fuel. Finally, because his work inaugurated a series of publications and research that would continue throughout the eighteenth and early nineteenth centuries, and marked the birth of a new science, 'caminology', which the *Encyclopédie méthodique* defined as follows: 'science des cheminées, ou art de construire des cheminées de manière qu'elles économisent le combustible et qu'elles ne fument pas' ('science of fireplaces, or the art of constructing fireplaces in such a way that they save fuel and do not smoke').[3]

The purpose of this chapter is to try to retrace the modalities and essential milestones of the constitution of this practical knowledge on domestic fire, which illustrates one of those 'cumulative processes of knowledge [...] nourished by borrowings, analogies and multiple circulations'.[4] These nourish and allow invention, but also to try to highlight the tools and methods used by the actors involved in this technological research, whom I will call "fire mechanics", i.e. specialists in domestic heating systems, trying to sketch a profile of this group of practitioners. Finally, I will show how the efforts of the

---

1 Jandot O., *Les délices du feu. L'homme, le chaud et le froid à l'époque moderne* (Ceyzérieu: 2017).
2 G*** [Gauger Nicolas], *La mécanique du feu, ou l'art d'en augmenter les effets et d'en diminuer la dépense* [...] (Paris, J. Estienne & J. Jombert: 1713).
3 *Encyclopédie méthodique. Physique, t. 2* (Paris, Veuve Agasse:1816) 208.
4 Hilaire-Pérez L. – Thébaud-Sorger M., "Inventer", in Hilaire-Pérez L. – Simon F. – Thébaud-Sorger M. (eds.), *L'Europe des sciences et des techniques. Un dialogue des savoirs, XVᵉ–XVIIIᵉ siècle* (Rennes: 2016) 34.

174   JANDOT

"fire mechanics" contributed to the emergence, at the turn of the nineteenth century, of an actual market in heating machines, giving sketches of the main actors and their commercial strategies.

1   The Birth of a New 'Science': the Gauger Moment (1713)

1.1   *Who Is Gauger?*

The personality of the author of *La mécanique du feu* is not well known. The work was published anonymously in 1713, with the only attribution 'G***', and if Gauger is well identified as the author of the work in the account given by the *Journal des sçavants*[5] at the time of its publication, and if his name appears on the 1749 reprint,[6] later authors, notably English ones, went so far as to deny even the reality of his existence. In 1820, an Englishman, Mickleham, author of a work on heating and ventilation, claimed that Gauger never existed, and that the author of *La mécanique du feu*, was in fact the Cardinal de Polignac, who published the work under the pseudonym Gauger.[7] If this assertion is obviously false, it contributes to blurring the knowledge that we can have of the character while recognizing his historical importance. Beyond these quarrels over attribution, the bio-bibliographical elements that can be gathered on Gauger as well as the analysis of his work allow us to understand the sudden emergence of the theme of domestic heating in technical literature.

The only certain biographical elements available on Gauger are a short biographical note included in the *Encyclopédie méthodique*,[8] proof of a certain recognition by his peers, and elements of a professional title that can be gleaned from some of his other publications. Born around 1680 and dying in 1730, Gauger is referred to as a 'physicien' ('physicist') in his biographical note. He is also referred to as 'avocat au parlement & censeur royal des livres' ('lawyer in parliament and royal censor of books'). Although he did not belong to any prominent institution, he seems to have acquired a certain reputation in Parisian learned circles, notably with the chevalier de Louville, by accurately repeating some of Newton's experiments before an audience of amateurs. He is the author of four published works: *La mécanique du feu* (1st ed. 1713,[9]

---

5   *Le journal des sçavants* (4th December 1713) 654–656.

6   Gauger Nicolas, *La mécanique du feu* [...] (Paris, C.A. Jombert: 1749).

7   Bosc E., *Encyclopédie générale de l'architecte-ingénieur. Traité complet théorique et pratique du chauffage et de la ventilation des habitations particulières et des édifices publics* [...] (Paris: 1875) 49.

8   *Encyclopédie méthodique. Physique*, t. 3 (Paris, Veuve Agasse: 1819) 254–255.

9   G*** [Gauger Nicolas], *La mécanique du feu* [...] (Paris, J. Estienne & J. Jombert: 1713).

FIRE MECHANICS: INVENTORS AND PROMOTERS OF HEATING SYSTEMS   175

2nd ed. 1749[10]), two works on thermometers and barometers (1710[11] & 1722[12])
and a work on the refrangibility of light rays (1728).[13]

1.2   *The Originality of Gauger's Thought Process*

This diversion through the scanty details of Gauger's biography are necessary
to try to understand how, at a given moment in history, technical thought was
to take over a field of study that it had hitherto supremely ignored. And the
analysis of Gauger's work also allows us to see the contributions of theoretical physics to technical reflection. For it is indeed as a physicist that Gauger
analyses the 'mechanics of fire' in an innovative way, applying by analogy to an
eminently practical subject analytical tools that come from Newtonian physics
and geometry. Gauger's main concern is how to increase the heat output of
domestic fires and, above all, how to improve the diffusion of heat from the
hearth.

The book is divided into three parts. The first looks at ways to increase the
heat produced by combustion in the hearth by studying the way in which heat
is radiated and suggests improving the air supply to the fireplace. The second
focuses on devices to improve smoke evacuation. Indeed, a classic defect of
old chimneys was that they had an unfortunate tendency to smoke the interior of the house, a problem that gave rise to an abundance of literature, both
in the form of specific works and as chapters on the subject in architectural
treatises. The third part of the book contains a list of practical tips to guide
the action of builders. It constitutes a real instruction manual for building this
new type of fireplace, which, in Gauger's view, is superior in every respect to
the then existing models. The originality of Gauger's thinking consists mainly
in the use, for the first time, of tools from physics and mathematics in a reflection on the way in which the heat produced by combustion in the fireplace
diffuses into the volume of the inhabited space. As he himself points out in
the preface to the book, the remarkable efficiency of the fireplace model he
sets out to design can be proved both by experience and – more importantly –
by the mathematical demonstrations and physical reasons he sets out to give.

---

10   Gauger Nicolas, *La mécanique du feu* [...] (Paris, C.A. Jombert: 1749).
11   Gauger Nicolas, *Résolution du problème proposé dans le 'Journal de Trévoux' du mois
     de mars dernier* [1710] *pour la construction de nouveaux thermomètres et de nouveaux
     baromètres* (Paris, J. Quillau: 1710).
12   Gauger Nicolas, *Théorie de nouveaux thermomètres et de nouveaux baromètres de toutes
     sortes de grandeurs,* [...] (Paris, Quillau et Desaint: 1722).
13   Gauger Nicolas, *Lettres de M. Gauger,* [...] *sur la différente réfrangibilité des rayons de la
     lumière et l'immutabilité de leurs couleurs,* [...] *avec le plan de son Traité de la lumière et des
     couleurs* (Paris, Simart: 1728).

Transferring and adapting concepts from optics (notably that of Newton, of which he appears to be a good connoisseur), Gauger analyses the diffusion of heat in terms of radiation, by an analogy between light rays and heat rays. For him, these rays of heat can be direct, when they come directly and immediately from the fire, or indirect, when they are reflected by a body that sends them back. The improvement of the calorific value of the fireplace is therefore achieved both by improving the combustion (maintained by a continuous air flow) and, above all, by changing the traditional geometry of the fireplace. The addition of a curved plate of sheet metal or copper plate at the bottom of the hearth should improve the diffusion of heat by optimising the radiation, as illustrated by the plates in the book.

1.3   Reception and Posterity of La mécanique du feu

The fundamentally innovative aspect of the work, as well as the fact that it deals with a universal problem (the mediocrity of the calorific performance of fireplaces emerges every harsh winter, when cold and frost penetrate to the very heart of dwellings), undoubtedly explains its rapid and wide distribution. First published in Paris in the late summer of 1713 by Jombert & Estienne, the book was then published in three other editions in 1714 in Holland, the first two by Henri Schelte[14] and David Mortier,[15] the third by an anonymous printer (probably a pirate edition).[16] The following year, the book was translated into German and English and published in Hannover[17] and London[18] respectively. The book seems to have been a real success amongst English and German-speaking audiences, as there were several reprints in the following years.[19] The book was also republished in French in 1749.[20]

The success of the book is explained by the radically new approach proposed by Gauger. Whereas architects had traditionally been concerned only

---

14   G*** [Gauger Nicolas], *La mécanique du feu* [...] (Amsterdam, Henri Schelte: 1714).

15   G*** [Gauger Nicolas], *La mécanique du feu* [...] (Amsterdam, David Mortier: 1714).

16   G*** [Gauger Nicolas], *La mécanique du feu* [...] (Cosmopoli: 1714).

17   G*** [Gauger Nicolas], *La mécanique du feu, oder Kunst die Würkung des Feuers zu vermehren, und die Kosten davon zu verringern* [...] (Hannover, Nicolaus Förstern: 1715).

18   Gauger Nicolas, *Fires improv'd: being a new method of building chimneys, so as to prevent their smoaking: in which A Small Fire, shall warm a Room better than a much Larger made the Common Way* [...] (London, J. Senex & E. Curll: 1715).

19   Gauger Nicolas, *The mechanism of fire made in chimneys* [...] (London, R. Bonwicke, T. Goodwin, J. Walthoe, M. Wotton, S. Manship, J. Nicholson, R. Wilkin, B. Tooke, R. Smith & T. Ward: 1716); Gauger Nicolas, *Fires Improved* [...] (London, J. Senex & E. Curll: 1736); G*** [Gauger Nicolas], *La mécanique du feu, oder Kunst Die Würckung des Feuers zu vermehren* [...] (Hannover, Nicolaus Förstern: 1717).

20   Gauger Nicolas, *La mécanique du feu* [...] (Paris, C.A. Jombert: 1749).

FIRE MECHANICS: INVENTORS AND PROMOTERS OF HEATING SYSTEMS 177

with the aesthetic aspect of fireplaces (technical considerations were generally limited, when present, to the question of improving smoke evacuation), Gauger focused on the question of the efficiency of this heating instrument. By studying the heat diffusion process from a physics perspective, it opens a completely new field of thinking. Therefore all subsequent works on the question of domestic heating, which will multiply in the second part of the eighteenth century and especially in the nineteenth century, will never fail to underline the founding aspect of Gauger's work. As the chemist Guyton de Morveau pointed out in 1802: 'C'est à Gauger [...] que l'on doit le premier système, le plus complet de vues et d'expériences, sur la circulation de la chaleur' ('It is to Gauger [...] that we owe the first and most complete system of views and experiments on the circulation of heat').[21] In so doing, Gauger inaugurated a new field of reflection and experimentation aimed at improving domestic comfort and saving fuel.

2 Modalities of the Constitution of Knowledge on Domestic Heating

2.1 *The Search for Practical Solutions to a Universal Problem*
Historians of material culture have rightly emphasised the turning point represented by the eighteenth century as the entry into 'the age of comfort'.[22] The search for thermal comfort is part of this general process in Western civilisation, but it was also stimulated by the search for economy in a context of latent energy crises and the continuous rise in the price of wood throughout the century. The combination of these two phenomena led to a sudden increase in the number of publications about domestic heating from the middle of the century. The technical solutions proposed by the various authors were more or less innovative, but they all started from a shared observation: the mediocre calorific performance of the traditional fireplace. Benjamin Franklin estimated that more than five-sixths of the heat produced by the fire in the fireplace disappeared instantly up the chimney flue, without contributing in any way to the heating of the room.[23] The concern for a more economical management of wood resources, both from a strict individual economy point of view and in the

---

21 Guyton de Morveau L.-B., "Description d'un poële sur les principes de la cheminée suédoise, avec bouches de chaleur", *Annales de chimie* 41 (1802) 89.

22 DeJean J., *The age of comfort: when Paris discovered casual – and the modern home began* (New York: 2009). See also Crowley J.E., *The invention of comfort. Sensibilities and design in early modern Britain and early America* (Baltimore: 2000).

23 Franklin Benjamin, "Description des nouveaux chauffoirs de Pensylvanie", in *Œuvres de M. Franklin, t. 2* (Paris, Quillau/Esprit: 1773), 89–90.

search for the common good, is therefore an important motivation for these innovators. Because it concerned a universal problem (everyone has a fireplace at home and can generally see its imperfections), the search for innovative technical solutions mobilised a broad and heterogeneous spectrum of actors. Soldiers, clergymen, lawyers, doctors, architects, important scientists (such as Franklin or Benjamin Thompson, Count Rumford) or simple amateurs – sometimes stimulated by academic prize contests – produced papers, memoirs, brochures and specialist books presenting innovations that promised to improve domestic comfort and reduce fuel costs.[24]

2.2 Similar Tools, Methods and Practices

The study of these works allows us to draw a group portrait of these fire mechanics whose tools, methods and practices are relatively similar. According to a cumulative process of knowledge, they generally rely on the existing literature on the subject (including Gauger of course) and make an assessment of the solutions proposed. Then the moment arrives to demonstrate the innovation they have developed, generally by successive trials. The constitution of this shared knowledge on domestic fire is inseparable from the publication of personal research, individual experiments and the demonstration of the new technical solutions they have led to. This technical literature is abundant, but often small and difficult to identify. The technical book, most often illustrated with figures, convenient milestones in the history of technology, is just the tip of the iceberg. Indeed, a large part of these technical discoveries was published in a less visible way in periodicals, not always specialist, which makes their identification difficult and random. For example, in an extensive bibliography on the subject compiled by a German clergyman in 1802, three quarters of the texts listed appeared in periodicals.[25]

Whatever the authors, the tools used to approach the problem are more or less the same. They are more a matter of empirical and concrete knowledge acquired through experience than of a real science. The modifications made to traditional fireplaces concern both the improvement of the air supply to the fireplace in order to promote combustion and the improvement of smoke evacuation, but also, and above all, the improvement of heat distribution in

---

24 For more details and references, see Jandot O., "The invention of thermal comfort in eighteenth-century France", in Stobart J. (ed.), *The Comfort of Home in Western Europe, 1700–1900* (London: 2020) 73–92.

25 Grimoldi A., "La littérature technique sur le chauffage au XVIII\ siècle. Entre France et Allemagne", in Bienvenu G. – Monteil M. – Rousteau-Chambon H. (eds.), *Construire ! Entre Antiquité et époque moderne. – Actes du 3ᵉ congrès francophone d'histoire de la construction, Nantes, 21–23 juin 2017* (Paris: 2019) 473–482.

FIRE MECHANICS: INVENTORS AND PROMOTERS OF HEATING SYSTEMS   179

the room. There is therefore a great deal of discussion about the geometry and dimensioning of the hearth, the positioning of the air inlet, the size of flues and the shape of jambs. The question of heating is then to be indissociably integrated with that of the ventilation of the buildings, heat being perceived as a fluid which circulates because of the thermal exchanges between the cold air coming from outside and the hot air produced by the fire.[26] By positioning a candle at different points in a heated room and observing how air currents tilt the flame, some inventors even attempted a kind of mapping of indoor air flows.[27] Surprisingly, while all of them boast of the unparalleled performance of the artefact they have modified or developed, very few are concerned with quantifying the indoor temperatures that their inventions achieved, as thermometric data is extremely rare. The administration of evidence is therefore based more on the rhetoric of the discourse than on objective data.

2.3   The Question of Heat, between Science and Technology

This eminently practical approach to improving the heating performance of the fireplace does not prevent a more theoretical approach to the problem. For if the second half of the eighteenth century saw an increase in technical reflection on the question of domestic heating, it was also a time of intense reflection and bitter debate on the questions of combustion and heat within the scientific community. This key moment in the history of the birth of modern chemistry saw a confrontation between the proponents of the phlogiston theory elaborated by the German physician and chemist Georg Ernst Stahl (1660–1734) and those of the caloric theory elaborated by Lavoisier and Laplace from the 1780s.[28] Although each of these two fields (technical innovation and fundamental research) has its own logic and mobilises different actors, there are sometimes bridges between the two which suggest that science and technology should not be opposed too schematically, which can be illustrated with two cases: Clavelin and Benjamin Thompson, Count Rumford.

The character of Clavelin is extremely mysterious. Referred to as 'citoyen Clavelin', he seems to have belonged to a religious order dissolved by the

---

26   Simonnet C., *Brève histoire de l'air* (Versailles: 2014), 99–115.

27   See for example Franklin Benjamin, "Description des nouveaux chauffoirs de Pensylvanie [...]" in *Œuvres de M. Franklin, t.* 2 (Paris, Quillau/Esprit: 1773) 82–84 and above all Cointeraux François, *L'économie des ménages* (Paris, Vezard & Le Normant: 1793) 16–17 and plate I.

28   Mazauric S., *Histoire des sciences à l'époque moderne* (Paris: 2009) 294–299; Locqueneux R., *Préhistoire et histoire de la thermodynamique classique (Une histoire de la chaleur)* (Paris: 1996) 31–51; Locqueneux R., *Sur la nature du feu aux siècles classiques. Réflexions des physiciens et des chimistes* (Paris: 2014) 183–214.

decree of the Assemblée nationale constituante on 13 February 1790.[29] He is known only by an unpublished work, of which are preserved – to the best of my knowledge – only summaries by Hallé and Jumelin, members of the Bureau de consultation des arts et métiers. In August 1793, a man named Petit introduced him to the Bureau de consultation, which was intended to 'récompenser les services rendus au corps social quand leur importance et leur durée méritent ce témoignage de reconnaissance' ('reward services rendered to the social body when their importance and duration merit such recognition').[30] Clavelin then submitted a manuscript consisting of thirteen in-folio notebooks illustrated with painted plates that brought together the fruit of twenty years of solitary experiments on 'the statics of air and fire'.[31] This work immediately aroused the admiration of the two rapporteurs responsible for examining it. In October 1794, the Bureau de consultation therefore decided to award him the maximum reward provided for in its statutes and promised to fund his printing.[32] But the troubled context of those years did not allow this publication to come about. In the end, Clavelin's work only came to public attention through the publication in scholarly journals of the summary report that Hallé and Jumelin wrote after reading the manuscript.[33] According to a contemporary opinion, Clavelin's work sheds a completely new light on the question of the 'statics of air and fire' as applied to the field of housing. The thousands of experiments that Clavelin carried out attempted to study all the factors that come into play in the heating of dwellings. They deal with the circulation of air, the comparative qualities of different types of fuels, the diffusion of heat in the volume of a room, and the physical principles at work in combustion. Clavelin's patient and persistent research is based on a purely experimental approach. He set up different 'laboratories' to conduct his experiments, inventing 'instruments' to quantify the phenomena he studied. His approach is marked by a constant concern for quantification. He used a watch, thermometers, scales, and a kind of anemometer he had made. He even designed a

---

29 Conservatoire national des arts et métiers (hereafter Cnam), Paris. Archives historiques, N 331/1: Lettre de Petit (25th August 1793). Clavelin is presented as a 'savant qui étoit membre d'un ordre distingué par leurs lumières' ('scholar who was a member of an order distinguished by their lights').

30 Ballot C. (ed.), *Procès-verbaux du Bureau de consultation des arts et métiers* (Paris: 1913) 16 (quote from Article 1 of the decree of 3 August 1790 which founded the institution).

31 Cnam, Paris. Archives historiques, N 331: Preliminary reports & N 333: Final synthesis report.

32 Ballot C. (ed.), *Procès-verbaux du Bureau de consultation des arts et métiers* 127.

33 *Magasin encyclopédique, ou Journal des sciences, des lettres et des arts* 5 (1795) 306–341 & *Annales de chimie ; ou Recueil de mémoires concernant la chimie et les arts qui en dépendent* [...] 33 (1800) 172–216. The following development and quotes are based on this source.

fireplace with adjustable dimensions (height and depth of the firebox, dimensions of the flue) in order to determine the ideal size of the firebox through hundreds of tests. As the authors of the report point out, its work establishes 'plusieurs vérités nouvelles' ('several new truths') based on 'preuves incontestables' ('incontrovertible evidence').

Much more famous than Clavelin, Benjamin Thompson, Count Rumford, is also an example of a lifetime devoted to the study of heat. Rumford's particularity is to have constantly mixed in his research both theoretical and very practical aspects that belong more to technology, without it being really possible to discern whether these are two parallel fields of study (and which, like parallels, never intersect) or whether Rumford's theoretical thinking may have influenced the inventions of which he was the author. In his *Mémoires sur la chaleur*,[34] Rumford published in 1804 the fruit of twenty years of theoretical research. The book is an important milestone in science history. But at the same time, he published the six volumes of his *Essais politiques, économiques et philosophiques*[35] in which he presented, amongst other things, his research on heating and cooking instruments. It is particularly in his fourth essay that he presented the model of the fireplace to which he gave his name.[36] By reducing the size of the hearth of an existing fireplace by lining it with bricks, Rumford claimed to reduce fuel consumption by more than fifty percent and to improve heat diffusion through the oblique angle of the jambs. This essay is therefore fully in line with the publications of the 'fire mechanics'.

3   The Emergence of a Market for Heating Devices

3.1   *Revolution in Home Heating*
The publication of Clavelin's and Rumford's work is also part of a very specific context, that of the last decade of the eighteenth century and the first years of the nineteenth, a period of major rupture in the history of France, the country in which the threads of this research on domestic heating – which were born in a vast area of the Western world extending from Pennsylvania (Franklin) to Bavaria (Rumford) – actually intertwined. For if the period that began in 1789

---

34   Rumford B.T., *Mémoires sur la chaleur* (Paris: 1804).

35   Rumford B.T., *Essais politiques, économiques et philosophiques* [...] (Genève – Paris: 1799–1806).

36   Rumford Benjamin Thomson, "Quatrième essai: Des cheminées et de leurs foyers. Différens moyens de les perfectionner pour épargner le combustible, rendre les maisons plus saines et plus agréables, et empêcher les cheminées de fumer" in Rumford Benjamin Thomson, *Essais politiques, économiques et philosophiques* [...], t. 1 (Genève, G.J. Manget: 1799) 309–386.

marked an obvious break in French history, it was also the moment of a revolution in the field of home heating. While the use of the stove, a much more efficient means of heating than the fireplace, was for a long time the subject of real reluctance on the part of the French population, its use tended to become more common in the last decades of the eighteenth century.[37] The growing demand for comfort and the search for more efficient and energy-saving heating devices (in a context of increasing wood prices throughout the eighteenth century) explains the questioning of this 'royauté des cheminées' ('royalty of fireplaces'), as Daniel Roche calls it.[38] This domestic revolution did not, of course, have the brutality or speed of the political revolution that was shaking France at the time. But a process was underway, which gradually changed the face of interiors as well as age-old gestures and habits. No doubt with a certain amount of exaggeration, the inventor and architect François Cointeraux noted in 1793 that 'dans les grandes villes [...] on voit une foule de poëliers, de faïenciers, de magasins remplis de poêles [...] de toutes sortes de formes et de matières' ('in the big cities [...] one sees a crowd of stove makers, earthenware makers, shops filled with stoves [...] of all kinds of shapes and materials').[39] The fact is that from these years onwards, we see the names of inventors appearing who, unlike those of previous decades, no longer simply present the fruit of their research and propose a sort of construction guide that everyone will be free to implement if they wish. A market for heating appliances is gradually taking shape. To meet the demand, a handful of inventors or craftsmen began to mass produce appliances which they marketed with a great deal of publicity, trying to prove the exceptional calorific performance of their invention in an increasingly competitive market. The archives of the Conservatoire National des Arts et Métiers[40] as well as the specialist periodicals of the time or the first dictionaries of inventions make it possible to draw up a list of these manufacturers, most of them Parisian, active during the Revolution and the First Empire.[41] Manufacturers of energy-saving fireplaces,[42] calorifiers,[43]

---

37 On this particular aspect and on the reasons for this reluctance, see Jandot O., "Le feu caché. Introduction du confort thermique et métamorphoses de l'économie des sens (France, 1700–1850)" in Duperron N. – Jouves-Hann B. – Métraux M.G. – Paulin M.-A. – Poulain B. (eds.) *Décoration intérieure et plaisir des sens (1700–1850)* (Rome: 2022).

38 Roche D., *Histoire des choses banales* (Paris: 1997) 138–145.

39 Cointeraux François, *L'économie des ménages* 41.

40 Cnam, Paris. Archives historiques, série N.

41 *Dictionnaire chronologique et raisonné des découvertes, inventions [...] en France, dans les sciences, la littérature, les arts, l'agriculture, le commerce et l'industrie de 1789 à la fin de 1820* (Paris: 1822–1824) 17 vol.

42 *Dictionnaire chronologique et raisonné des découvertes, inventions*, vol. III, 96–114.

43 *Dictionnaire chronologique et raisonné des découvertes, inventions*, vol. II, 313–322.

FIRE MECHANICS: INVENTORS AND PROMOTERS OF HEATING SYSTEMS   183

energy-saving or fumivorous furnaces,[44] and stoves can now be counted by the dozen.[45] Amongst these was Voyenne, head of the company 'Voyenne et compagnie, mécaniciens' ('Voyenne and company, mechanics)', which, according to one of its advertising brochures, 'tient fabrique et magasin de poêles à courant d'air et à vapeur, établis d'après un procédé nouveau et infiniment économique, au prix le plus modéré' ('runs a factory and shop for air current and steam stoves, established according to a new and infinitely economical process, at the most moderate price') and undertakes to deliver in less than a month throughout France.[46] But we also find Thilorier, inventor of a fumivorous stove and various heating devices,[47] Curaudau, inventor of economical fireplaces the principle of which consists in multiplying the smoke evacuation pipes behind the decorative glass that traditionally adorns the overmantel in order to improve the diffusion of heat through the installation of heat registers[48] and above all two ambitious entrepreneurs who are trying to conquer a booming market: Desarnod and Ollivier.

3.2   *Entrepreneurial Logic and Commercial Strategy*

The first of the two, Joseph-François Desarnod, was an architect from Lyon who, from the early 1780s, embarked on a real industrial venture: the mass production of a heating appliance largely inspired by the Pennsylvania fireplace (also known as 'Franklin stove') described by the American inventor in a brochure printed in Philadelphia in 1745 and translated into French in 1773 on the occasion of the publication in Paris of his main works.[49] The appliance consists of a set of perfectly interlocking cast iron plates that form a sort of parallelepiped that fits into the fireplace. After designing the device from 1783 onwards, the main difficulty encountered by Desarnod was to find a forge capable of producing his prototype with accuracy and precision. Indeed, for it to be fully effective, the various parts of the device had to fit together perfectly, which was a real technical challenge at the time. This is why he travelled around France during the years 1784–1786 meeting various forge masters and having tests carried out at the forges of Villersexel,[50] in eastern France (January 1785),

---

44   *Dictionnaire chronologique et raisonné des découvertes, inventions*, vol. VII, 396–407.
45   *Dictionnaire chronologique et raisonné des découvertes, inventions*, vol. XIII, 525–542.
46   Cnam, Paris. Archives historiques, N 34.
47   Cnam, Paris. Archives historiques, N 29 & N 161.
48   Cadet de Vaux A.-A., "Des cheminées de M. Curaudeau", *Bibliothèque des propriétaires ruraux, ou journal d'économie rurale et domestique* 8 (1805) 169–174. See also Cnam, Paris. Archives historiques, N 26 & N 30.
49   Franklin Benjamin, "Description des nouveaux chauffoirs de Pensylvanie", in *Œuvres de M. Franklin*, t. 2 (Paris, Quillau/Esprit: 1773), 81–118.
50   Villersexel, currently located in the department of Haute-Saône.

then Dampierre,[51] in the Perche (1786).[52] Desarnod invested several tens of thousands of livres in the venture. In 1788, he owed almost 25,000 livres to one of his creditors who took legal action against him to get paid.[53] Once the manufacturing challenge had been met, the next step was to ensure the marketing of the device. Now settled in Paris, to convince his potential buyers, he sought and obtained the backing of two prestigious institutions during the year 1788: the Royal Academy of Sciences and the Royal Society of Medicine. The advertising campaign could then begin. In 1789, he published a small booklet of about sixty pages, illustrated with plates, sold for the modest sum of 36 sols.[54] The brochure presents the properties and advantages of these 'nouveaux foyers économiques et salubres' ('new economical and healthy hearths'), reproduces the approvals of the Academy of Sciences and the Society of Medicine and also contains installation and maintenance instructions for future purchasers. The device, delivered in kit form, can be ordered from the author, 18 rue Caumartin in Paris. From there, Desarnod would become a major player in the emerging heating devices market. He had his devices approved by all the official bodies designed to encourage inventors born with the Revolution (Lycée des Arts, Bureau de consultation des arts et métiers, etc.) and took part in competitions and exhibitions that promoted the products of industry, which he never failed to point out in the numerous leaflets he published. In order to satisfy a wide range of consumers, it offers its devices in three sizes and, for each of them, in three types of finish, from the simplest (glossy with lead) to the most luxurious (black varnished, painted in gold). Prices ranged from 260 to 1000 livres, a considerable investment for the time. It also provides after-sales service by selling spare parts individually if necessary.[55] Fifty years before Godin, the most famous stove maker of the nineteenth century in France, Desarnod was an example of an ambitious entrepreneur with particularly innovative business methods for his time.

---

51   Dampierre-sur-Blévy, currently located in the commune of Maillebois, department of Eure-et-Loir.

52   University of Pennsylvania, Philadelphia. Rare Book & Manuscript Library, LJS 180: *Mémoire sur les procédés à suivre pour construire des chauffoirs et sur les changemens à faire dans la manière de les couler en fonte*, after 1786.

53   Archives départementales de la Haute-Marne, 2E69, Fonds de la famille Diderot-Caroillon de Vandeul. Famille Caroillon. Papiers de famille. Claude-Xavier Caroillon-Destillières et sa famille. Procès avec Desarnod (1788–1790).

54   Desarnod François-Joseph, *Mémoire sur les foyers économiques et salubres de M. le docteur Franklin, et du Sieur Désarnod, architecte à Lyon* [...] (Lyon – Paris, Dessenne et Gattey, Royer, Bailli: 1789).

55   Cnam, Paris. Archives historiques, N 162.

Louis-François Ollivier is also one of those entrepreneurs who were able to capture the spirit of the times and adapt their production to market developments. In 1777, he inherited his father's earthenware factory located at 73 rue de la Roquette, in the faubourg Saint Antoine, in Paris.[56] Inventor of a 'calorifère salubre' ('salubrious calorifier') in 1785, he was also the depositary of the first French patent in 1791. Supporter of the Revolution's ideas, he was also known for having offered the Convention Nationale a stove in the shape of a Bastille in 1792. During the same years (1793–1794), he published a catalogue to promote the stove models he produced in his factory. Praising the beauty, utility and solidity of his productions, he proposes no less than eighteen models of stoves, that are also beautiful decorative elements.[57] While he produced stoves in series, he also offered his customers to produce specific models on request, by modifying the shape, design, size and colour if they so wish. For this purpose, the catalogue presents the models in coloured plates systematically backed up by a black and white plate which the customer can colour himself to match the stove to the colours of the room in which it is to be installed. Like Desarnod, he was keen to make his production known and to have its quality recognised by the various institutions responsible for encouraging the industry.[58] Like him too, he was a well-known manufacturer on the Paris market whose excellent products are recognised by specialists.[59]

3.3 *Evaluating Performance, Comparing Inventions*

In about fifteen years (circa 1790–circa 1805), a new branch of the industry was thus born, or at least developed considerably: that of the manufacturers of heating appliances. These inventors and promoters of heating systems are clearly a group of professionals operating in a competitive market. Each specialises in a type of production (fireplace, stove, furnace, etc.), which they constantly improve and promote, especially at national industry exhibitions. Desarnod was the only exhibitor of heating appliances at the first exhibition in 1798.[60] But at the second exhibition, in 1801, he was joined by Thilorier and

---

56 Plinval de Guillebon R. de, "Une manufacture de faïence à Paris, de François Genest à Louis François Ollivier (1734–1808)", *Sèvres* 27 (2018) 44–63.

57 *Collection de dessins des poëles de formes antique et moderne, de l'invention et de la Manufacture du sieur Ollivier, rue de la Roquette, fauxbourg Saint-Antoine* (Paris: ca. 1794).

58 Cnam, Paris. Archives historiques, N 25.

59 *Bulletin de la Société d'Encouragement pour l'Industrie Nationale* 24 (1805) 300.

60 *Catalogue des produits industriels qui ont été exposés au Champ de Mars pendant les trois derniers jours complémentaires de l'An VI* [...] (Paris, Imprimerie de la République: 1798) 6.

186   JANDOT

Ollivier.[61] At the fourth exhibition, in 1806, there were still new manufacturers such as Curaudau, Voyenne or Huson & Verdier; these latter were stove makers specialising in the production of 'grands poêles en faïence imitant la porcelaine' ('large earthenware stoves imitating porcelain').[62] These exhibitions were also a means of obtaining awards (gold or silver medals, jury mentions, etc.) which would be real commercial persuasions later on.[63]

Faced with this proliferation of devices whose designers emphasised their exceptional performance, the Minister of the Interior asked the Conservatoire national des arts et métiers in 1807 to carry out comparative tests to try to get a clearer picture. The government's interest in a subject as common as domestic heating appliances is because energy is a highly political issue. The development of industrial uses of wood as fuel (in the form of wood or charcoal) increased the tension on a market destroyed by the Revolution.[64] The state had everything to gain by finding ways to reduce the domestic use of wood so as not to hinder industrial development, and this includes the development of more efficient heating and cooking equipment. During the winter of 1807–1808, members of the Conservatoire tested eight models of stoves and eight models of fireplaces supplied by a dozen manufacturers. The appliances were successively installed in the former calefactory of the monks of the priory of Saint-Martin-des-champs, in which the Conservatoire is located, a vast room of 400 m^2 and 560 m^3. The performance of each appliance was evaluated for eight hours by taking regular temperature readings at different locations in the room and feeding the appliances with an equal amount of fuel. A summary table compiling the main results of the experiments shows that the reputation of the devices produced by Ollivier, Desarnod, Curaudau or Voyenne was not usurped. With 40 kg of wood, their appliances managed to increase the indoor temperature by more than 12 degrees on average over eight hours, while some others performed poorly.[65]

---

61   *Seconde exposition publique des produits de l'industrie française* [...] *qui seront exposées dans la grande Cour du Louvre, pendant les cinq jours complémentaires de l'an 9* [...] (Paris, Imprimerie de la République: 1801) 7, 23 & 28. See also: *Procès-verbal des opérations du jury nommé par le ministre de l'intérieur pour examiner les produits de l'industrie française mis à l'exposition des jours complémentaires de la dixième année de la République* (Paris, Imprimerie de la République : 1802) 63–64.

62   *Catalogue des produits de l'industrie française qui seront exposés, pendant les derniers jours de septembre 1806* [...] (Paris, Imprimerie impériale : 1806) 5, 14, 34 & 44.

63   See for example *Rapport du jury sur les produits de l'industrie française, présenté à S. E. M. de Champagny, ministre de l'intérieur* (Paris, Imprimerie impériale: 1806), 166–168.

64   Woronoff D. (ed.), *Révolution et espaces forestiers* (Paris: 1989).

65   Cnam, Paris. Archives historiques, N 45.

4　Conclusion

At the beginning of the nineteenth century, barely a hundred years after the publication of Gauger's pioneering work and after a century of reflection and technical innovation, efficient heating instruments finally existed. This long eighteenth century is therefore the scene of the gradual emergence of technical knowledge about domestic heating and constitutes a turning point in the history of domestic comfort. This painstaking technical research is the result of both a search for comfort and economy. They make it possible to find solutions to an age-old problem: that of fighting the cold, which until then had been solved in imperfect ways, with fairly inefficient heating methods. Thanks to these fire mechanics, the house ceased to be a fragile, smoky shelter pierced by icy draughts and became a home. This is a key moment in the history of sensitivities that will modify the thresholds of tolerance to cold, define new standards of comfort, and make it possible to assert new requirements. This process can be compared with the progress made at the same time in France in the field of lighting and, to a lesser extent, in techniques relating to the domestic use of water.[66] Technology is gradually taking over the home and the construction or improvement of the house requires the intervention of specialist trades. From the end of the 1820s onwards, specialist manuals for professionals will appear which will order the knowledge accumulated during the previous century, such as Ardenni's, which was constantly re-edited and expanded throughout the nineteenth century.[67] Of course, at the time, this progress only concerned a small part of the population. The structures of everyday life, as Fernand Braudel calls them, change only slowly and gradually and comfort is then a privilege.[68] But the way to increasingly effective control of thermal environments through more and more sophisticated techniques was now open.

66　Castelluccio S., *L'éclairage, le chauffage et l'eau aux XVIIe et XVIIIe siècles* (Montreuil: 2016).

67　Ardenni P., *Manuel du poêlier-fumiste ou Traité complet et simplifié de cet art* [...] (Paris: 1828).

68　Braudel F., *Civilization and capitalism: 15th–18th Century. 1. The Structures of everyday life: the limits of the possible* (London: 1983).

Bibliography

Primary Sources

Annales de chimie ; ou Recueil de mémoires concernant la chimie et les arts qui en dépendent [...] 33 (Paris: 1800) 172–216.

Ardenni P., *Manuel du poêlier-fumiste ou Traité complet et simplifié de cet art* [...] (Paris: 1828).

Ballot C. (ed.), *Procès-verbaux du Bureau de consultation des arts et métiers* (Paris: 1913).

Bosc E., *Encyclopédie générale de l'architecte-ingénieur. Traité complet théorique et pratique du chauffage et de la ventilation des habitations particulières et des édifices publics* [...] (Paris: 1875).

Bulletin de la Société d'Encouragement pour l'Industrie Nationale 24 (Paris: 1805) 300.

Cadet de Vaux A.-A., 'Des cheminées de M. Curaudeau', *Bibliothèque des propriétaires ruraux, ou journal d'économie rurale et domestique* 8 (Paris: 1805) 169–174.

Catalogue des produits de l'industrie française qui seront exposés, pendant les derniers jours de septembre 1806 [...] (Paris, Imprimerie impériale : 1806).

Catalogue des produits industriels qui ont été exposés au Champ de Mars pendant les trois derniers jours complémentaires de l'An VI [...] (Paris, Imprimerie de la République: 1798).

Cointeraux François, *L'économie des ménages* (Paris, Vezard & Le Normant: 1793).

Collection de dessins des poëles de formes antique et moderne, de l'invention et de la Manufacture du sieur Ollivier, rue de la Roquette, fauxbourg Saint-Antoine (Paris: ca. 1794).

Desarnod François-Joseph, *Mémoire sur les foyers économiques et salubres de M. le docteur Franklin, et du Sieur Désarnod, architecte à Lyon* [...] (Lyon – Paris, Dessenne et Gattey, Royer, Bailli: 1789).

Dictionnaire chronologique et raisonné des découvertes, inventions [...] *en France, dans les sciences, la littérature, les arts, l'agriculture, le commerce et l'industrie de 1789 à la fin de 1820* (Paris: 1822–1824).

Encyclopédie méthodique. Physique, t. 2 (Paris, Veuve Agasse: 1816).

Encyclopédie méthodique. Physique, t. 3 (Paris, Veuve Agasse: 1819).

Franklin Benjamin, 'Description des nouveaux chauffoirs de Pensylvanie', in *Œuvres de M. Franklin, t. 2* (Paris, Quillau/Esprit: 1773), 81–118.

G*** [Gauger Nicolas], *La mécanique du feu* [...] (Amsterdam, David Mortier: 1714).

G*** [Gauger Nicolas], *La mécanique du feu* [...] (Amsterdam, Henri Schelte: 1714).

G*** [Gauger Nicolas], *La mécanique du feu* [...] (Cosmopoli: 1714).

G*** [Gauger Nicolas], *La mécanique du feu, oder Kunst Die Würckung des Feuers zu vermehren* [...] (Hannover, Nicolaus Förstern: 1717).

G*** [Gauger Nicolas], *La mécanique du feu, oder Kunst die Würkung des Feuers zu vermehren, und die Kosten davon zu verringern* [...] (Hannover, Nicolaus Förstern: 1715).

FIRE MECHANICS: INVENTORS AND PROMOTERS OF HEATING SYSTEMS   189

G*** [Gauger Nicolas], *La mécanique du feu* [...] (Paris, J. Estienne & J. Jombert: 1713).

Gauger Nicolas, *Fires improv'd: being a new method of building chimneys, so as to prevent their smoaking: in which A Small Fire, shall warm a Room better than a much Larger made the Common Way* [...] (London, J. Senex & E. Curll: 1715).

Gauger Nicolas, *Fires Improved* [...] (London, J. Senex & E. Curll: 1736).

Gauger Nicolas, *The mechanism of fire made in chimneys* [...] (London, R. Bonwicke, T. Goodwin, J. Walthoe, M. Wotton, S. Manship, J. Nicholson, R. Wilkin, B. Tooke, R. Smith & T. Ward: 1716).

Gauger Nicolas, *Théorie de nouveaux thermomètres et de nouveaux baromètres de toutes sortes de grandeurs,* [...] (Paris, Quillau et Desaint: 1722).

Gauger Nicolas, *La mécanique du feu* [...] (Paris, C.A. Jombert: 1749).

Gauger Nicolas, *Lettres de M. Gauger,* [...] *sur la différente réfrangibilité des rayons de la lumière et l'immutabilité de leurs couleurs,* [...] *avec le plan de son Traité de la lumière et des couleurs* (Paris, Simart: 1728).

Gauger Nicolas, *Résolution du problème proposé dans le 'Journal de Trévoux' du mois de mars dernier* [1710] *pour la construction de nouveaux thermomètres et de nouveaux baromètres* (Paris, J. Quillau: 1710).

Guyton de Morveau L.-B., 'Description d'un poële sur les principes de la cheminée suédoise, avec bouches de chaleur', *Annales de chimie* 41 (1802) 79–105.

Le journal des sçavants (4th December 1713).

Magasin encyclopédique, ou Journal des sciences, des lettres et des arts 5 (1795) 306–341.

Procès-verbal des opérations du jury nommé par le ministre de l'intérieur pour examiner les produits de l'industrie française mis à l'exposition des jours complémentaires de la dixième année de la République (Paris, Imprimerie de la République : 1802).

Rapport du jury sur les produits de l'industrie française, présenté à S. E. M. de Champagny, ministre de l'intérieur (Paris, Imprimerie impériale : 1806).

Rumford B.T., *Essais politiques, économiques et philosophiques* [...] (Genève – Paris : 1799–1806).

Rumford B.T., *Mémoires sur la chaleur* (Paris : 1804).

Seconde exposition publique des produits de l'industrie française [...] *qui seront exposées dans la grande Cour du Louvre, pendant les cinq jours complémentaires de l'an 9* [...] (Paris, Imprimerie de la République : 1801).

Secondary Sources

Bienvenu G. – Monteil M. – Rousteau-Chambon H. (eds.), *Construire ! Entre Antiquité et époque moderne. – Actes du 3ᵉ congrès francophone d'histoire de la construction, Nantes, 21–23 juin 2017* (Paris: 2019).

Braudel F., *Civilization and capitalism: 15th–18th Century. 1. The Structures of everyday life: the limits of the possible* (London: 1983).

Castelluccio S., *L'éclairage, le chauffage et l'eau aux XVIIe et XVIIIe siècles* (Montreuil: 2016).

Crowley J.E., *The invention of comfort. Sensibilities and design in early modern Britain and early America* (Baltimore: 2000).

DeJean J., *The age of comfort: when Paris discovered casual – and the modern home began* (New York: 2009).

Duperron N. – Jouves-Hann B. – Métraux M.G. – Paulin M.-A. – Poulain B. (eds.) *Décoration intérieure et plaisir des sens (1700–1850)* (Rome: 2022).

Hilaire-Pérez L. – Simon F. – Thébaud-Sorger M. (eds.), *L'Europe des sciences et des techniques. Un dialogue des savoirs, XVe–XVIIIe siècle* (Rennes: 2016).

Jandot O., *Les délices du feu. L'homme, le chaud et le froid à l'époque moderne* (Ceyzérieu: 2017).

Locqueneux R., *Sur la nature du feu aux siècles classiques. Réflexions des physiciens et des chimistes* (Paris: 2014).

Locqueneux R., *Préhistoire et histoire de la thermodynamique classique (Une histoire de la chaleur)* (Paris: 1996).

Mazauric S., *Histoire des sciences à l'époque moderne* (Paris: 2009).

Plinval de Guillebon R. de, "Une manufacture de faïence à Paris, de François Genest à Louis François Ollivier (1734–1808)", *Sèvres* 27 (2018) 44–63.

Roche D., *Histoire des choses banales* (Paris: 1997).

Simonnet C., *Brève histoire de l'air* (Versailles: 2014).

Stobart J. (ed.), *The Comfort of Home in Western Europe, 1700–1900* (London: 2020).

Woronoff D. (ed.), *Révolution et espaces forestiers* (Paris: 1989).

CHAPTER 7

'The Manner of Conducting Fire': Firing Architectural Terracotta in the Modern Era, between Know-How, Wood Shortage, and Innovations

Cyril Lacheze

Abstract

Architectural terracotta in modern France was produced through an operating chain where an imposing kiln held a central place. This was no high technology, but required a specific know-how as to the "manner of conducting fire." The oven could be a masonry oven fuelled with wood, or an "open" coal kiln built with the very bricks that were to be fired; both these structures made use of relatively stable techniques from the Middle Ages to the beginning of the twentieth century. The appearance of smoke and the colour of the walls remained the main indicators to handle the firing process. The kilns construction underwent some slow changes aimed at improving the circulation of fire in the firing chamber, but the difficulty of controlling fire, coupled with the economic catastrophe represented by a failed firing, often pushed tile makers to invoke supernatural powers to protect their kilns.

Beyond these technical aspects, due to the growing wood shortage that affected modern Europe, tile makers using wood-fired ovens (50 to 100 cubic meters per firing) were subject to increasing pressure as to fuel supplies. Preservation measures were taken by various authorities as of the sixteenth century, but many establishments still found themselves in great difficulty until the mid-eighteenth century. The first tests of coal firing (or more rarely with peat or other types of fuel) were attempted at this time, with mixed results due to low availability of the chosen fuel, an incomplete technical mastery, but also the resistance to culture change. Another possibility was to radically innovate the oven construction, in order to reduce its wood consumption. If this dynamic gave birth in the mid-nineteenth century to new technical solutions, some still in use today, this movement of innovation was ongoing since at least the end of the eighteenth century: some inventors thus proposed the first models of juxtaposed or superimposed kilns, recovering heat from a fire to preheat the nearby oven. These first tests, remaining for the most part without future, did not question the prevalence of the craftsmen's know-how: the superimposed kiln of the baron de La Tour d'Aigues, in

© KONINKLIJKE BRILL BV, LEIDEN, 2025 | DOI:10.1163/9789004521766_009

1787, was thus tested by 'people of the countryside', who still had to rely on their senses to guide the firing, not through sight but the 'considerable noise [of] the air column'.

Keywords

architectural terracotta – know-how – wood – coal – scarcity – innovation

The production of architectural terracotta – namely of bricks, roofing and paving tiles, and some other related products such as pipes – requires a furnace or kiln, which is an expensive apparatus, hard to use, but central to the operational chain. Even if one narrows the focus to early modern France, as is the case with this paper, the existence of several types of kiln is to be acknowledged. None of these models represented a particularly advanced technology for the time, but their use required a specific know-how, which is called the 'manner of conducting fire (*manière de conduire le feu*)' by Droz, a lawyer, author of a memoir on the topic in the mid-eighteenth century.[1] From a technical standpoint, producing architectural terracotta thus relied on the skilful management of fire. Yet fire management also played a crucial role at another level, namely fuel supplies. Wood was by far the most used fuel for architectural terracotta kilns in early modern France. To make repeat firings, it was necessary to manage wood supplies efficiently, both at the company level and on a larger regional scale, which was all the more necessary given the dearth of such supplies. The use of alternative fuels, such as peat or coal, was a possibility, but not necessarily obvious at the time, given technical, logistical, economic or cultural obstacles. In this chapter, I thus intend to explore this twofold management of fire, which has, as we shall see, a particularly intricate history. This research is part of a larger project that addresses the production of architectural terracotta in France from the thirteen to the nineteenth century.[2]

As is customary in the history of medieval and early modern technology, the research is based on an open corpus, calling on both handwritten or printed – and archaeological sources, in this case mainly concerning a geographical area

---

1 Droz, "Mémoire sur la manière de perfectionner les tuileries", *Mémoires et observations recueillies par la Société Œconomique de Berne* (1765) 314.

2 Lacheze C., *L'art du briquetier, XIIIᵉ–XIXᵉ siècles. Du régime de la pratique aux régimes de la technique*, PhD dissertation (Paris 1 Panthéon-Sorbonne University: 2020). Lacheze C., *L'art du briquetier, XVIᵉ–XIXᵉ siècle. Du régime de la pratique aux régimes de la technique* (Paris: 2023).

extending from the Center of France to Burgundy, via the Paris region. For the specific question of how to conduct fire, of a fundamentally practical nature and therefore rarely described anyway, the majority of sources for the early modern period dates from the eighteenth century, because of both the better preservation of recent documents and an increasingly present administration, including at the local level, generating more documentation. However, the problems of wood scarcity were also becoming increasingly apparent at this time, hence the greater presence of regulations and reflections by administrators and craftsmen concerning the use of fuel. This also implies that solutions found to this energy crisis in the second half of the eighteenth century were, for many, effectively employed only in the following decades. In order to take this 'time-lag'[3] into account, we will extend the chronological period discussed until the middle of the nineteenth century, a time when, despite some innovations previously mentioned, most of the dynamics of the previous century still remained valid.

Before dealing with fire management itself, we need to address the place that the kiln held on the one hand in the production process of architectural terracotta, and on the other hand within the factory. Two types of factories will be discussed in this chapter, corresponding to two different types of furnace. The main one is the tile factory, which could also produce bricks, consisting of a set of buildings and a kiln, common throughout France. These infrastructures were usually arranged around a central sand-covered area, which could reach up to 500 m². The clay was mined a few hundred metres away, mixed in one or more pits partially filled with water. The products were then shaped on a table using simple wooden moulds, possibly reinforced with iron. They were unmoulded on the sanded central area, where they dried for a few days in the sun, then moved into halls surrounding the square to complete drying off the rain.[4] When products were almost completely dry, they were placed in the kiln for firing. A normal kiln, in brick (for the inside) and stone (for the outside), essentially consisted of a cube open at the top, of a few metres on each side, including two chambers: the bottom, divided into two or three heating corridors, contained the fire; the hot gases, moving up, passed through a perforated floor; and then through the main chamber measuring from 20 to 50 m³, where

---

3 This expression is used with reference to the following paper, which clearly demonstrates the importance of taking into account the lifespan of objects – or techniques – and the time lags that this may entail: Adams W.H., "Dating Historical Sites: The Importance of Understanding Time Lag in the Acquisition, Curation, Use, and Disposal of Artifacts", *Historical Archaeology* 37.2 (2003) 38–64.

4 Grésillon B. – Lambert O. – Mioche P., *De la terre & des hommes. La tuilerie des Milles d'Aix-en-Provence (1882–2006)* (Mirabeau: 2007) 15–16.

the products were placed. Such a kiln could typically accommodate one to five tens of thousands of products, equivalent to two or three weeks' worth of moulding work.[5] The factory could be supplemented by a dwelling and a few fields, vineyards or ancillary production spaces, linked to generalised multiple activities: the lease of a tile factory in Le Bignon-Mirabeau (Loiret, central France), in 1775, thus described it as 'a tile kiln, a hall, two workshops, a house and garden [...] such as tile-makers are accustomed to enjoy'.[6]

The other type of factory, notably employed in the Netherlands and in Belgium, and consequently to an extent in the North of France under the name of the 'Flemish kiln', was exclusively used for brick production in large quantities. In this configuration, a team of itinerant brick-makers set up near a construction site, moulded bricks 'in the field' without building permanent installations, and put them to dry on the spot. This same team, or another more specialist one, then built a kiln with the bricks to be fired, comprising at least 50,000 bricks, often 200,000 to 500,000, and sometimes more than a million for more than 6 m on a side (including in height). The fuel was then coal, placed horizontally between rows of bricks. The kiln had to be ignited before its construction was complete, so that the bottom bricks could harden and thus support the weight of those above.[7] This type of kiln being specific to the extreme north of France throughout the early modern era, we will above all be interested in fire management in the masonry-clad kilns and fixed factories previously discussed, and we will explicitly indicate when we refer to these 'Flemish' kilns.

1   Firing Know-How, Core of the Technical Process

Firing is the most obvious critical point in the production process of architectural terracotta, given the difficulty of mastering it and the complete loss of both infrastructure and load in the event of an error: a rise in temperature that was too slow or too fast could indeed not be rectified, and, in the second case, the furnace could melt with its charge, possibly causing a fire.[8] Consequently,

---

5   Gardner J.S. – Eames E., "A tile kiln at Chertsey Abbey", *Journal of the British Archaeological Association* 17 (1954) 24–42.

6   French National Archives – MC/ET LVIII 470, Paris notaries. Lease of a tile factory, February 1, 1775.

7   Thuillier F., "Le four à briques en meule d'époque moderne de Bruille-lez-Marchiennes (Nord)", *Revue du Nord* 17 (2012) 467–476.

8   Dufaÿ B., "Les tuileries parisiennes du faubourg Saint-Honoré d'après les fouilles des jardins du Carrousel (1985–1987)", in Van Ossel P., *Les Jardins du Carrousel à Paris. Fouilles 1989–1990. Volume 3* (Paris: 1991) 475–488.

it was generally carried out by the 'master' of the tile factory, who had in theory the required know-how, a specialization sometimes specified in engagement contracts when the tile-maker was only a tenant; ordinary workers were usually confined to the other tasks. It was however necessary to monitor the fire and supply wood constantly for several days, involving the use of at least one helper, whether an assistant, an apprentice, or in the worst case a child, as in Auxon (Aube, Champagne) where this dangerous practice seemed usual in 1847: if he fell asleep or simply wasn't careful enough, the baking failed, and the load or even the whole kiln was lost.[9]

The know-how of fire management actually started with loading the products into the kiln, since they had to be placed so that they all heat up as evenly as possible and without collapsing on top of each other. This arrangement is known through some technical treatises,[10] but above all from archaeological discoveries of not completely emptied kilns, black traces corresponding to spacing left visible for aesthetic purposes in some regions (notably in Sologne, central France), archaeomagnetic analyses, and even traces of drips and sticking on glazed products. These clues suggest an orthogonal layout of the pieces or at 30° and 60° angles, with enough space between the pieces to fit the fingers through.[11] Ethnographic observations, from a few installations in France still using this type of kiln, and in particular at Bezanleu (Seine-et-Marne, Parisian region) and L'Hôme-Chamondot (Orne, Normandy) confirm in part this way of loading, with possible variations in composition and orientation according to the rows of products.[12] On the 22 rows of the kiln at Bezanleu, the first sixteen rows were aligned above each arch of the heating corridors, the next five at an angle, and the last row parallel to the arches. If the load

---

9   Aube departmental archives – 5M 397, Unsanitary, inconvenient and dangerous establishments. Auxon, 1847; Kleinmann D., "Les briqueteries en Chinonais et leurs marques de fabrication (I)", *Bulletin de la Société des Amis du Vieux Chinon* 8.6 (1982) 854; Peirs G., *La brique. Fabrication et traditions constructives* (Paris: 2004) 38–45. Gouzouguec S., "Le métier de tuilier à Paris au XVᵉ siècle: approche sociale et organisation du travail dans la juridiction de Saint-Germain-des-Prés", *Archéologie médiévale* 36 (2006) 236.

10   Diderot Denis – d'Alembert Jean (eds.), *Recueil de planches, sur les sciences, les arts libéraux et les art méchaniques, avec leur explication. Tome 1* (Paris, Le Breton: 1762); Duhamel du Monceau Henri-Louis – Fourcroy de Ramecourt Charles-René – Gallon Jean-Gaffin, *Descriptions des arts et métiers, faites ou approuvées par Messieurs de l'Académie royale des sciences. L'art du tuilier et du briquetier* (Paris, Saillant & Nyon: 1763).

11   Carette M. – Derœux D., *Carreaux de pavement médiévaux de Flandre et d'Artois (XIIIᵉ–XIVᵉ siècles)* (Arras: 1985) 39–41.

12   Bernstein D. – Champetier J.-P. – Sartre J., *L'histoire de la brique en France. Approche exploratoire. Rapport de recherche* (Paris: 1984) 64–90. Anonymous, *Les Briqueteries de Langeais et Cinq-Mars* (Langeais: 1987); Mandion R., "La Tuilerie de Bezanleu... la tradition. Un métier de diable et de bon Dieu", *Notre Département: la Seine-et-Marne* 17 (1991) 26–27.

included different products, it started with bricks, then roofing tiles and finally paving tiles. In order to fill the kiln as effectively as possible, the quantity of products moulded during the previous weeks had to be calculated according to the loading capacity, a tile-maker from Champagne thus testifying to calculating the layout of the pieces in his kiln to optimize 'surfaces and volumes', and moulding a few extra products at the last minute if he had fallen a bit short.[13]

Firing physically corresponds to melting the surface of the minerals constituting the paste of the product, so that they mix during the solidification corresponding to the cooling phase.[14] This process is punctuated by a series of particularly critical changes of state: disappearance of residual water between 110 and 200 °C, end of plasticity between 400 and 700 °C, then release of carbon dioxide up to 900 °C. The firing can continue even beyond that point, up to 1,200 °C. The cooling is marked by a sudden contraction of the material around 570 °C.[15] The colour that the finished product takes on is determined by paste constitution, firing atmosphere and temperature: siliceous clays bake from brick-red to brown, and limestones from orange-red to green and yellow; elements added during baking could be used to give a specific colour, for example bundles of green wood for a black surface.[16]

In practice, once the products were placed in the kiln, the goal of the firing process was to consistently raise the temperature uniformly in time and space, since bumpy firing, with reducing phases within an overall oxidative firing, could result in fragile products, identifiable by their greyish interior,[17] not to mention the risk of losing control of the firing. The firing having to be homogeneous, but hot gases tending to go to already hot places inside the firing chamber, the draft was rectified by adjusting the covering of tiles or other terracotta overhanging the load and 'closing' the kiln.[18] The 'small fire' first allowed the kiln to be brought to 200 or 300 °C to make the remaining water disappear, over a period of at least two days and up to a week, depending on the factory. After starting the fire with stick bundles, ideally in the evening to take advantage of the cold air improving the draft, the fuel was switched to

---

13 Czmara J.-C., *Les métiers du feu en Champagne méridionale. La terre* (Saint-Cyr-sur-Loire: 2009) 51.

14 Peirs, *La brique* 28.

15 Haussonne M., *Technologie céramique générale. Faïences, grès, porcelaines*, vol. 2 (Paris: 1954) 312–316.

16 Campbell J., *L'art et l'histoire de la brique. Bâtiments privés et publics du monde entier* (Paris: 2004) 17.

17 Haussonne, *Technologie céramique générale* 334–337.

18 Charpail M.-P., *Une tuilerie à Bezanleu* (Saint-Cyr-sur-Morin: 1995) 42. Dufaÿ B., "Les tuileries parisiennes ..." 263.

kindling, then small logs of about 40 cm in diameter burning slowly, all placed at the bottom of the heating corridors using a wooden pole. The total evaporation of the water was visually detectable, with the release of whitish steam giving way to a blackish smoke marking the start of the actual baking.[19] The firing process then switched to 'full fire': for a day or two the kiln's mouth was partially bricked up to stabilize the draft, and the fuel returned to stick bundles igniting instantly, in order to increase it. During this phase, flames crossed the entire firing chamber, and were visible above the kiln, heated to its maximum temperature of about 1,000 °C.

The ability to determine the end of 'full fire', before the kiln was in danger of overheating, was one of the most crucial skills of the tile-maker. Observation of the flame and the smoke, as well as the colour of the load and of the kiln itself, which turned white at high temperature, gave purely practical and sensory indications, thus difficult to quantify and interpret for the historian.[20] However, there were two complementary systems. The first, still within the field of know-how, was based on the observation of height indicators placed on top of the load, exploiting the phenomenon of collapsing caused by the shrinkage of products during baking: the total height of the load was supposed to decrease by 'four to six inches', or 8 to 12 cm, according to an observer in 1817, or by 'the height of a flat brick' at L'Hôme-Chamondot, where witness bricks are still used today in 'traditional' firing.[21] The second system was that of 'timers' changing state at a given temperature. Some producers used objects made by themselves, or even recovered, for example glass bottles. However, at the end of the nineteenth century standardized products were marketed, being small numbered clay cones curving at a defined temperature.[22] The idea was to place three indicators on the load: one curving a few degrees before the desired temperature to warn of the imminence of the optimum moment, one curving at exactly the right temperature, and one curving at a slightly higher temperature

---

19   Auger F. – Lépine J., "Evolution et fonctionnement d'une briqueterie de Sologne aux XIXe et XXe siècles: Montevray à Nouan-le-Fuzelier (Loir-et-Cher)", in Heude B. (dir.), *Les briqueteries-tuileries de Sologne. Cher, Loir-et-Cher, Loiret. Histoires d'hommes et d'établissements qui ont fait de la Sologne un pays de briques* (Lamotte-Beuvron 2012) 43.

20   Peirs, *La brique* 38–45.

21   French National Archives – F[12] 2380, Manufacture of tiles and bricks. Draft of a treatise on the art of tile making by Dutroncy, 1817; Bernstein – Champetier – Sartre, *L'histoire de la brique en France* 64–90. Public communication by Laurent Fontaine, L'Hôme-Chamondot factory owner and tile-maker, 12 September 2019.

22   Val-de-Marne departmental archives – 70J 51, Gournay brick factory. Studies, chemical analysis of bricks, 1883–1950; Bernstein D. – Bruel A. – Filatre P. – Champetier J.-P., *Histoire d'un matériau de construction: la brique en France* (n.p.: 1989) 42; Armogathe J.-R., *Bernard Palissy. Mythe et réalité* (Agen: 1990) 62–63.

to call for an emergency stop. However, despite this apparent practicality and precision, craftsmen preferred to rely on their experience rather than on these timers. The system indeed only worked properly in optimal conditions, and they sometimes observed a contradiction between timer indications and the temperature which they could estimate from experience through fire colour and shrinkage.[23] For this type of kiln, the fundamental know-how therefore remained identical at the turn of the nineteenth and twentieth centuries to that of the early modern era.

Once 'full fire' was considered achieved, by whatever means, all openings were sealed, and the kiln left to cool for a week before the products were deemed cold enough to be unloaded. A few days being necessary after the firing to clean the kiln and recover the ashes, it was therefore theoretically possible to carry out a firing every 20 days, a rate actually documented at Laines-aux-Bois (Aube, Champagne) in 1846.[24] It is worth noting that, given its impressive, dangerous and hard to control nature, the firing process found a particular echo in religious and magical thinking, among the neighbouring population but also among the tile-makers themselves. The Bezanleu firing area was for example called the 'devil's pit', and a calvary can be found on some kilns, as in the masonry of the Guilly kiln (Loiret, central France).[25] At L'Hôme-Chamondot tile factory, when the full fire is stopped by blocking the opening with bricks, a calvary is (still nowadays) drawn on one of the bricks, a meal is shared between the workers, and a prayer is recited: 'Do what you can, I did what I could', emphasizing the master's inability to further influence the process, and placing his kiln 'in the hands of God'.

Even fewer are the clues to 'Flemish' kiln processes, as these installations, being temporary and only regional on the scale of France, are less documented in written sources, less often found in archaeological excavations, and have not been the subject of in-depth ethnographic study. In any case, these firings too required great technical ability, this time not only from the master, but from the whole team, called a 'hand', of a dozen workers. In addition to the quantity of bricks to be produced and handled (several hundreds of thousands per kiln), and the size of the kiln, working conditions were particularly tough and dangerous, since the kiln was built and fired simultaneously. As previously

---

23   Bourry É., *Traité des industries céramiques* (Paris: 1897) 392–399; Arnaud D. – Franche G., *Manuel de céramique industrielle* (Paris: 1906) 368–379. Haussonne, *Technologie céramique générale* 322–330.

24   Aube departmental archives – 5M 404, Unsanitary, inconvenient and dangerous establishments. Laines-aux-Bois, 1845–1846. Peirs, *La brique* 38–45.

25   Inventory of architectural heritage (Mérimée database) – IA00013325, Guilly, tile factory, 19th century. Mandion, "La Tuilerie de Bezanleu" 26. Charpail, *Une tuilerie à Bezanleu* 42.

‘THE MANNER OF CONDUCTING FIRE’   199

mentioned, as soon as the height exceeded one metre, the lower bricks could not withstand the pressure of the upper ones if kept raw, hence they were baked as the kiln was assembled.[26] The most complete description of the firing technique of such a kiln for the early modern era is that made by Fourcroy de Ramecourt in the *Descriptions des Arts et Métiers*, published under the direction of the Paris Academy of Sciences in the middle of the eighteenth century, taking as an example a kiln of 500,000 bricks, or about 10 m on a side.[27] In this example, seven rows of bricks constituted the base of the kiln, and were surmounted by the first layer of coal. The fire was lit with small wood as soon as this base was completed, and took about ‘eight to ten hours’ to reach the coal, at which time the base was obstructed and the coal layer took over as fuel. Meanwhile, workers laid rows of 10,000 bricks each, at the rate of three to five rows per day, and laying a layer of coal every three rows of bricks. It is necessary to underline here the physical side of this pace. Bricks could have been laid to dry on the field very far away from the kiln, and this one quickly became very high; as a result, most of the workers formed a chain to bring the bricks, and only two ‘loaders’ were at the top of the kiln to lay them. They then had to place a 3 or 4 kg brick every three seconds by rough estimate, bending down and getting up each time. Because of the heat, it was impossible to stay on the kiln for more than half an hour, so the shifts had to rotate regularly. The situation was even worse in the morning, since the loading stopped during the night: the fire then caught up with the last row of bricks, and the first shift could not remain in place for more than five minutes. Setting up layers of coal was just as gruelling, as each weighed roughly six tons.

Besides this hardship, the team leader (or two specialist leaders) had to have a particular know-how on two specific points, related to the arrangement of fuel in the very middle of the products, and to the absence of masonry. First of all, it was necessary to ensure homogeneous firing, as in a masonry-clad kiln. Here, however, that meant analysing the process according to the observation of the smoke, and adding or removing coal accordingly, in a counter-intuitive way: by adding coal to close the interstices between bricks, and thus slow down the fire. The other problem was that, despite these efforts, the inside often baked faster than the outside, and the kiln consequently sank faster in the middle, thus putting pressure on the outer edges. To avoid collapse, it was therefore necessary to know how to build solid though flexible edges by a particular arrangement of bricks, and to add intermediate ‘false rows’ of bricks during

---

26   Peirs, *La brique* 30.
27   Duhamel du Monceau – Fourcroy de Ramecourt – Gallon, *Descriptions des arts et métiers* 20–55.

loading in the middle of the kiln, to keep it level. Indeed, archaeological examples with some rows still in place, as at Bruille-lez-Marchiennes (Nord, near Flanders, sixteenth–seventeenth centuries), show at least for the base of the structure bricks placed obliquely or on different sides, even within the same row, and especially at the edges.[28] Comparisons have been made with Albania, Greece and Turkey, where this type of production is still practised nowadays on a small scale, in a family context, according to need, with collective oral transmission, and kilns no larger than 20,000 bricks and 2.50 m in height.[29] The bricks are arranged in such a way that one can 'pass the hand' between them to leave a circulation space for hot gases; the temperature is estimated by the colour of bricks near the entrance of the heating corridors; firing of at least 12 hours is considered complete when no more smoke rises although the kiln is still aflame; and the end of the two-day cooling is estimated by simply placing a hand on the wall, which must be cold. Unsurprisingly, as per the eighteenth century sources, prayers and other magical protective gestures similar to those previously described might also take place.

Whatever the technical option considered, firing, given the difficulty of evenly distributing the heat, systematically gave rise to losses, or at least to poorly baked products, in an estimated proportion of almost 30% of failures for tiles in a masonry oven, whether burnt, underbaked, cracked or warped, and 15% of bricks in a 'Flemish' kiln.[30] However, defective bricks and tiles, the latter usually tested by cantilevering them between two bricks and climbing on them,[31] were rarely thrown away but rather sold at a lower price as inferior quality, or reused in kiln construction.[32] Archaeological clues suggest the same

---

28   Thuillier, "Le four à briques en meule" 469–471.

29   Fenet A., "L'apport des fours à briques traditionnels de la région d'Apollonia (Albanie) à la compréhension des techniques antiques", in Boucheron P. – Broise H. – Thébert Y. (eds.), *La brique antique et médiévale. Production et commercialisation d'un matériau. Actes du colloque international, Saint-Cloud, 16–18 novembre 1995* (Rome: 2000) 103–111.

30   Duhamel du Monceau – Fourcroy de Ramecourt – Gallon, *Descriptions des arts et métiers* 20–55; Anonymous, "Médaille d'or de 500 francs, pour le perfectionnement, dans le département du Haut-Rhin, de la fabrication des briques, d'après la méthode flamande", *Bulletin de la Société industrielle de Mulhouse* 12 (1839) 496–498; Czmara, *Les métiers du feu* 11–12 ; Grizeaud J.-J. (dir.), *Aude (11), Carcassonne, Les Jardins de Grèzes. Découvertes sur le site de Grèzes: un établissement rural gaulois, des vestiges du Bas-Empire et d'une tuilerie du XVIIe s. Rapport final d'opération* (Nîmes: 2013) 131–136; Lebrun G., *Hauts-de-France, Oise. Passel "le Vivier". La briqueterie médiévale. Rapport de fouille. Volume 6* (Glisy: 2017) 17–18.

31   Nègre V., *L'ornement en série. Architecture, terre cuite et carton-pierre* (Paris: 2006) 26.

32   Richard F., "Tuileries et briqueteries en Saône-et-Loire", *Annales de Bourgogne* 62 (1990) 28.

'THE MANNER OF CONDUCTING FIRE' 201

conclusion, the rubbish pits of tile factories containing few whole pieces, even defective ones, and poor quality products being found in place on roofs, walls or floors.[33] Bricks were typically classified according to at least three qualities corresponding to differentiated uses: well-fired bricks for facings; underbaked or cracked, making a dull sound, for interior walls; and over-fired, which may correspond to sets of fused bricks called 'toads' and separated with a crowbar, for foundations.[34] Eighteenth century masons recognized four qualities of brick in Paris, according to their colour, and six in Toulouse (Haute-Garonne, southern France) in 1753, with a price differential of 50% between the best and the worst. Their marketing could possibly have been adapted accordingly: in the nineteenth century, Marseille (Bouches-du-Rhône, French Riviera) tile-makers sold the good tiles locally and exported the others.[35] Thus, presence, in acceptable proportions, of non-optimal products in the firing result was not due to a lack of know-how, but was an integral part of the process and could be valued by selling these products at a lower cost, thus reaching more potential markets.

Kiln firing thus appears as an essential technical node in the architectural terracotta production process, concentrating both the greatest technicality and the greatest risks, both from a technical and economic point of view. The exact learning processes are unfortunately unknown due to lack of data, but oral transmission from a tiler to one or a few individuals serving as his helper(s) is very likely, or from experienced members to new ones in the case of a 'Flemish' team. However, as mentioned in the introduction, the firing itself was only one aspect of 'fire management' for this production. This in fact also increasingly involved the management of fuel resources, which had become chronically scarce at the end of the modern period – a concern that went far beyond the sole framework of this specific production.

---

33   Pallot É., "La toiture vernissée de l'église de Brou, Bourg-en-Bresse. Le contexte d'une restitution", *Monumental* 15 (1996) 8; Gillon P., "Le pavement du XIIIe siècle du collatéral nord du chœur de l'abbatiale de Saint-Maur-des-Fossés (Val-de-Marne)", in Chapelot J. – Chapelot O. – Rieth B. (eds.), *Terres cuites architecturales médiévales et modernes en Île-de-France et dans les régions voisines* (Caen: 2009) 118.

34   Peirs, *La brique* 31–32.

35   Seine-et-Marne departmental archives – 190J 246–247, By tile factory, customer account records, 1866–1878. Leroy V., *Notice sur les constructions des maisons à Marseille au XIXe siècle* (n.p., n.n.: 1847; reprint, Edisud – Aix-en-Provence: 1989) 98; Nègre, *L'ornement en série* 24–26.

2 Managing the Fuel, from Scarcity to Innovation

Indeed, the average wood consumption of a masonry-clad kiln could be estimated at 2 to 2.5 cubic metres per hour, that is a total of 50 to 100 cubic meters for one firing. A tile factory thus consumed the kindling of around 17 ha of forest annually.[36] The tile-maker could use for his supply collection (legal or not), acquisition of standing timber, purchase from a forester (punctual, regular or even in associations), or from resellers, for example innkeepers or specialist merchants.[37] However, wood consumption for various branches of production, construction or domestic use since the Middle Ages had led to a crisis of the forest cover, reaching its climax at the end of the early modern era: the French forest covered 60% of the territory in the middle of the medieval time, 25% in the sixteenth century, and 12.5% in 1825.[38] Burgundian reports of woods inspections around tile factories, as well as convictions of tile-makers for illegally collecting wood in the forests, effectively testify to the attention paid at least occasionally to this matter from the sixteenth century. Wood origin could be traced, and the establishment of new workshops was controlled, based on a 1669 ordinance on forests, a 1723 royal declaration, and the 1827 Forest Code, prohibiting establishment of tile factories 'on the edge of [royal] forests, or half a league [approximately 2 km] from here' without a royal letter of permission.[39] Some establishments were in difficulty or even

---

36 Daumas M., *Histoire générale des techniques. Tome 2. Les premières étapes du machinisme: XVᵉ–XVIIIᵉ siècle* (Paris: 1964) 43–44 ; Collective, *Éclats d'histoire. 10 ans d'archéologie en Franche-Comté. 25000 ans d'héritages* (Besançon: 1995) 81–85.

37 French National Archives – MC/ET CV 509, Paris notaries. Sale of firewood to Guillaume Martin, tile-maker, April 30, 1611; Dietrich Philippe-Frédéric de, *Description des gîtes de minérai et des bouches à feu de la France. Tome troisième* (Paris, Didot jeune: 1800) 322–324; Chapelot J., *Potiers de Saintonge. Huit siècles d'artisanat rural. Exposition, Musée national des arts et traditions populaires, 1975–1976* (Paris: 1975) 52–53; Chapelot O., *La construction sous les ducs de Bourgogne Valois: l'infrastructure (moyens de transport, matériaux de construction)* vol. 3, PhD dissertation (Paris 1 Panthéon-Sorbonne University: 1975), 375–389. Gouzouguec S., "L'artisanat de la terre cuite architecturale en Île-de-France (XIIIᵉ–XVᵉ siècles) : l'apport des sources écrites", in Collective, *Actes des journées archéologiques Île-de-France 1998* (Paris: 2000) 93.

38 Delsalle P., *La France industrielle aux XVIᵉ–XVIIᵉ–XVIIIᵉ siècles* (Gap: 1993) 19–23; Chapelot J., "Une famille d'artisans de la terre cuite à Genouillé (Vienne) (seconde moitié du XVᵉ–début du XVIᵉ siècle", *Archéologie médiévale* 35 (2005) 163–164; Chalvet M., *Une histoire de la forêt* (Paris: 2011) 136.

39 Côte-d'Or departmental archives – B 12226, Parliament of Burgundy. Acquisition of a tile factory in Aubigny, 1584; B 12262, Judgment concerning the heating of a tile factory, 1649; Louis XIV, *Édict portant règlement général pour les eaux et forests. Vérifié en Parlement le 13 aoust 1669* (Paris, Frédéric Léonard: 1669) 89; Gallon de, *Conférence de l'ordonnance*

'THE MANNER OF CONDUCTING FIRE' 203

forced to close from the middle of the sixteenth century, for example the tile factory at Moutiers-Saint-Jean (Côte-d'Or, Burgundy) in the 1650s.[40] Similarly, some authorization requests for setting up tile factories were rejected for lack of wood in the middle of the eighteenth century, for example at Cadalen in 1748 or Fronton in 1760 (Tarn and Haute-Garonne, both in southern France).[41] A survey of kiln-using factories carried out in 1788 was therefore particularly oriented on this question, because of the 'sufficiently known wood dearth'.[42] However, this wood shortage concerned above all larger trees, which were not those used in tile kilns, since a large part of the fire was ensured by bundles of small wood; similarly, not all wood species were necessarily sought after.[43]

In reality, societal parameters also had an impact. First of all, the lack of fuel did not necessarily mean that surrounding land had been totally deforested, but could correspond to a refusal by forest owners to sell wood, tile-makers getting their supplies within a short radius (3 km; because of weight and transport difficulties), thus alternatives were limited.[44] Conversely, a tile factory could also make it possible to value the otherwise unexploited kindling, and establishments were even created specifically for this purpose from the Middle Ages to the nineteenth century.[45] Moreover, and in spite of recurrent and fairly

---

de Louis XIV du mois d'Août 1669. Sur le fait des eaux et forêts, avec les Édits, Déclarations, Coutumes, Arrêts, Règlemens, & autres Jugemens, rendus avant & en interprétation de ladite ordonnance, depuis l'an 1115 jusqu'à présent. Contenant les loix forestières de France, vol. 1 (Paris, au Palais: 1752) 677–678; vol. 2, 455, 467, 472, 741–742, 752, 758–759; Baudrillart J.-J., Code forestier, suivi de l'ordonnance réglementaire et d'une table des matières (Paris: 1827) 42.

40 Côte-d'Or departmental archives – 8H 154, Benedictine Abbey of Moutiers-Saint-Jean. Tilery, 1536–1731. Mordefroid J.-L., "Les chartreux franc-comtois et la terre cuite au XVIIIe siècle. Approche historique et archéologique des ateliers de potiers et des tuileries-briqueteries de Nermier, Bon-Lieu et La Frasnée", Travaux de la Société d'émulation du Jura (1995) 106–108.

41 French National Archives – F[12] 680, Survey of factories and manufactures employing fire in every generality. Montauban generality, 1788–1789; Lesur A., Les poteries et les faïences françaises, vol. 1 (Paris: 1957) 253, 471.

42 French National Archives – F[12] 680, Survey of factories and manufactures employing fire in every generality. Aix generality, 1788–1789.

43 Haussonne, Technologie céramique générale 343.

44 Roupnel G., La ville et la campagne au XVIIe siècle. Étude sur les populations du Pays Dijonnais (Paris: 1955) 301 ; Michelin R., "Une tuilerie de la Bresse louhannaise à l'époque préindustrielle : la tuilerie de Châteaurenaud (1821–1837)", Mémoires de la Société d'Histoire et d'Archéologie de Chalon-sur-Saône 56 (1987) 98–99.

45 Aube departmental archives – 5M 406, Unsanitary, inconvenient and dangerous establishments. Maraye-en-Othe, 1836. Combes A., "Agriculture du Midi. II: Arrondissement de Castres (Tarn)", Journal d'agriculture pratique, de jardinage et d'économie domestique (1839) 168; Thomas J.-B., Traité général de statistique, culture et exploitation des bois, vol. 2

usual accusations of abuse that affected tile-makers as well as forge masters for example, administrative officers or ordinary villagers often had to authorize tile-makers to take wood for their kiln, out of necessity for their products.[46] Finally, deforestation did not affect all regions equally, some retaining abundant forests even at the end of the eighteenth century: the tile-makers of Verdun-sur-le-Doubs (Saône-et-Loire, Burgundy) in 1780 and Mazères (Ariège, southern France) in 1785 thus encountered no difficulties in supply, and some had their own forest plots, as in Verrières-le-Buisson (Essonne, Parisian region) in 1761.[47]

In any case, regulations restricted access rights to wood already in the twelfth century, especially in the royal domain.[48] It was not so much a matter for lords of preserving the forest as of financially profiting from the dynamism of these activities, by forcing tile-makers and other craftsmen to buy the wood instead of collecting it for free, or by taxing it. This logic was even more asserted in the sixteenth and especially seventeenth centuries, when the forest had clearly become, for its owners, a capital asset to develop and exploit in a rational way, with royalties to be paid to the lord, as at the Mesnil-Saint-Père tile factory in 1488, bound to the Montiéramey abbey (Aube, Champagne), or in the royal forest of Pontailler-sur-Saône in 1623 (Côte-d'Or, Burgundy). The same logic was also valid if the owner was a municipality, for example in Aniane in 1601 (Hérault, French Riviera).[49] The situation changed however in the eighteenth

---

(Paris: 1840), 93–95; Gourcy C. de, *Second voyage agricole en Belgique, en Hollande et dans plusieurs départements de la France* (Paris: 1850) 244; Gouzouguec, "L'artisanat de la terre cuite architecturale" 93.

46   Amouric H. – Démians d'Archimbaud G., "Potiers de terre en Provence-Comtat Venaissin au Moyen Âge", in Barral i Altet X. (ed.), *Artistes, artisans et production artistique au Moyen Âge. 2: Commande et travail. Colloque international, 2–6 mai 1983* (Paris: 1988) 609; Chalvet, *Une histoire de la forêt*, 135.

47   French National Archives – F¹² 1498, Application for establishment of a brick factory by Delpech, priest, 1785–1786; MC/ET XXXV 709, Paris notaries. Wood sale by Jeanne Sortet, widow of Remi Hanel, tile-maker, September 28, 1761; Saône-et-Loire departmental archives – 2F 385, Palatin de Dyo, lord of Bresse-sur-Grosne. Tile factory, 1780; Martin L. – Fournier S., *Villevieille – RN 202 à Moriez (Alpes-de-Haute-Provence). Rapport final d'opération. Fouille archéologique* (Nîmes: 2007) 22.

48   French National Library – Chappée 24 70, Lease of the Sublaines tile factory, July 14, 1496; Chapelot J., "Le droit d'accès à l'argile et à la pierre des tuiliers-chauniers d'Écoyeux (Charente-Maritime) aux XVᵉ–XVIᵉ siècles. L'apport des sources judiciaires", in Benoit P. – Braunstein P., *Mines, carrières et métallurgie dans la France médiévale. Actes du colloque de Paris, 19, 20, 21 juin 1980* (Paris: 1983) 120–121.

49   Côte-d'Or departmental archives – B 5711, Pontailler. Tiles revenue, 1623–1624; Mondière Melchior, *L'Abrégé des voitures par l'abrégé des sinuositez & navigation des eaux mortes des valons, paluds, terres inondées & jonction des rivières navigables joignant les deux*

'THE MANNER OF CONDUCTING FIRE'   205

century, when the wood shortage began to become a real concern. Some lords then tried to go back on old, rather permissive, rights, and more severe regulations appeared, prohibiting, for example, the creation of new earthenware kilns in Rouen in 1725 (Seine-Maritime, Normandy).[50] After the Revolution, the administration kept exercising significant vigilance over forests, with a more proactive policy of consumption control or even reforestation. Absence of growth in wood consumption became an increasingly crucial criterion for obtaining authorizations to set up tile factories, with extreme cases of total prohibition, such as the closure by decree of 'any permanent pottery, tile and lime factory' in the Hautes-Alpes department in 1808 (Alps).[51] The local authorities did the same if necessary, the municipal council of Saint-Michel-de-Lanès (Aude, southern France), for example, refusing access to communal woods to a tile factory 'existing for a century' in 1805.[52]

To respond to these difficulties, or to take advantage of local opportunities, use of alternative fuels occurred. Peat could be used when it was available, a known practice in Holland, but also in the nineteenth century in Villadin (Essonne, Parisian region) in 1834, or in Criquebœuf (Calvados, Normandy)

---

      mers (Paris, Melchior Mondière: 1627) 7; Chauffour Jacques de, *Instruction sur le faict des Eaues et Forests. Contenant en abregé les moyens de les gouverner & administrer suivant les Ordonnances des Roys tant anciennes que nouvelles, Arrests & Reglemens sur ce donnéz, & autres observations accoustumées* (Rouen, David du Petit Val: 1642) 495, 501–502; Lesur, *Les poteries et les faïences françaises* vol. 2, 795. Vayssettes J.-L., *Les potiers de terre de Saint-Jean-de-Fos* (n.p.: 1987) 146–149; Lefebvre S., "Les enclaves en forêt de Soignes, XVIᵉ–XVIIIᵉ siècles", in Collective, *Herbeumont. Cinquième congrès de l'association des cercles francophones d'histoire et d'archéologie de Belgique* (n.p.: 1999) 470–471; Chalvet, *Une histoire de la forêt* 111–112.

50   Seine-et-Marne departmental archive – J 515, Saint-Ouen-en-Brie. Lease for a tile factory, 1746; Courtépée Claude – Béguillet Edmé, *Description générale et particulière du duché de Bourgogne, précédée de l'abrégé historique de cette province,* vol. 2 (Dijon, L.N. Frantin: 1777) 354; Rostand A., "Reconnaissance des anciens droits d'usage en forêt des potiers de Saussemesnil (Manche)", *Annales de Normandie* 7.2 (1957); Carel P. – Rioland S., "Les faïenceries du faubourg Saint-Sever, implantation et développement. XVIIᵉ–XIXᵉ siècle. Prospection / inventaire", *Haute-Normandie Archéologique* 4 (1994) 91.

51   French National Archives – F14 4282, Hautes-Alpes. Serres pottery. Decree of October 18, 1810 relating to the closure of pottery and tile factories in the department, 1810; 4408, Meurthe. Transformation of Fremonville tile factory into an earthenware factory, 1802–1803. Essonne departmental archives – 5M 1/1, Unsanitary, inconvenient and dangerous establishments. Angervilliers, 1824.

52   French National Archives – F12 2380, Complaint of Roussel, brick-maker in Saint-Michel-de-Lanès, 1805.

in 1840,[53] and lignite was tried in the middle of the same century.[54] At the turn of the nineteenth century, the question of the use of coal was increasingly raised, especially since it has a significantly higher caloric value than that of vegetable fuels.[55] Coal production increased seven or eight times during this century, but in very specific areas. In almost all cases, this fuel remained less common than wood, even rarefied. English coal was only readily available in port cities (notably Rouen in Normandy), and northern and Flanders coal did not spread until the 1850s. Therefore, except for local consumption, coal had generally to be transported over considerable distances, supply then being conditioned by transport capacities.[56] Real development of coal use for architectural terracotta firing in France was therefore generally later, despite an effort in this direction from the end of the eighteenth century. For example, an attempt at Aniane (Hérault, French Riviera) in 1780 proved unsuccessful, in particular for lack of transport routes, and, in the Yonne department (Burgundy), coal did not develop until the 1840s.[57]

The first specific examples of effective coal use for architectural terracotta firing in France, mainly on the initiative of producers affected by the rise in wood prices, were located in the south of the country, in 1746 in Alès (Gard) with a request for a privilege with an unknown result, and in the Tarn department in 1763 with Carmaux coal, also mentioned without success in 1775 for use in the

---

53   Essonne departmental archive – 8S 1, Statistics of the mineral industry. Tables of quarries, plaster and lime kilns, tile and brick factories, 1835. Dietrich, *Description des gîtes de minérai* 129; Schnitzler J.-H., *De la création de la richesse, ou Des intérêts matériels en France. Tome premier* (Paris: 1842) 149; Termote J., "Production et utilisation de la brique à l'abbaye des Dunes, Coxyde", in Derœux D. (ed.), *Terres cuites architecturales au Moyen Âge. Musée de Saint-Omer, colloque du 7–9 juin 1985* (Arras: 1986) 66–68; Bernouis P. – Dufournier D. – Lecherbonnier Y., *Céramique architecturale en Basse-Normandie. La production de briques et de tuiles, XIXᵉ–XXᵉ siècles* (Cabourg: 2006) 120. Gérard C., "Les tourbières du département de l'Aube au XIXᵉ siècle", *La Vie en Champagne*, 59 (2009) 47–48.

54   Duhamel J.-P.-F.-G., "Rapport sur un Mémoire du Cⁿ Riboud, concernant des matières bitumeuses et charbonneuses trouvées dans le département de l'Ain", *Procès-verbaux des Séances de l'Académie* 2 (1800–1804) 608–610. Schnitzler, *De la création de la richesse* 145–146.

55   Mestiviers B. (ed.), *Briques et tuiles* (Paris: 1985) 33.

56   Woronoff D., *Histoire de l'industrie en France du XVIᵉ siècle à nos jours* (Paris: 1998) 117–119; Belhoste J.-F., "L'essor de l'usage industriel du charbon en France au XVIIIᵉ siècle", in Benoit P. – Verna C. (eds.), *Le charbon de terre en Europe occidentale avant l'usage industriel du coke* (Turnhout: 1999) 187–190; Bernouis – Dufournier – Lecherbonnier, *Céramique architecturale en Basse-Normandie* 120–121.

57   Daumas M., *L'archéologie industrielle en France* (Paris: 1980) 53; Vayssettes, *Les potiers de terre de Saint-Jean-de-Fos* 146–149; Delor J.-P., *L'industrie de la tuile et de la brique au nord de l'Yonne* (Dijon: 2005) 32–33.

'THE MANNER OF CONDUCTING FIRE' 207

Toulouse (Haute-Garonne) area.[58] Examples are hardly more numerous in the 1770s: in 1772 in Franche-Comté with Ronchamp coal (Haute-Saône, eastern France),[59] in 1777 in Toulouse (Haute-Garonne, southern France) where the Noubel brothers were allowed to open two brick factories using coal,[60] and the same year in Montélimar (Drôme, southern France). In the latter case, coal loads had to pass through a dozen tolls over the 120 km route, greatly reducing its economic interest.[61] This fuel was introduced into the faïence factories of Rouen only in 1781, by the Irishman William Sturgeon.[62] Coal use was therefore above all conditioned by the existence of a viable supply source, and if possible of the right type of coal.[63] Thus, according to the general survey of 1788 of kilns, the 196 tile and brick factories of the Montauban generality (southern France) still did not use coal, even though deforestation was almost complete and the price of wood had tripled in half a century. The few inclinations to use coal were generally described as failures, and good management of wood stocks was sometimes enough to avoid shortages, for example in the Moissac sub-delegation.[64] On the other hand, in some regions such as the Languedoc province (southern France) and despite requests from the administration in this regard, producers 'didn't know how to' (or perhaps didn't want to) use coal, which initially had the reputation of making brittle products.[65]

58 Tarn departmental archive – C 364, Use of Carmaux stone coal for bricks manufacture, 1763; C 1005, Request by the Albi community for the prescription to be made to brick-makers, dyers and others to use coal and stop burning wood, 1787–1790. Venel Gabriel-François, *Instructions sur l'usage de la houille, plus connue sous le nom impropre de charbon de terre, pour faire du feu; sur la maniere de l'adapter à toute sorte de feux; & sur les avantages, tant publics que privés, qui résulteront de cet usage* (Avignon, Gabriel Regnault: 1775) 441–459; Lesur, *Les poteries et les faïences françaises* vol. 1 (1957), 13–14; Beltran A. – Griset P., *Histoire des techniques aux XIXᵉ et XXᵉ siècles* (Paris: 1990) 8–9.

59 Haute-Saône departmental archive – C 558, Ronchamp coal mines. Coal supply for tile, brick and lime kilns, 1772.

60 French National Archives – F¹² 1498, Memory of the Noubel brothers, merchants in Toulouse, to establish two bricky factories on the city suburb, 1777. Nègre, *L'ornement en série* 31–32.

61 French National Archives – F¹² 1498, Proposal of de Chabrignac to establish in Montélimart three lime kilns, a tile factory and a silk spinning mill supplied with Forèz coal, 1777. Lesur, *Les poteries et les faïences françaises* vol. 2, 832.

62 Carel – Rioland, "Les faïenceries du faubourg Saint-Sever" 92.

63 Haussonne, *Technologie céramique générale* vol. 2, 339–340.

64 French National Archives – F¹² 680, Survey of factories and manufactures employing fire in every generality. Montauban generality, 1788–1789.

65 French National Archives – F¹² 680, Survey of factories and manufactures employing fire in every generality. Province of Languedoc, 1788–1789. Caron F., *La dynamique de l'innovation. Changement technique et changement social (XVIᵉ–XXᵉ siècle)* (Paris: 2010) 42.

Further north, the situation seemed less worrying, due in particular to a less significant shortage of wood, although real in places.[66] In Touraine (Loire Valley), coal use only developed in the town of Langeais, and only in the middle of the nineteenth century.[67] Moreover, if it was therefore not always used, coal was nevertheless more abundant than in the southern regions, and could be used if necessary: it was, for example, widely used in Anjou lime kilns, although rarely in local tile factories. The only two brick factories of the Rennes generality (Ille-et-Vilaine, Brittany) used coal in 1788, that of Croisic bringing its own from Nantes (Loire-Atlantique).[68] In the Paris region, where coal was used as early as 1769 at the Port-à-l'Anglais (Essonne) brick factory,[69] a privilege to fire with coal was requested in 1785 by a private individual who claimed to import the practice from 'abroad'; the government was careful not to grant him an exclusive privilege, and contented itself with granting him a bonus, the enterprise not seeming to have succeeded in the end.[70] The municipality also tried to ban wood in the following years, with an ordinance that remained unenforced in 1787, and in 1788 supported coal use by the manufacturer Von-Esch.[71] For the same reason, coal use also developed late in Flanders for masonry-clad kilns (its use being implicit for 'Flemish' kilns, cf. *supra*), despite the proximity of northern and Belgium coal basins. Only in 1782 a tile-maker from Hainaut asked to be authorized to set up four coal-fired tile factories in 'the Provinces of Hainaut, Flanders, Artois, Picardy'.[72] The 1788 survey indeed confirmed the abundance of wood and the general lack of interest of tile-makers in coal in Flanders and Picardy: the intendant of Flanders was thus surprised that, in general, the sixteen 'brick and tile factories ha[d] not yet adopted the use of stone coal',[73] and only one brickyard out of twenty-four used coal in that of

---

66 French National Archives – F¹² 680, Survey of factories and manufactures employing fire in every generality. Tours generality, 1788–1789.

67 Indre-et-Loire departmental archive – 97J 28, Langeais county. Tile and brick factories, 2002.

68 French National Archives – F¹² 680, Survey of factories and manufactures employing fire in every generality. Rennes generality, 1788–1789.

69 Le Roux T., *Les nuisances artisanales et industrielles à Paris. 1770–1830*, vol. 1, PhD dissertation (Paris 1 Panthéon-Sorbonne University: 2007) 221.

70 French National Archives – F¹² 2380, Application of coal to plaster, lime, brick and tile firing, by Jazet, 1785.

71 Le Roux, *Les nuisances artisanales et industrielles à Paris*, vol. 1, 221.

72 French National Archives – F¹² 2380, Application of coal to brick firing, by Verchain, 1782.

73 Coal or 'stone coal (*charbon de terre*)', as opposed to charcoal or 'wood coal (*charbon de bois*)'.

'THE MANNER OF CONDUCTING FIRE' 209

Soissons.[74] Real interest in coal therefore rather developed in the nineteenth century, from the 1830s in the Paris region with Sarcelles (Val-d'Oise) in 1837, in the 1840s in Yonne (Burgundy), or even in 1875 in Lachapelle-aux-Pots (Oise, Picardy).[75] These establishments were then able to benefit in particular from Charleroi coal (Belgium), specifically used in 1861 in Andilly (Val-d'Oise, Parisian region), yet 250 km away.[76]

If it were not so simple to modify fuel at the turn of the nineteenth century, supply difficulties were also a powerful driver of innovation. Instead of modifying the type of fuel, another possibility was to reduce its consumption in the furnace. The general idea pursued in this direction appeared at the end of the eighteenth century: it was to recover the heat released by the full fire phase in one kiln to 'preheat' another kiln, thus partly or totally avoiding the 'small fire' phase in this second kiln, or at least spare some fuel for it. A few superimposed kilns appeared at the end of the eighteenth and the beginning of the nineteenth century. Generally, two kilns were stacked, both were loaded with products, and the fire set normally in the lower kiln. Once this fire was finished, the top kiln was already hot, and there was no need for a small fire in this one. The oldest known example in French records seems to be the one of the agronomist de La Tour d'Aigues.[77] This author suggested that individuals make their own bricks and tiles with this kiln 'in their spare time', especially in winter

---

74 French National Archives – F^{12} 680, Survey of factories and manufactures employing fire in every generality. Flanders and Soisson generalities, 1788–1789.

75 Yvelines departmental archive – 5M 3, Industrial statistics. State of manufactures by district. Correspondence and inquiries, 1806–1812; Brongniart A., *Traité des arts céramiques ou des poteries considérées dans leur histoire, leur pratique et leur théorie*, vol. 1 (Paris: 1844) 345–347 ; Gentil A. – Musmacque A., "La Céramique, n°301, février 1913. Excursion des membres de l'Union céramique et chaufournière de France, à Beauvais et dans le pays de Bray, les 22–23 mai 1912", *Bulletin du Groupe de Recherches et d'Études de la Céramique du Beauvaisis* 13 (1991) 182; Dauphin J.-L. – Delor J.-P., *De tuile et de brique. Contribution à l'étude de l'artisanat tuilier et de l'habitat traditionnel dans le nord de l'Yonne* (Villeneuve-sur-Yonne: 1994) 16–17; Baduel D., *Briqueteries et tuileries disparues du Val-d'Oise* (Saint-Martin-du-Tertre: 2002) 242; Baduel D. – Gentili F. – Warmé N., "Les fours de la briqueterie du Petit Merisier (XVIIIe–XIXe siècle) à Sarcelles (Val-d'Oise)", *Bulletin du Groupe de recherches et d'études de la céramique du Beauvaisis* 26 (2005) 12, 34.

76 Val-d'Oise departmental archive – 7M 99, Unsanitary, inconvenient and dangerous establishments. Andilly, 1861. Baduel, *Briqueteries et tuileries disparues du Val-d'Oise* 34.

77 De La Tour d'Aigues, "Description d'un four, dans lequel on peut cuire des briques, des tuiles, & toutes sortes de poterie très-économiquement", *Mémoires d'agriculture, d'économie rurale et domestique* (1787) 1–8; Cubells M., "Un agronome aixois au XVIIIe siècle: le président de la Tour d'Aiguës, féodal de combat et homme des Lumières", *Annales du Midi: revue archéologique, historique et philologique de la France méridionale* 96.165 (1984) 31–59.

when their activities were reduced. He seemed to use it effectively, with 'country people' as workforce. This type of kiln was particularly known for faïence production, for example at the national factory of Sèvres. In fact, a second documented superimposed kiln corresponds to the patent of a faïence maker, Elzéar Bonnet of Apt (Vaucluse, French Riviera), which he used in particular to produce paving tiles.[78] Thanks to the superimposition, firing in the upper kiln consumed half the wood as in the lower kiln fired before. Another set of such furnaces was that of Brest and Lorient arsenals (Brittany), described a few times by students of the Paris Mines School on compulsory study trips between 1829 and 1852.[79] The Lorient ones were a little particular, as the lower part was used for lime production, carried out between the two phases of the brick kiln fire on top. Finally, the factory of Cheuvreuse and Bouvert in Metz (Moselle, Lorraine) also seems to have used a superimposed kiln around the 1830s.[80]

A simpler variant consisted in juxtaposing kilns instead of superimposing them: it involved building several kilns next to each other, and connecting them by heat pipes. It was possible to open or close these pipes as needed, to allow the heat from a firing kiln to preheat one or more of the other kilns.[81] The first French patent for this type of kiln was that of Champion, Fabre and Janier-Dubry in 1830.[82] This solution had an increasing success from the end of the 1830s, with in particular Lamy's patent in 1838, which was not fundamentally different from the previous one, and could group together an indefinite number of kilns (theoretically from two to twenty according to the patent, and

---

78   French National Institute of Industrial Property – 1BA1101, Bonnet, economical faïence kiln, 1806. Terras A., "Description d'un Four à briques inventé par M. Bonnet, faïencier à Apt, département de Vaucluse, mars 1811", *Bulletin de la Société d'Encouragement pour l'Industrie Nationale* 10.81 (1811) 65–69.

79   Library of the Paris Mines School – J 1829 (23), Travel diary of Garella, Reverchon and Bineau vol. 4 (1829) 66–69; J 1831 (30), Travel diary of de Hennezel, Senez and Foy, vol. 3 (1831); J 1852 (143), Travel diary of Dormoy (1852) 27–32. Challeton de Brughat F., *L'art du briquetier*, vol. 1 (Paris: 1861) 156–157. Thuillier G., "Une source documentaire à exploiter: les 'Voyages métallurgiques' des élèves-ingénieurs des Mines", *Annales. Histoire, Sciences humaines*, 17.2 (1962) 302–307.

80   Library of the ceramic museum of Sèvres – U 18, Documents and notes by Alexandre Brongniart, 19th century.

81   Some also proposed to recover heat from other types of kilns or machines, for example a steam engine. French National Institute of Industrial Property – 1BA9118, Meltzer, bricks firing, 1841.

82   French National Institute of Industrial Property – 1BA6040, Champion, Fabre and Janier-Dubry, manufacture and firing of tiles and bricks, 1830.

'THE MANNER OF CONDUCTING FIRE'   211

four in Lamy's factory at Pont-sur-Yonne in Burgundy).[83] A number of similar designs were patented over the next few years,[84] kilns being quickly arranged according to a circular plan by some inventors,[85] with the development of the concept of 'continuous firing' of the set of kilns, even if each kiln taken individually actually had to be stopped regularly. This type of kiln is regularly documented both in patents and in practice until at least the 1860s, generally without noticeable changes in design; we can however note the Lieutard variant in 1853, with a single central hearth firing permanently, the flow of heat being regulated individually towards each kiln arranged around it.[86]

However, a logical development of the juxtaposed kiln, that can already be glimpsed with the Lieutard kiln, consisted in completely suppressing the small fire while keeping only the large one. This solution was particularly suitable for firing very large quantities of products in order to supply the construction sites of developing cities, and when coal was available, from the middle of the nineteenth century. It involved juxtaposing around twenty kilns open to each other, forming a ring around a chimney, and place the ignited coal in one of the kilns, which then corresponded to the 'full fire'. Opposite this kiln in the ring, by a system of hatches, a kiln communicated with the fresh air from outside, and the nearby kiln was connected to the chimney, with a tight separation between the two. Because of the draft, the air thus traversed the

---

83   Yonne departmental archive – 5M 10/25, Unsanitary, inconvenient and dangerous establishments. Pont-sur-Yonne, 1838. French National Institute of Industrial Property – 1BA7017, Lamy, united and dependent brick kilns, 1838.

84   French National Institute of Industrial Property – 1BA8261, Delpierre, production and firing with continuous kiln of bricks and tiles, 1840; 1BA8452, Lebrun, continuous brick and tile kilns, 1840; 1BA8578, Boistel, various upgrades to a brick kiln, 1839; 1BA12554, Cheuvreuse and Bouvert, clay preparation machine and heating kiln, 1838.

85   French National Institute of Industrial Property – 1BA7357, Maille, brick kiln, 1839.

86   Seine-et-Marne departmental archive – 5 Mp 193/2, Unsanitary, inconvenient and dangerous establishments. Fresnes-sur-Marne, 1862. French National Institute of Industrial Property – 1BB7814, Daleth, brick and tile kin, 1848; 1BB8327, Breuille and Breuille, brick kiln, 1849; 1BB8386, Thackeray, drain pipes, 1849; 1BB8807, Breuille, brick kiln, 1849; 1BB13774, Palm, brick and tile kilns, 1852; 1BB16791, Gastellier, brick kiln, 1853; 1BB16991, Lieutard, brick kiln, 1853; 1BB20361, Grimberghs, brick, tile and pipe kiln, 1854; 1BB21652, Muller and Gilardoni, brick and tile kilns, 1854. Thirion, Mastaing de, "Four à cuire la briques, les poteries, tuyaux de drainage, etc.", Le Génie industriel. Revue des inventions françaises et étrangères, annales des progrès de l'industrie agricole et manufacturière 19 (1860) 216–218; Cartier C., "De la brique de campagne à la brique vernissée", L'archéologie industrielle en France 39 (2001) 15. Thomas J., Fours à chaux, Tuileries, Briqueteries en Touraine (Tours: 2005) 109–113; Campbell, L'art et l'histoire de la brique 210. Delor, L'industrie de la tuile et de la brique 180. Paul-Cavallier G. – Hamon M., D'argile et d'hommes. Carnet de voyage au cœur de Terreal (Paris: 2005) 22.

whole ring, first cooling already fired brick batches, passing to the 'full fire' level, then continuing by preheating batches closest to the chimney. This also allowed to load and unload products in these two kilns away from the heat; when the process was completed, all the elements were shifted by one, and the fire thus traversed the whole of the ring without interruption, over at least one or two weeks. Carl Arnold, a German mason, seems to have built such a kiln in 1839 in Fürstenwalde, near Berlin, without patenting it.[87] These kilns are however clearly linked to the first patent concerning them, that of Friedrich Hoffmann, filed in 1858 in Germany and the following year in France.[88] The first Hoffmann kiln built in France seems to have been at the Sachot tile factory in Montereau-Fault-Yonne (Seine-et-Marne, Parisian region) in 1865, and they then only appeared occasionally until around 1880, before a very progressive diffusion. An 1870 source indicated a total of 540 Hoffmann kilns in operation or under construction worldwide, including 295 in Prussia alone, 70 in Britain, 51 in Austria, 38 in the rest of Germany and 30 in Italy. France came next, with 19 kilns, preceding 10 kilns in the United States.[89] However, we were only able to find definite testimonies for barely a dozen kilns of this type in France before 1885, for example, in addition to that of Montereau previously mentioned, those of the Oustau factory in Aureilhan in 1873 (Hautes-Pyrénées, southern France), Septveilles in 1879 and Mortcerf in 1881 (Seine-et-Marne, Parisian region), or Lachapelle-aux-Pots in 1885 (Oise, Picardy). The famous example of the Le Creusot-Montceau ecomuseum (Saône-et-Loire, Burgundy) was only built in 1892.[90] In any case, they were generally variants of the Hoffmann patent, build by specialist contractors with patent licences.

---

87 Lehmann J., *Die Obere Havel und der Finowkanal* (Erfurt: 2015) 145.

88 French National Institute of Industrial Property – 1BB43230 Hoffmann and Licht, system of annular kilns with continuous action, 1859; Hoffmann F., Licht A., "Four à feu continu pour la cuisson des briques", *Le Génie industriel. Revue des inventions françaises et étrangères, annales des progrès de l'industrie agricole et manufacturière* 30 (1865) 296–300.

89 Bois V., "Rapport fait sur le four annulaire à action continue de M. Frédérich Hoffman Hoffman, ingénieur à Berlin, représenté, à Paris, par M. Bourry, 80, rue Taitbout", *Bulletin de la Société d'Encouragement pour l'Industrie Nationale*, 69.17 (1870) 273.

90 Gentil – Musmacque, "La Céramique", 182. Singer C., Williams T. (eds.), *A history of technology*, vol. 5 (Oxford: 1958) 667–668; Plancke R.-C., "En 1909 à Mortcerf, le centenaire des établissements céramiques Houbé", *Notre département, la Seine-et-Marne* 9 (1989) 17–18; Cartier J., "La tuile mécanique, une technologie du XIXe siècle", *Monumental* 15 (1996) 28; Bonhôte J. – Cranga Y., "Les usines et la villa Oustau. Un pôle patrimonial de l'industrie céramique en Midi-Pyrénées", *L'archéologie industrielle en France* 39 (2001) 30–32; Bonnot T. – Notteghem P., "La Briqueterie. Écomusée du Creusot-Montceau", *L'archéologie industrielle en France* 39 (2001) 38. Cartier, "De la brique de campagne" 11–13; Peirs, *La brique* 296. Paul-Cavallier – Hamon, *D'argile et d'hommes* 23–24; Bernouis – Dufournier – Lecherbonnier, *Céramique architecturale en Basse-Normandie* 35. Rambaud I., *La Seine-et-Marne industrielle. Innovations, talents, archives inédites* (Lyon: 2010) 27; Cochet G., "La

'THE MANNER OF CONDUCTING FIRE' 213

Parallel to these circular furnaces, another possibility to use a continuous full fire was to make mobile, not the fire, but the load. The first corresponding patent seems to have been the one filed in France by Testu in 1847: the idea was to keep the fire in a fixed place and to pass the products in front of it, rising in temperature when they approached it, and cooling when they moved away.[91] This first patent placed the products in one big column inside a 6.50 m high chimney, with the fire 2 m high, probably out of habit compared to masonry-clad kilns where the products are stacked, and to allow hot gases, naturally rising, to preheat the products. This arrangement was still very impractical, requiring a mobile carriage and a powerful jack at the base of the column to remove the fired products. A much more effective solution was proposed in 1852 by the engineer Péchiné and the tile-maker Colas, namely to use the exact same technical principle, but arranging the 'chimney' almost horizontally forming a kind of inclined gallery, the products here again entering from the highest end, on trolleys, to benefit from the natural draft of the hot gases.[92] This model was several times improved in the following years, the two men forming the Lyon Ceramic Products General Company to exploit it and selling at least one operating licence, and another variant patented and built by Demimuid in Commercy (Meuse, Lorraine) in 1853.[93] This type of kiln, however, had limited immediate success: Borie, who held the French patent for modern hollow bricks, had for example apparently tried a Demimuid kiln, before opting for juxtaposed kilns, because his clay was baking too slowly.[94] Thus, if this type of kiln, known as a tunnel-kiln and with gas as fuel, is currently

---

briqueterie-tuilerie du 'moulin de Septveilles'", *Bulletin de la Société d'Histoire et d'Archéologie de l'Arrondissement de Provins* 167 (2013) 65–66; Chapelot J., "La tuilerie au début du XXe siècle d'après les cartes postales anciennes. Images de la fin d'un artisanat et de la puissance de l'industrie", in Thuillier F. (dir.), *Les terres cuites architecturales en France du Moyen Âge à l'époque contemporaine. Recherches sur les tuileries et les productions tuilières* (Drémil-Lafage: 2019) 425–426; Orefici L., Nauleau J.-F., "Les tuileries d'Arthon (Chaumes-en-Retz, Loire-Atlantique). Un site majeur à découvrir", in Thuillier F. (dir.), *Les terres cuites architecturales en France du Moyen Âge à l'époque contemporaine. Recherches sur les tuileries et les productions tuilières* (Drémil-Lafage: 2019) 316.

91 French National Institute of Industrial Property – 1BB5269, Testu, brick firing with continuous firing, 1847.

92 French National Institute of Industrial Property – 1BB13383, Péchiné and Colas, continuous firing kiln for tiles and bricks, 1852.

93 French National Institute of Industrial Property – 1BB15598, Demimuid, tile and brick kiln, 1853; 1BB16011, Péchiné and Colas, continuous firing kiln for tiles and bricks, 1853. Saône-et-Loire departmental archive – 6U 535, Society of Saône-et-Loire perfected tile works, 1855.

94 Anonymous, "Rapport sur les briques creuses dites tubulaires de M. P. Borie et sur la machine qui sert à les fabriquer à l'usine de la rue de la Muette, à Paris", *Bulletin de la Société d'Encouragement pour l'Industrie Nationale* 56 (1857) 673–683.

the most used in Europe, it still remained very theoretical at the end of the nineteenth century, the circular kiln being the most popular of the continuous kilns, and non-continuous masonry-clad kilns actually remaining the most common until the turn of the twentieth century.

The 'manner of conducting fire' in the context of the production of architectural terracotta in France during a large early modern period thus corresponds not only to a purely technical process, but to a much larger set of dynamics and interactions, in particular social, economic and innovative. Both technically and economically, fire management was the key point in the process and know-how in tile and brick factories, up until the nineteenth century. Much more complex than extracting clay or moulding products, it was really the 'core' of the business, and a failed firing, with or without destroying the kiln, could ruin a business. Consequently, whether we consider small perennial tile factories or mobile teams of 'Flemish' brick-makers, the mastery of know-how, namely how to arrange products inside the kiln, and know how to 'read' the fire to understand more or less intuitively the processes at work, was what 'made' the master tile-maker. However, due to time constraints, he also had to rely at times on less qualified assistants and apprentices, thus allowing a transmission of know-how but also generating a risk. On a larger scale, the increasing wood crisis brought with it a variety of consequences going far beyond mere technical issues. Wood became a merchandise that owners could sell at a good price, taking advantage of the crisis to make a profit. Tile-makers sometimes could not do anything about this and had to close their business – perhaps a less serious problem than one might imagine, given that these people in any case evolved in a world of multiple activities and could therefore often practice other crafts, tile-making being sometimes from the start an ancillary activity. When they could, they tried to adapt, with limited possibilities: in particular, coal was not always suitable, for logistical reasons, but also out of cultural distrust. The crisis finally forced innovation, by pushing the development of kilns consuming less fuel: a battery of interconnected kilns to recover energy was a particularly interesting solution. Through development, this pressure also led, in the middle of the nineteenth century, to the making of new, more clearly innovative solutions, departing from the way of thinking about traditional kilns. The circular Hoffmann-type kiln, still quite similar to previous juxtaposed kilns, was the most popular during the last quarter of the century, although adopted much more slowly in France than in other industrialized countries, starting with Germany. The other possibility, the tunnel kiln, was on the other hand radically 'alien' in its operation: consequently, even though it appeared at the same time or even a little before the circular kiln, and it seems more specifically in France, it had even less success. Being at the core of

'THE MANNER OF CONDUCTING FIRE' 215

current industrial systems, its adoption did not take place until the twentieth century, with a further evolution of fuel from coal to gas.

Bibliography

Primary Sources

French National Library – Chappée 24 70, Lease of the Sublaines tile factory, 14 July 1496.

French National Archives – F12 680, Survey of factories and manufactures employing fire in every generality, 1788–1789.

French National Archives – F12 1498, Memory of the Noubel brothers, merchants in Toulouse, to establish two bricky factories on the city suburb, 1777. Proposal of de Chabrignac to establish in Montélimart three lime kilns, a tile factory and a silk spinning mill supplied with Forèz coal, 1777. Application for establishment of a brick factory by Delpech, priest, 1785–1786.

French National Archives – F12 2380, Application of coal to brick firing, by Verchain, 1782. Application of coal to plaster, lime, brick and tile firing, by Jazet, 1785. Manufacture of tiles and bricks. Complaint of Roussel, brick-maker in Saint-Michel-de-Lanès, 1805. Draft of a treatise on the art of tile making by Dutroncy, 1817.

French National Archives – F14 4282, Hautes-Alpes. Serres pottery. Decree of October 18, 1810 relating to the closure of pottery and tile factories in the department, 1810.

French National Archives – 4408, Meurthe. Transformation of Fremonville tile factory into an earthenware factory, 1802–1803.

French National Archives – MC/ET XXXV 709, Paris notaries. Wood sale by Jeanne Sortet, widow of Remi Hanel, tile-maker, 28 September 1761.

French National Archives – MC/ET LVIII 470, Paris notaries. Lease of a tile factory, February 1, 1775. French National Archives – MC/ET CV 509, Paris notaries. Sale of firewood to Guillaume Martin, tile-maker, 30 April 1611.

Aube departmental archives – 5M 397, Unsanitary, inconvenient and dangerous establishments. Auxon, 1847.

Aube departmental archives – 5M 404, Unsanitary, inconvenient and dangerous establishments. Laines-aux-Bois, 1845–1846.

Aube departmental archives – 5M 406, Unsanitary, inconvenient and dangerous establishments. Maraye-en-Othe, 1836.

Côte-d'Or departmental archives – B 5711, Pontailler. Tiles revenue, 1623–1624.

Côte-d'Or departmental archives – B 12226, Parliament of Burgundy. Acquisition of a tile factory in Aubigny, 1584.

Côte-d'Or departmental archives – B 12262, Judgment concerning the heating of a tile factory, 1649.

Côte-d'Or departmental archives – 8H 154, Benedictine Abbey of Moutiers-Saint-Jean. Tilery, 1536–1731.

Essonne departmental archives – 5M 1/1, Unsanitary, inconvenient and dangerous establishments. Angervilliers, 1824.

Essonne departmental archive – 8S 1, Statistics of the mineral industry. Tables of quarries, plaster and lime kilns, tile and brick factories, 1835.

Haute-Saône departmental archive – C 558, Ronchamp coal mines. Coal supply for tile, brick and lime kilns, 1772.

Indre-et-Loire departmental archive – 97J 28, Langeais county. Tile and brick factories, 2002.

Saône-et-Loire departmental archives – 2F 385, Palatin de Dyo, lord of Bresse-sur-Grosne. Tile factory, 1780.

Saône-et-Loire departmental archive – 6U 535, Society of Saône-et-Loire perfected tile works, 1855.

Seine-et-Marne departmental archive – J 515, Saint-Ouen-en-Brie. Lease for a tile factory, 1746.

Seine-et-Marne departmental archives – 190J 246–247, By tile factory, customer account records, 1866–1878.

Seine-et-Marne departmental archive – 5 Mp 193/2, Unsanitary, inconvenient and dangerous establishments. Fresnes-sur-Marne, 1862.

Tarn departmental archive – C 364, Use of Carmaux stone coal for bricks manufacture, 1763.

Tarn departmental archive – C 1005, Request by the Albi community for the prescription to be made to brick-makers, dyers and others to use coal and stop burning wood, 1787–1790.

Val-de-Marne departmental archives – 70J 51, Gournay brick factory. Studies, chemical analysis of bricks, 1883–1950.

Val-d'Oise departmental archive – 7M 99, Unsanitary, inconvenient and dangerous establishments. Andilly, 1861.

Yonne departmental archive – 5M 10 / 25, Unsanitary, inconvenient and dangerous establishments. Pont-sur-Yonne, 1838.

Yvelines departmental archive – 5M 3, Industrial statistics. State of manufactures by district. Correspondence and inquiries, 1806–1812.

Inventory of architectural heritage (Mérimée database) – IA00013325, Guilly, tile factory, 19th century.

French National Institute of Industrial Property – 1BA1101, Bonnet, economical faïence kiln, 1806.

French National Institute of Industrial Property – 1BA6040, Champion, Fabre and Janier-Dubry, manufacture and firing of tiles and bricks, 1830.

'THE MANNER OF CONDUCTING FIRE'

French National Institute of Industrial Property – 1BA7017, Lamy, united and dependent brick kilns, 1838.

French National Institute of Industrial Property – 1BA7357, Maille, brick kiln, 1839.

French National Institute of Industrial Property – 1BA8261, Delpierre, production and firing with continuous kiln of bricks and tiles, 1840.

French National Institute of Industrial Property – 1BA8452, Lebrun, continuous brick and tile kilns, 1840.

French National Institute of Industrial Property – 1BA8578, Boistel, various upgrades to a brick kiln, 1839.

French National Institute of Industrial Property – 1BA9118, Meltzer, bricks firing, 1841.

French National Institute of Industrial Property – 1BA12554, Cheuvreuse and Bouvert, clay preparation machine and heating kiln, 1838.

French National Institute of Industrial Property – 1BB7814, Daleth, brick and tile kin, 1848.

French National Institute of Industrial Property – 1BB5269, Testu, brick firing with continuous firing, 1847.

French National Institute of Industrial Property – 1BB8327, Breuille and Breuille, brick kiln, 1849.

French National Institute of Industrial Property – 1BB8386, Thackeray, drain pipes, 1849.

French National Institute of Industrial Property – 1BB8807, Breuille, brick kiln, 1849.

French National Institute of Industrial Property – 1BB13383, Péchiné and Colas, continuous firing kiln for tiles and bricks, 1852.

French National Institute of Industrial Property – 1BB13774, Palm, brick and tile kilns, 1852.

French National Institute of Industrial Property – 1BB15598, Demimuid, tile and brick kiln, 1853.

French National Institute of Industrial Property – 1BB16011, Péchiné and Colas, continuous firing kiln for tiles and bricks, 1853.

French National Institute of Industrial Property – 1BB16791, Gastellier, brick kiln, 1853.

French National Institute of Industrial Property – 1BB16991, Lieutard, brick kiln, 1853.

French National Institute of Industrial Property – 1BB20361, Grimberghs, brick, tile and pipe kiln, 1854.

French National Institute of Industrial Property – 1BB21652, Muller and Gilardoni, brick and tile kilns, 1854.

French National Institute of Industrial Property – 1BB43230 Hoffmann and Licht, system of annular kilns with continuous action, 1859.

Library of the Paris Mines School – J 1829 (23), Travel diary of Garella, Reverchon and Bineau, vol. 4 (1829) 66–69.

Library of the Paris Mines School – J 1831 (30), Travel diary of de Hennezel, Senez and Foy, vol. 3 (1831).

Library of the Paris Mines School – J 1852 (143), Travel diary of Dormoy (1852) 27–32.

Library of the ceramic museum of Sèvres – U 18, Documents and notes by Alexandre Brongniart, 19th century.

Secondary Sources

Adams W.H., "Dating Historical Sites: The Importance of Understanding Time Lag in the Acquisition, Curation, Use, and Disposal of Artifacts", *Historical Archaeology* 37.2 (2003) 38–64.

Amouric H. – Démians d'Archimbaud G., "Potiers de terre en Provence-Comtat Venaissin au Moyen Âge", in Barral i Altet X. (ed.), *Artistes, artisans et production artistique au Moyen Âge. 2 : Commande et travail. Colloque international, 2–6 mai 1983* (Paris: 1988) 601–623.

Anonymous, "Médaille d'or de 500 francs, pour le perfectionnement, dans le département du Haut-Rhin, de la fabrication des briques, d'après la méthode flamande", *Bulletin de la Société industrielle de Mulhouse* 12 (1839) 496–498.

Anonymous, "Rapport sur les briques creuses dites tubulaires de M. P. Borie et sur la machine qui sert à les fabriquer à l'usine de la rue de la Muette, à Paris", *Bulletin de la Société d'Encouragement pour l'Industrie Nationale* 56 (1857) 673–683.

Anonymous, *Les Briqueteries de Langeais et Cinq-Mars* (Langeais: 1987).

Armogathe J.-R., *Bernard Palissy. Mythe et réalité* (Agen: 1990).

Arnaud D. – Franche G., *Manuel de céramique industrielle* (Paris: 1906).

Auger F. – Lépine J., "Evolution et fonctionnement d'une briqueterie de Sologne aux XIXe et XXe siècles : Montevray à Nouan-le-Fuzelier (Loir-et-Cher)", in Heude B. (dir.), *Les briqueteries-tuileries de Sologne. Cher, Loir-et-Cher, Loiret. Histoires d'hommes et d'établissements qui ont fait de la Sologne un pays de briques* (Lamotte-Beuvron 2012) 23–52.

Baduel D., *Briqueteries et tuileries disparues du Val-d'Oise* (Saint-Martin-du-Tertre: 2002).

Baduel D. – Gentili F. – Warmé N., "Les fours de la briqueterie du Petit Merisier (XVIIIe–XIXe siècle) à Sarcelles (Val-d'Oise)", *Bulletin du Groupe de recherches et d'études de la céramique du Beauvaisis* 26 (2005) 11–40.

Baudrillart J.-J., *Code forestier, suivi de l'ordonnance réglementaire et d'une table des matières* (Paris: 1827).

Belhoste J.-F., "L'essor de l'usage industriel du charbon en France au XVIIIe siècle", in Benoit P. – Verna C. (eds.), *Le charbon de terre en Europe occidentale avant l'usage industriel du coke* (Turnhout: 1999) 187–198.

Beltran A. – Griset P., *Histoire des techniques aux XIXe et XXe siècles* (Paris: 1990).

Bernouis P. – Dufournier D. – Lecherbonnier Y., *Céramique architecturale en Basse-Normandie. La production de briques et de tuiles, XIXe–XXe siècles* (Cabourg: 2006).

Bernstein D. – Champetier J.-P. – Sartre J., *L'histoire de la brique en France. Approche exploratoire. Rapport de recherche* (Paris: 1984).

Bernstein D. – Bruel A. – Filatre P. – Champetier J.-P., *Histoire d'un matériau de construction: la brique en France* (n.p.: 1989).

Bois V., "Rapport fait sur le four annulaire à action continue de M. Frédérich Hoffman Hoffman, ingénieur à Berlin, représenté, à Paris, par M. Bourry, 80, rue Taitbout", *Bulletin de la Société d'Encouragement pour l'Industrie Nationale*, 69.17 (1870) 263–275.

Bonhôte J. – Cranga Y., "Les usines et la villa Oustau. Un pôle patrimonial de l'industrie céramique en Midi-Pyrénées", *L'archéologie industrielle en France* 39 (2001) 30–36.

Bourry É., *Traité des industries céramiques* (Paris: 1897).

Bonnot T. – Notteghem P., "La Briqueterie. Écomusée du Creusot-Montceau", *L'archéologie industrielle en France* 39 (2001) 38–47.

Brongniart A., *Traité des arts céramiques ou des poteries considérées dans leur histoire, leur pratique et leur théorie* (Paris: 1844).

Campbell J., *L'art et l'histoire de la brique. Bâtiments privés et publics du monde entier* (Paris: 2004).

Carel P. – Rioland S., "Les faïenceries du faubourg Saint-Sever, implantation et développement. XVIIe–XIXe siècle. Prospection / inventaire", *Haute-Normandie Archéologique* 4 (1994) 89–98.

Caron F., *La dynamique de l'innovation. Changement technique et changement social (XVIe–XXe siècle)* (Paris: 2010).

Cartier C., "De la brique de campagne à la brique vernissée", *L'archéologie industrielle en France* 39 (2001) 8–18.

Cartier J., "La tuile mécanique, une technologie du XIXe siècle", *Monumental* 15 (1996) 26–31.

Chapelot J., *Potiers de Saintonge. Huit siècles d'artisanat rural. Exposition, Musée national des arts et traditions populaires, 1975–1976* (Paris: 1975).

Chapelot J., "Le droit d'accès à l'argile et à la pierre des tuiliers-chauniers d'Écoyeux (Charente-Maritime) aux XVe–XVIe siècles. L'apport des sources judiciaires", in Benoit P. – Braunstein P., *Mines, carrières et métallurgie dans la France médiévale. Actes du colloque de Paris, 19, 20, 21 juin 1980* (Paris: 1983) 117–168.

Chapelot J., "Une famille d'artisans de la terre cuite à Genouillé (Vienne) (seconde moitié du XVe–début du XVIe siècle", *Archéologie médiévale* 35 (2005) 159–165.

Chapelot J., "La tuilerie au début du XXe siècle d'après les cartes postales anciennes. Images de la fin d'un artisanat et de la puissance de l'industrie", in Thuillier F. (dir.), *Les terres cuites architecturales en France du Moyen Âge à l'époque contemporaine. Recherches sur les tuileries et les productions tuilières* (Drémil-Lafage: 2019) 395–501.

Chapelot O., *La construction sous les ducs de Bourgogne Valois : l'infrastructure (moyens de transport, matériaux de construction)* (Ph.D. dissertation, Paris 1 Panthéon-Sorbonne University: 1975).

Carette M. – Derœux D., *Carreaux de pavement médiévaux de Flandre et d'Artois (XIIIe–XIVe siècles)* (Arras: 1985).

Challeton de Brughat F., *L'art du briquetier* (Paris: 1861).

Chalvet M., *Une histoire de la forêt* (Paris: 2011).

Charpail M.-P., *Une tuilerie à Bezanleu* (Saint-Cyr-sur-Morin: 1995).

Chauffour Jacques de, *Instruction sur le faict des Eaues et Forests. Contenant en abregé les moyens de les gouverner & administrer suivant les Ordonnances des Roys tant anciennes que nouvelles, Arrests & Reglemens sur ce donnéz, & autres observations accoustumées* (Rouen, David du Petit Val: 1642).

Cochet G., "La briqueterie-tuilerie du 'moulin de Septveilles'", *Bulletin de la Société d'Histoire et d'Archéologie de l'Arrondissement de Provins* 167 (2013) 55–75.

Collective, *Éclats d'histoire. 10 ans d'archéologie en Franche-Comté. 25000 ans d'héritages* (Besançon: 1995).

Combes A., "Agriculture du Midi. II: Arrondissement de Castres (Tarn)", *Journal d'agriculture pratique, de jardinage et d'économie domestique* (1839) 163–174.

Courtépée Claude – Béguillet Edmé, *Description générale et particulière du duché de Bourgogne, précédée de l'abrégé historique de cette province* (Dijon, L.N. Frantin: 1777).

Cubells M., "Un agronome aixois au XVIIIe siècle: le président de la Tour d'Aiguës, féodal de combat et homme des Lumières", *Annales du Midi: revue archéologique, historique et philologique de la France méridionale* 96.165 (1984) 31–59.

Czmara J.-C., *Les métiers du feu en Champagne méridionale. La terre* (Saint-Cyr-sur-Loire: 2009).

Daumas M., *Histoire générale des techniques.* Tome 2: *Les premières étapes du machinisme: XVe–XVIIIe siècle* (Paris: 1964).

Daumas M., *L'archéologie industrielle en France* (Paris: 1980).

Dauphin J.-L. – Delor J.-P., *De tuile et de brique. Contribution à l'étude de l'artisanat tuilier et de l'habitat traditionnel dans le nord de l'Yonne* (Villeneuve-sur-Yonne: 1994).

De La Tour d'Aigues, "Description d'un four, dans lequel on peut cuire des briques, des tuiles, & toutes sortes de poterie très-économiquement", *Mémoires d'agriculture, d'économie rurale et domestique* (1787) 1–8.

Delor J.-P., *L'industrie de la tuile et de la brique au nord de l'Yonne* (Dijon: 2005).

Delsalle P., *La France industrielle aux XVIe–XVIIe–XVIIIe siècles* (Gap: 1993).

Diderot Denis – d'Alembert, Jean (dirs.), *Recueil de planches, sur les sciences, les arts libéraux et les art méchaniques, avec leur explication. Tome 1* (Paris, Le Breton: 1762).

Dietrich Philippe-Frédéric de, *Description des gîtes de minérai et des bouches à feu de la France. Tome troisième* (Paris, Didot jeune: 1800).

Droz, 'Mémoire sur la manière de perfectionner les tuileries', *Mémoires et observations recueillies par la Société Œconomique de Berne* (1765) 281–322.

Dufaÿ B., "Les tuileries parisiennes du faubourg Saint-Honoré d'après les fouilles des jardins du Carrousel (1985–1987)", in Van Ossel P., *Les Jardins du Carrousel à Paris. Fouilles 1989–1990*, vol. 3 (Paris: 1991) 461–492.

Duhamel J.-P.-F.-G., "Rapport sur un Mémoire du C[n] Riboud, concernant des matières bitumeuses et charbonneuses trouvées dans le département de l'Ain", *Procès-verbaux des Séances de l'Académie* 2 (1800–1804) 608–610.

Duhamel du Monceau Henri-Louis – Fourcroy de Ramecourt Charles-René – Gallon Jean-Gaffin, *Descriptions des arts et métiers, faites ou approuvées par Messieurs de l'Académie royale des sciences. L'art du tuilier et du briquetier* (Paris, Saillant & Nyon: 1763).

Fenet A., "L'apport des fours à briques traditionnels de la région d'Apollonia (Albanie) à la compréhension des techniques antiques", in Boucheron P. – Broise H. – Thébert Y. (eds.), *La brique antique et médiévale. Production et commercialisation d'un matériau. Actes du colloque international, Saint-Cloud, 16–18 novembre 1995* (Rome: 2000) 103–111.

Gallon de, *Conférence de l'ordonnance de Louis xiv du mois d'Août 1669. Sur le fait des eaux et forêts, avec les Édits, Déclarations, Coutumes, Arrêts, Règlemens, & autres Jugemens, rendus avant & en interprétation de ladite ordonnance, depuis l'an 1115 jusqu'à présent. Contenant les loix forestières de France* (Paris, au Palais: 1752).

Gardner J.S. – Eames E., "A tile kiln at Chertsey Abbey", *Journal of the British Archaeological Association* 17 (1954).

Gentil A. – Musmacque A., "La Céramique, n°301, février 1913. Excursion des membres de l'Union céramique et chaufournière de France, à Beauvais et dans le pays de Bray, les 22–23 mai 1912", *Bulletin du Groupe de Recherches et d'Études de la Céramique du Beauvaisis* 13 (1991) 167–183.

Gérard C., "Les tourbières du département de l'Aube au XIXe siècle", *La Vie en Champagne*, 59 (2009) 46–49.

Gillon P., "Le pavement du XIIIe siècle du collatéral nord du chœur de l'abbatiale de Saint-Maur-des-Fossés (Val-de-Marne)", in Chapelot J. – Chapelot O. – Rieth B. (eds.), *Terres cuites architecturales médiévales et modernes en Île-de-France et dans les régions voisines* (Caen: 2009) 103–122.

Gourcy C. de, *Second voyage agricole en Belgique, en Hollande et dans plusieurs départements de la France* (Paris: 1850).

Gouzouguec S., "L'artisanat de la terre cuite architecturale en Île-de-France (XIIIe–XVe siècles) : l'apport des sources écrites", in Collective, *Actes des journées archéologiques Île-de-France 1998* (Paris: 2000) 92–100.

Gouzouguec S., "Le métier de tuilier à Paris au XVe siècle : approche sociale et organisation du travail dans la juridiction de Saint-Germain-des-Prés", *Archéologie médiévale* 36 (2006) 213–240.

Grésillon B. – Lambert O. – Mioche P., *De la terre & des hommes. La tuilerie des Milles d'Aix-en-Provence (1882–2006)* (Mirabeau: 2007).

Grizeaud J.-J. (ed.), *Aude (11), Carcassonne, Les Jardins de Grèzes. Découvertes sur le site de Grèzes : un établissement rural gaulois, des vestiges du Bas-Empire et d'une tuilerie du XVIIe siècle. Rapport final d'opération* (Nîmes: 2013).

Haussonne M., *Technologie céramique générale. Faïences, grès, porcelaines*, vol. 2 (Paris: 1954).

Hoffmann F., Licht A., "Four à feu continu pour la cuisson des briques", *Le Génie industriel. Revue des inventions françaises et étrangères, annales des progrès de l'industrie agricole et manufacturière* 30 (1865) 296–300.

Kleinmann D., "Les briqueteries en Chinonais et leurs marques de fabrication (1)", *Bulletin de la Société des Amis du Vieux Chinon* 8.6 (1982) 847–855.

Lacheze C., *L'art du briquetier, XIIIe–XIXe siècles. Du régime de la pratique aux régimes de la technique* (Ph.D. dissertation, Paris 1 Panthéon-Sorbonne University: 2020).

Lacheze C., *L'art du briquetier, XVIe–XIXe siècle. Du régime de la pratique aux régimes de la technique* (Paris: 2023).

Le Roux T., *Les nuisances artisanales et industrielles à Paris. 1770–1830* (Ph.D. dissertation, Paris 1 Panthéon-Sorbonne University: 2007).

Lebrun G., *Hauts-de-France, Oise. Passel 'le Vivier'. La briqueterie médiévale. Rapport de fouille. Volume 6* (Glisy: 2017).

Lefebvre S., "Les enclaves en forêt de Soignes, XVIe–XVIIIe siècles", in Collective, *Herbeumont. Cinquième congrès de l'association des cercles francophones d'histoire et d'archéologie de Belgique* (n.p.: 1999) 470–477.

Lehmann J., *Die Obere Havel und der Finowkanal* (Erfurt: 2015).

Leroy V., *Notice sur les constructions des maisons à Marseille au XIXe siècle* (n.p., n.n.: 1847; reprint, Edisud – Aix-en-Provence: 1989).

Lesur A., *Les poteries et les faïences françaises* (Paris: 1957).

Louis XIV, *Édict portant règlement général pour les eaux et forests. Vérifié en Parlement le 13 aoust 1669* (Paris, Frédéric Léonard: 1669).

Mandion R., "La Tuilerie de Bezanleu … la tradition. Un métier de diable et de bon Dieu", *Notre Département : la Seine-et-Marne* 17 (1991) 21–28.

Martin L. – Fournier S., *Villevieille – RN 202 à Moriez (Alpes-de-Haute-Provence). Rapport final d'opération. Fouille archéologique* (Nîmes: 2007).

Mestiviers B. (ed.), *Briques et tuiles* (Paris: 1985).

Michelin R., "Une tuilerie de la Bresse louhannaise à l'époque préindustrielle: la tuilerie de Châteaurenaud (1821–1837)", *Mémoires de la Société d'Histoire et d'Archéologie de Chalon-sur-Saône* 56 (1987) 93–122.

Mondière Melchior, *L'Abrégé des voitures par l'abrégé des sinuositez & navigation des eaux mortes des valons, paluds, terres inondées & jonction des rivières navigables joignant les deux mers* (Paris, Melchior Mondière: 1627).

Mordefroid J.-L., "Les chartreux franc-comptois et la terre cuite au XVIIIe siècle. Approche historique et archéologique des ateliers de potiers et des tuileries-briqueteries de Nermier, Bon-Lieu et La Frasnée", *Travaux de la Société d'émulation du Jura* (1995) 73–118.

Nègre V., *L'ornement en série. Architecture, terre cuite et carton-pierre* (Paris: 2006).

Orefici L. – Nauleau J.-F., "Les tuileries d'Arthon (Chaumes-en-Retz, Loire-Atlantique). Un site majeur à découvrir", in Thuillier F. (dir.), *Les terres cuites architecturales en France du Moyen Âge à l'époque contemporaine. Recherches sur les tuileries et les productions tuilières* (Drémil-Lafage: 2019) 299–334.

Pallot É., "La toiture vernissée de l'église de Brou, Bourg-en-Bresse. Le contexte d'une restitution", *Monumental* 15 (1996) 78–89.

Paul-Cavallier G. – Hamon M., *D'argile et d'hommes. Carnet de voyage au cœur de Terreal* (Paris: 2005).

Plancke R.-C., "En 1909 à Mortcerf, le centenaire des établissements céramiques Houbé", *Notre département, la Seine-et-Marne* 9 (1989) 17–21.

Peirs G., *La brique. Fabrication et traditions constructives* (Paris: 2004).

Rambaud I., *La Seine-et-Marne industrielle. Innovations, talents, archives inédites* (Lyon: 2010).

Richard F., "Tuileries et briqueteries en Saône-et-Loire", *Annales de Bourgogne* 62 (1990) 25–42.

Rostand A., "Reconnaissance des anciens droits d'usage en forêt des potiers de Saussemesnil (Manche)", *Annales de Normandie* 7.2 (1957) 221–224.

Roupnel G., *La ville et la campagne au XVIIe siècle. Étude sur les populations du Pays Dijonnais* (Paris: 1955).

Schnitzler J.-H., *De la création de la richesse, ou Des intérêts matériels en France. Tome premier* (Paris: 1842).

Singer C. – Williams T. (eds.), *A history of technology* (Oxford: 1958).

Termote J., "Production et utilisation de la brique à l'abbaye des Dunes, Coxyde", in Derœux D. (ed.), *Terres cuites architecturales au Moyen Âge. Musée de Saint-Omer, colloque du 7–9 juin 1985* (Arras: 1986) 60–71.

Terras A., "Description d'un Four à briques inventé par M. Bonnet, faïencier à Apt, département de Vaucluse, mars 1811", *Bulletin de la Société d'Encouragement pour l'Industrie Nationale* 10.81 (1811) 65–69.

Mastaing de T., "Four à cuire la briques, les poteries, tuyaux de drainage, etc.", *Le Génie industriel. Revue des inventions françaises et étrangères, annales des progrès de l'industrie agricole et manufacturière* 19 (1860) 216–218.

Thomas J., *Fours à chaux, Tuileries, Briqueteries en Touraine* (Tours: 2005).

Thomas J.-B., *Traité général de statistique, culture et exploitation des bois* (Paris: 1840).

Thuillier F., "Le four à briques en meule d'époque moderne de Bruille-lez-Marchiennes (Nord)", *Revue du Nord* 17 (2012) 467–476.

Thuillier G., "Une source documentaire à exploiter : les 'Voyages métallurgiques' des élèves-ingénieurs des Mines", *Annales. Histoire, Sciences humaines,* 17.2 (1962) 302–307.

Vayssettes J.-L., *Les potiers de terre de Saint-Jean-de-Fos* (n.p.: 1987).

Venel Gabriel-François, *Instructions sur l'usage de la houille, plus connue sous le nom impropre de charbon de terre, pour faire du feu ; sur la maniere de l'adapter à toute sorte de feux ; & sur les avantages, tant publics que privés, qui résulteront de cet usage* (Avignon, Gabriel Regnault: 1775).

Woronoff D., *Histoire de l'industrie en France du XVI e siècle à nos jours* (Paris: 1998).

CHAPTER 8

John Smeaton's Fire Engine Trials

Andrew M.A. Morris

Abstract

The early history of the steam engine is dominated by James Watt's invention of the separate condenser in 1765. Unfortunately, this achievement has overshadowed the no less significant improvements made around the same period by fellow engineer John Smeaton, during a series of trials on a prototype steam engine he built at his home near Leeds, UK. This chapter will give an account of Smeaton's trials and the improvements he made, as well as exploring in more detail the methods he used. Since his youth, Smeaton had been perfecting ways of improving scientific instruments, experimental apparatuses and works of engineering. The steam engine trials showcase this approach at its most mature, when Smeaton was at the pinnacle of his career. I argue that Smeaton played a key role in the transformation of engineering from a tradition-based practice that used ad hoc trial-and-error to solve problems, into a methodologically systematic activity based on a universally applicable problem-solving approach based on parameter variation, that tests hypotheses, improves optimization and provides general rules for the functioning of machines.

Keywords

open science – experimentation – optimization – Industrial Enlightenment – fire engine – efficiency

Smeaton's career as an instrument maker and engineer was marked by the development of a single, widely applicable method of investigation. Although Smeaton was not the first or the only person to work with this method, he wrote on it extensively, describing his approach in detail via books and articles. This paper will focus on Smeaton's experiments to improve the Newcomen fire engine. Thanks to his experiments and improvements, Smeaton improved the efficiency of the engine by up to 100% (estimations vary), although this

© KONINKLIJKE BRILL BV, LEIDEN, 2025 | DOI:10.1163/9789004521766_010

impressive result was overshadowed by the separate condenser, and other innovations, introduced by James Watt (1736–1819) just a few years later.[1]

Smeaton's writings contribute to the eighteenth-century drive to share technical know-how, which contrasted with earlier traditions of craftsman secrecy. Works such as John Theophilus Desaguliers's *Course of Experimental Philosophy* (1734), or Bernard Forest de Bélidor's *Architecture hydraulique* (1737), provided invaluable accounts of different machines and rules for building those machines.[2] Despite earning his living first as an instrument maker, then as a civil engineer, Smeaton nevertheless disseminated the results of his research, happy to satisfy the 'avidity that the public showed' for research that was useful, but which also piqued one's curiosity.[3]

Not only was Smeaton willing to share his research, he also explicitly strove to draw general conclusions from the results of his experiments – conclusions which could be applied more widely than the individual case under consideration. This will become clearer in the sections that follow, but it is a clear indication that Smeaton not only tolerated his work being shared, rather he actively tried to make his research useful to others. For this reason, Smeaton can be viewed as a quintessential figure of the Industrial Enlightenment, trying to live up to the ideal of sharing knowledge for the betterment of humanity.[4] Before the eighteenth century, technical knowledge remained a guarded secret in a mainly oral tradition, available only to members of certain guilds, or via training and apprenticeship in a trade.[5] Smeaton attempted to bring these techniques and methods out into the open, to the benefit of all. As we will see, he wanted to put practical ideas and methods to the test, using a rigorous

---

1 Cardwell D., *The Fontana History of Technology* (London: 1994) 140.
2 Desaguliers John Theophilus, *A Course of Experimental Philosophy Volume 1* (London, John Senex: 1734); Bélidor Bernard Forest de, *Architecture hydraulique, ou L'art de conduire, d'élever et de ménager les eaux pour les différens besoins de la vie*, vol. 1 (Paris, Charles-Antoine Jombert: 1737).
3 Smeaton John, *A Narrative of the Building and a Description of the Construction of the Edystone Lighthouse with Stone* (London, H. Hughs: 1791) 7.
4 Mokyr J., *The Gifts of Athena. Historical Origins of the Knowledge Economy* (Princeton: 2002), and Jones P.M., *Industrial Enlightenment. Science, Technology and Culture in Birmingham and the West Midlands 1760–1820* (Manchester: 2008), on the Industrial Enlightenment, a period when the growth of knowledge began to have a significant influence on early industrial activity.
5 In addition to this oral tradition, a market for technical literature also developed. See Lefèvre W., *Minerva Meets Vulcan: Scientific and Technological Literature – 1450–1750* (Cham: 2021), https://doi.org/10.1007/978-3-030-73085-7, and Hilaire-Pérez L. et al., (eds.), *Le livre technique avant le XXᵉ siècle. À l'échelle du monde* (Paris: 2017) for a more detailed discussion of this literature.

method drawn from the natural sciences. In this way he actively participated in the codification and verification of practical – and often tacit – knowledge.[6] This was part of the move towards 'open science' (or 'open technology') that made previously hard to access know-how much more available to the enlightened public.[7] The secrecy of tacit knowledge was not entirely replaced by open science, however, but by a hybrid system of patents, where the technical know-how itself was made public, but the right to use it was sometimes restricted.[8]

The first part of this paper will trace the development of Smeaton's method from his earliest experiments on electricity, his famous waterwheel and windmill experiments, and his trials of different mortar recipes for the Eddystone Lighthouse. The second part of this paper will look at Smeaton's mature use of this method in his fire engine trials.

Twentieth-century commentators in the history and philosophy of science have tended to downplay the role and importance of experimentation in the sciences, giving more attention to theory building. Experimentation was sometimes seen as an afterthought, merely necessary to verify or falsify hypotheses.[9] Towards the end of the century, this view was gradually replaced by a more nuanced approach to experiment which has sometimes been called the 'New Experimentalism'.[10] Following the increase in interest in different types of experimentation, commentators began to pay more attention to the different ways in which experiments were used in the history of the sciences, in particular their exploratory dimension, as opposed to their traditional role testing hypotheses.[11]

---

6   Mokyr J., "Intellectual Origins of Modern Economic Growth", *Journal of Economic History* 65 (2005) 285–351, and Valleriani, M. (ed.), *The Structures of Practical Knowledge* (Cham: 2017) https://doi.org/10.1007/978-3-319-45671-3.

7   Hilaire-Pérez L., "Une tendance récente: le cultural turn de l'histoire des techniques", in *Regards Croisés Sur l'historiographie Française Aujourd'hui* (Paris: 2020) 143–152, proposes the term 'technique ouverte'. See also, Mokyr, "Intellectual Origins of Modern Economic Growth".

8   Mokyr, *The Gifts of Athena* 37.

9   The most important work in this tradition is Popper K., *The Logic of Scientific Discovery* (London: 1934).

10   See, for example Hacking I., *Representing and Intervening* (Cambridge: 1983); and Ackermann, R., "The New Experimentalism", *The British Journal for the Philosophy of Science* 40, no. 2 (1989) 185–190, https://doi.org/10.1093/bjps/40.2.185.

11   On exploratory experimentation, see Steinle F., *Exploratory Experiments. Ampère, Faraday, and the Origins of Electrodynamics*, trans. Levine. A (Pittsburgh: 2016). On the relation between Smeaton's work and exploratory experimentation, see Morris, A.M.A., "English Engineer John Smeaton's Experimental Method(s): Optimisation, Hypothesis

1 Electrical Experiments

Smeaton's first published work was a very short letter that appeared in the *Gentleman's Magazine*, a periodical that printed accounts of amateur research.[12] He stated that he began his experiments in 1745, at the age of 21.[13] The letter recounting his electrical experimentation was published two years later, in 1747. The experiments that interest us in this letter concern the ignition of alcohol using electricity. The first person to succeed in igniting a flammable material using electricity was Christian Friedrich Ludolff (1701–1763) at a meeting of the Berlin Academy of Sciences in January 1744.[14] Briefly, the experimental set-up – shown in Fig. 8.1 – involved a friction machine that generated static electricity by rubbing a spinning piece of glass. The charge was transferred into a prime conductor – usually a heavy rod of metal such as a gun barrel, suspended with silk – which was then put into contact with alcohol in a spoon, often using another, smaller metal object. The person holding the smaller metal object was insulated from the ground using wax. When the charge was large enough, the shock would ignite the alcohol. In Smeaton's case, he used a piece of wire, since this experiment was part of a broader attempt to determine the velocity of the charge moving through a wire.

As soon as this experiment was first carried out, others attempted to replicate it, including Smeaton. Despite succeeding, Smeaton had difficulty replicating the experiment reliably. So, he began a course of experiments to improve his success rate. First, Smeaton needed to ensure that his failure to replicate the experiment was not caused by different environmental conditions – he was attempting the experiment in winter, so he needed to be sure that it wasn't the cold, damp atmosphere that was a problem.[15] It was already common knowledge that moisture hampered attempts to generate electricity.[16]

Having established that external factors were not preventing him from igniting the alcohol, Smeaton set about attempting to optimize his experimental set-up in the hopes of generating a more powerful shock. First, Smeaton

---

Testing and Exploratory Experimentation", *Studies in History and Philosophy of Science Part A* 89 (2021) 283–294, https://doi.org/10.1016/j.shpsa.2021.07.004.

12   Kuist J.M., "A Collaboration in Learning: The *Gentleman's Magazine* and Its Ingenious Contributors", *Studies in Bibliography* 44 (1991) 302–317.

13   Smeaton John, "Experiments, with Some Queries on Electricity", *The Gentleman's Magazine* 17 (1747) 15.

14   Heilbron J.L., *Electricity in the 17th and 18th Centuries* (Berkeley: 1979) 272.

15   Smeaton, "Experiments, with Some Queries on Electricity" 15.

16   John Theophilus Desaguliers, *A Dissertation Concerning Electricity* (London, W. Innys and T. Longman: 1742) 6.

FIGURE 8.1　Illustration to the French edition of William Watson's *Expériences et observations pour server à l'explication de la nature et des propriétés de l'électricité* (Paris, Sébastien Jorry: 1748) plate 3. The friction machine on the right (of the bottom image) charges the horizontal rod (the prime conductor) held by the man on the left, who is holding a sword in his other hand, that he uses to ignite the liquid in the spoon. The man is insulated from the ground so that the charge is transmitted from the prime conductor, across his body and into the sword.

increased the quantity of matter in the wire, by adding an anvil to the circuit, but this didn't sensibly alter the force of the shock.[17] Second, he lengthened the wire, reasoning that a greater shock might be achieved if the wire was longer. Smeaton discovered that an 860-foot piece of wire produced a more powerful shock than a wire 450 feet long.[18]

Seeing that his hypothesis according to which the shock could be viewed as a moving body had been confirmed, Smeaton extended the analogy. He supposed that by increasing the magnitude of the charge before it travelled along the wire, as well as increasing the length of the wire itself, this would produce an even greater shock. This he did, as might be deduced from his very brief description, by using the anvil in conjunction with the prime conductor to increase the storage capacity of the conductor.[19] This addition to the experimental set-up did not increase the shock. Building on the idea that a longer

---

17　Ibidem, 15.
18　Ibidem.
19　Ibidem.

wire might increase the magnitude of the shock, Smeaton tested even longer wires, but found that the increase in magnitude of the shock trailed off beyond 860 feet.

Smeaton concluded this experiment by stating that the increase in shock was best achieved with 500 to 600 feet of wire.[20] He had carried out trials with both shorter and longer wires in order to establish the optimal length, and found that the shorter wires gave a smaller shock, and longer wires provided diminishing returns.

These experiments to optimize his electrical set-up were carried out by applying a simple method that combined hypothesis testing and parameter variation. Smeaton came up with potential alterations that could improve the performance of his electrical apparatus, then he tested them by varying different features of this apparatus and keeping other features the same. Thus, to test the hypothesis according to which more wire would give a greater shock, Smeaton compared a wire with more mass to a wire with less mass. Similarly, he tested whether a longer wire through which the charge moved could cause an increase in the magnitude of the shock, by comparing the performance of long and short wires.

By varying features in this way, Smeaton was able to optimize his apparatus and provide the largest shock possible. At the end of his letter, Smeaton provided a summary of his results. However, in stating these results he also generalised them:

> Does it [the electric matter] not move with an accelerated motion; since, if it moved with an equal one, as great an effect ought to be produced from a short wire as from a long one, if it be true that the effect is not increased by increasing the quantity of matter electrified.[21]

Whereas in his description of the experiments, Smeaton only wrote of comparing long and short wires, in his conclusion he states a more general rule that can be derived from those results: the charge *accelerates* along the wire, increasing the magnitude of the shock the longer the wire is. Since this effect tapered off beyond 500 or 600 feet, Smeaton also concluded that the accelerating charge came up against an increasing resistance in the wire, causing the increase in the magnitude of the shock to slow down once the wire became longer than 600 feet.[22]

---

20   Ibidem, 16.
21   Smeaton, "Experiments, with Some Queries on Electricity" 16.
22   Ibidem.

2 Mortar Trials

In the 1750s Smeaton transitioned from an instrument maker into a civil engineer. The two major projects that Smeaton undertook in the period when he made his name as an engineer are his research on waterwheels (and the construction of mills), and supervising the building of the Eddystone Lighthouse. Smeaton's waterwheel trial methodology is very similar to that employed in his fire engine trials, so I will discuss the lime mortar trials first, setting aside the fact that they were carried out about four years *after* the waterwheel trials.

Smeaton was contracted to supervise the construction of a new lighthouse on the Eddystone rocks off the coast of England, in 1756.[23] He decided to replace the wooden lighthouse design with a lighthouse made of stone, but to secure the blocks of stone he needed a mortar that would harden in the damp conditions of an offshore lighthouse.[24] To do this, he carried out a series of trials to find a mortar mixture that would harden in wet conditions, rolling different mortar mixtures into small balls and leaving them in water for up to three months, then testing for hardness with his fingers.[25]

In contrast to his electrical experiments, Smeaton's main source of hypotheses to test was practical know-how gathered from workmen's practices. The method of optimization was very similar, however.[26] He varied eight different parameters in this series of experiments: he compared well and poorly burnt lime (a crucial ingredient in the mixture), he varied the quantities of sand and tarras (a type of rock made from volcanic ash), he trialled different types of limestone used to make the lime (it was commonly claimed that a harder limestone produced a harder lime), he compared salt and fresh water (on the lighthouse salt water was freely available but fresh water was not), he compared limestone sourced in different locations, he varied the degree to which the mortar was beaten or mixed, and he also experimented with other materials such as shell lime and plaster of Paris.[27]

This method combined hypothesis testing and optimization via parameter variation. Smeaton took workmen's construction practices and transformed them into hypotheses to be tested. For example, he tested the claim

---

23 Turner T. – Skempton A.W., "John Smeaton", in *John Smeaton, FRS*, ed. A.W. Skempton (London: 1981) 14.

24 Gargiani R., *Concrete: From Archeology to Invention 1700–1769* (Lausanne: 2013) 310.

25 Smeaton, *A Narrative of the Building and a Description of the Construction of the Edystone Lighthouse with Stone* 104.

26 Morris, "English Engineer John Smeaton's Experimental Method(s)" 287–288.

27 Smeaton, *A Narrative of the Building and a Description of the Construction of the Edystone Lighthouse with Stone* 103–122.

N°	Water Lime with *Puzzolana.*	Lime Powder.	Pozzolana.	Common Sand.	N° of Cube Feet.	Expence per Cube Foot. *l. d.*
		Bushels.	Bushels.	Bushels.		
1	Edyftone Mortar —	2	2	—	2.32	3 8
2	Stone Mortar —	2	1	1	2.68	2 11½
3	—— 2d Sort —	2	1	2	3.57	1 7½
4	Face Mortar —	2	1	3	4.67	1 4
5	—— 2d Sort —	2	0½	3	4.17	1 1
6	Backing Mortar —	2	0½	3	4.04	0 11
	Water Lime with *Minion.*		Minion.			
7	Face Mortar —	2	2	1	3.22	1 5½
8	—— Calder Compofition —	2	1	2	3.57	1 1
9	Backing Mortar —	2	0½	3	4.17	0 10
10	—— 2d Sort —	2	0½	3	4.04	0 9½
	Common Lime with *Tarras.*		Tarras.			
11	Tarras Mortar —	2	1	—	1.67	4 0
12	—— increafed —	2	1	1	2.50	2 9
13	—— further —	2	1	2	3.45	2 0½
14	—— ftill further —	2	1	3	4.35	1 8
15	Tarras Backing Mortar —	2	0½	3	3.50	1 2½
16	—— 2d Sort —	2	0½	3	3.37	0 11½
	Common Lime with *Minion.*		Minion.			
17	Ordinary Face Mortar —	2	2	2	2.75	1 5½
18	—— 2d Sort —	2	1	3	4.34	0 8½
19	Ordinary Backing Mortar —	2	0½	3	4.05	0 8
20	—— 2d Sort —	2	0½	3	3.92	0 7½

FIGURE 8.2 The table of results for Smeaton's experiments on mortars from John Smeaton, *A Narrative of the Building and a Description of the Construction of the Edystone Lighthouse with Stone* (London, H. Hughs: 1791) 122

that harder limestone produced a better lime by mixing mortars made using lime from both harder and softer limestones, then compared those mortars. If the mortar made using the harder limestone was better, then the hypothesis was confirmed *and* the mortar was optimized for strength in water. By testing many workmen's practices by varying multiple parameters one after the other, Smeaton was able to select the best mortar recipe overall. Fig. 8.2 is the list of different mortar recipes.

Smeaton did not only seek to optimize his mortar, he also wanted to provide more general rules that might help others trying to make mortars. This meant

JOHN SMEATON'S FIRE ENGINE TRIALS

generalising his results beyond the specific constraints of the Eddystone Lighthouse. The most significant case of a generalised rule in these mortar trials involved finding an alternative to the maxim according to which 'the harder the limestone, the harder the lime'.[28] Smeaton had found that this did not hold for hydraulic mortars, but he was able to establish an alternative rule.

During his trials, Smeaton determined that a specific kind of limestone called Aberthaw 'blue lyas' produced lime that made up the hardest mortar. However, in order to turn this discovery into a general rule, Smeaton needed to find a feature that distinguished this kind of limestone from other less suitable kinds. To do this, he turned to his friend William Cookworthy (1705–1780), a chemist who lived in Plymouth where the Eddystone Lighthouse onshore work was being done.[29] Together, Smeaton and Cookworthy applied *aqua fortis* (nitric acid) to the limestone, and determined that limestones that went into harder lime mortars had higher levels of clay in them, which formed a residue when the rock was broken down with acid.[30] Thus instead of needing to procure the same specific limestone, future engineers could find their own convenient source of limestone that was suitable for hydraulic mortars, by simply comparing limestone samples using *aqua fortis*.

3 Waterwheel and Windmill Experiments

The most significant precursor to the fire engine trials in Smeaton's career are his waterwheel and windmill trials. They will provide us with a template to help understand the fire engine trials, since the only source material available to us concerning Smeaton's fire engine experiments, is what was published by John Farey Jr. (1791–1851), since the original manuscripts have unfortunately been lost.[31]

Smeaton decided to embark on a series of trials of waterwheels and windmills in 1752.[32] This may have been because he had been commissioned to

---

28 Smeaton, *A Narrative of the Building and a Description of the Construction of the Edystone Lighthouse with Stone* 103.

29 Gargiani, *Concrete: From Archeology to Invention 1700–1769* 314.

30 Smeaton, *A Narrative of the Building and a Description of the Construction of the Edystone Lighthouse with Stone* 108.

31 See Woolrich A.P., "John Farey and the Smeaton Manuscripts", *History of Technology* 10 (1985) 181–216.

32 For a detailed account of Smeaton's place in the history of waterwheel development, see Reynolds T.S., *Stronger than a Hundred Men. A History of the Vertical Waterwheel* (Baltimore: 1983) 223–233.

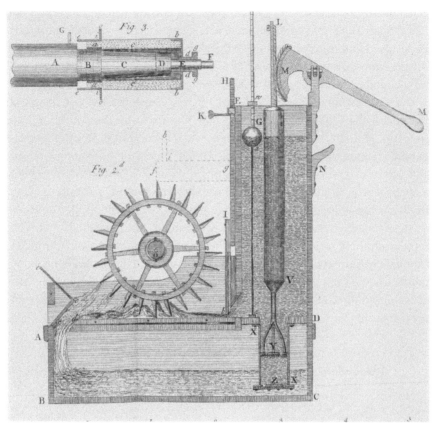

FIGURE 8.3 A technical drawing of the workings of Smeaton's waterwheel apparatus from John Smeaton, "An experimental enquiry concerning the natural powers of water and wind to turn mills, and other machines, depending on a circular motion", *Philosophical Transactions* (1759) 102, tab v

build a watermill, or because he wanted to learn how to build one.[33] His earliest mill design dates from the following year, i.e. 1753.[34] Smeaton used his skills as an instrument maker to construct a model waterwheel and a model windmill on which to carry out experiments [Fig. 8.3]. He then used his conclusions to inform his practical work, thereby fully testing and confirming those conclusions in practice, in the mid-1750s, before publishing his results in 1759.[35]

33   Smith N., "Scientific Work", in *John Smeaton, FRS*, ed. A.W. Skempton (London: 1981) 37.
34   Smith, "Scientific Work" 58.
35   Smeaton John, "An Experimental Inquiry Concerning the Natural Powers of Water and Wind to Turn Mills, and Other Machines, Depending on a Circular Motion", *Philosophical Transactions of the Royal Society* 51 (1759) 100–174, https://doi.org/10.1098/rstl.1759.0019.

JOHN SMEATON'S FIRE ENGINE TRIALS

In the waterwheel trials, he varied five main parameters. He varied the flow rate of the water in his model, by varying the height of the water in the reservoir (to the right of the wheel) and opening or closing a sluice (I) through which the water passed. This meant he could estimate the performance of the wheel in different locations, such as next to a small stream or on a more powerful river. Smeaton varied the number of paddles on the waterwheel – this had become the topic of some debate in the literature, featuring interventions by Henri Pitot (1695–1771), Bernard Forest de Bélidor (1698–1761) and John Theophilus Desaguliers (1683–1744).[36]

Significantly, Smeaton also compared the performance of undershot and overshot waterwheels. On an undershot wheel, the water flows under the wheel, pushing the paddles below the axle forward, thus turning the wheel. It is also known as an impulse wheel. On the other hand, water is guided over the top of an overshot wheel, with the weight of the water pulling the paddles down in order to turn the wheel. It is also known as a gravity wheel. It had been thought that both types of wheel were equally efficient:

> In reasoning without experiment, one might be led to imagine that, however different the mode of application is, yet that wherever the same quantity of water descends thro' the same perpendicular space the natural effective power would be equal.[37]

In other words, in both cases, it was the downward flow of water that either determined the velocity of the water striking the paddles (on an undershot wheel), or determined the height through which the water falls, turning the wheel (on an overshot wheel). Smeaton's model waterwheel design allowed him to easily switch between overshot and undershot operations, in contrast to a full-sized wheel, which would require complex alterations.

The second part of Smeaton's paper dealt with windmills, using the same approach.[38] It is worth noting that Smeaton had carried out a fact-finding mission to the Netherlands in 1755 where he was able to study Dutch windmill

---

36   Morris, "English Engineer John Smeaton's Experimental Method(s)" 286.

37   Smeaton, "An Experimental Inquiry Concerning the Natural Powers of Water and Wind to Turn Mills, and Other Machines, Depending on a Circular Motion" 124.

38   These trials have received less attention in the secondary literature. See Anderson J.D., *A History of Aerodynamics* (Cambridge: 1997); and Channell D.F., "The Emergence of Engineering Science", in *The Rise of Engineering Science*, ed. Channell D.F., (Cham: 2019) 75–109, https://doi.org/10.1007/978-3-319-95606-0_5, as well as Morris, "English Engineer John Smeaton's Experimental Method(s)".

TABLE III. Containing Nineteen Setts of Experiments on Windmill-Sa[ils] of various Structures, Positions, and Quantities of Surfaces.

The kind of fails made use of.	N°.	Angle at the extremities	Greatest angle.	Turns of the fails unloaded.	Turns of ditto at the maximum.	Load at the maximum.	Greatest load.	Product.	Quantity of surface.	Ratio of greatest velocity to the velocity at a maximum.	Ratio of greatest load to the load at maximum.	Ratio of surface to the product.
		°	°			lb.	lb.		Sq.In			
Plain fails at an angle of 55°.	1	35	35	66	42	7,56	12,59	318	404	10:7	10:6	10:7,9
Plain fails weather'd according to the common practice.	2	12	12		70	6,3	7,56	441	404		10:8,3	10:10,
	3	15	15	105	69	6,72	8.12	464	404	10:6,6	10:8,3	10:10,
	4	18	18	96	66	7,0	9.81	462	404	10:7,	10:7,1	10:10,
Weathered according to Maclaurin's theorem.	5	9	26½		66	7,0		462	404			10:11,
	6	12	29½		70½	7,35		518	404			10:12,
	7	15	32½		63¼	8,3		527	404			10:13
Sails weathered in the Dutch manner, tried in various positions.	8	0	15	120	93	4,75	5,31	442	404	10:7,7	10:8,9	10:11
	9	3	18	120	79	7,0	8,12	553	404	10:6,6	10:8,6	10:13
	10	5	20		78	7,5	8,12	585	404		10:9,2	10:14
	11	7½	22½	113	77	8,3	9,81	639	404	10:6,8	10:8,5	10:15
	12	10	25	108	73	8,69	10,37	634	404	10:6,8	10:8,4	10:15
	13	12	27	100	66	8,41	10,94	580	404	10:6,6	10:7,7	10:14
Sails weathered in the Dutch manner, but enlarged towards the extremities.	14	7½	22½	123	75	10,65	12,59	799	505	10:6,1	10:8,5	10:15
	15	10	25	117	74	11,08	13,69	820	505	10:6,3	10:8,1	10:16
	16	12	27	114	66	12,09	14,23	799	505	10:5,8	10:8,4	10:15
	17	15	30	96	63	12,09	14,78	762	505	10:6,6	10:8,2	10:15
8 fails being sectors of ellipses in their best positions.	18	12	22	105	64½	16,42	27,87	1059	854	10:6,1	10:5,9	10:12
	19	12	22	99	64½	18,06		1165	1146	10:5,9		10:10
	1.	2.	3.	4.	5.	6.	7.	8.	9.	10.	11.	12

FIGURE 8.4 The table of results for Smeaton's experiments on windmills from John Smeaton, "An experimental enquiry concerning the natural powers of water and wind to turn mills, and other machines, depending on a circular motion", *Philosophical Transactions* (1759) 144, tab. v

design (he also saw tarras being used in mortars, which would be relevant for his later research on the Eddystone Lighthouse project).[39]

In the windmill experiments, Smeaton varied the angles of the sails (called weathering – see Fig. 8.4); different angles were proposed by both theoreticians

---

39 The account of this trip is in Smeaton John, *John Smeaton's Diary of His Journey to the Low Countries 1755* (Leamington Spa: 1938).

JOHN SMEATON'S FIRE ENGINE TRIALS

and practitioners during the eighteenth century, such as French mathematician Antoine Parent (1666–1716), or Dutch mill builders.[40] He also varied the twist in the sail. Because the outer edge of the sail travelled much faster than the sail near the axle, the angle of the sail needed to be flatter towards the edge to avoid creating wind resistance to the circular motion.

With his model windmill apparatus, Smeaton was able to recreate changes in wind speed, so that he could develop an optimal design for different weather conditions. As with the waterwheel model, Smeaton could vary the load being lifted in order to model the performance of the machine when used in different ways. Finally, Smeaton tested the sail coverage, because '[m]any have imagined, that the more sail, the greater the advantage, and have therefore proposed to fill up the whole area'.[41]

As with his electrical experiments and mortar trials, Smeaton used parameter variation as a way to test hypotheses and optimize windmill design. He also wanted to provide general rules for the construction of waterwheels and windmills. To do this, he organised his data into tables and then derived general rules (called 'maxims') from that data. These maxims were simply mathematical relations between different features of a waterwheel or windmill, which could be used by anyone attempting to build their own. For example:

> That the quantity of water expended being the same, the effect is nearly as the square of its velocity.[42]

or

> The load at the maximum is nearly, but somewhat less than, as the square of the velocity of the wind, the shape and position of the sails being the same.[43]

These maxims are clearly approximative, and Smeaton conceived of them as rough rules of thumb rather than exact laws. Importantly, however, it was not simply a matter of reading off these rules from his tables of results – Smeaton actually needed to study his results carefully, and either carry out mathematical

---

40   Smeaton, "An Experimental Inquiry Concerning the Natural Powers of Water and Wind to Turn Mills, and Other Machines, Depending on a Circular Motion" 146–148.
41   Smeaton, "An Experimental Inquiry Concerning the Natural Powers of Water and Wind to Turn Mills, and Other Machines, Depending on a Circular Motion" 149.
42   Ibidem, 118.
43   Ibidem, 154.

operations (in the case of the waterwheel and windmill experiments), chemical analysis (in the case of mortars), or abstract from the concrete results (in the case of his experiments on electricity) in order to formulate general rules or maxims.

Before looking at Smeaton's fire engine trials in more detail, it will be worth distilling Smeaton's method down into a few principles. Smeaton himself provided a clear statement of his method in his paper on waterwheels and windmills:

> Many obvious and considerable improvements upon the common practice naturally offer themselves from a due consideration of the principles here established, as well as many popular errors show themselves in view: but as my present purpose extends, no farther than the laying down such general rules as will be found to answer in practice, I leave the particular application to the intelligent artist, and to the curious in these matters.[44]

Smeaton aimed to optimize the apparatus under consideration ('considerable improvements') through a critical examination of different known techniques for making electrical apparatuses, waterwheels, windmills or mortars ('common practice'). This Smeaton accomplished by rejecting erroneous rules and by establishing new maxims that could guide practitioners ('principles here established [...] such general rules as will be found to answer in practice'). The maxims established by Smeaton would permit others ('the intelligent artist' and 'the curious') to build their own apparatuses based on their own particular situation, without the need to design the object and try to calculate its optimal performance from scratch.

Although this method is based on the use of parameter variation, much like the electrical research of the early nineteenth century, which is the focus of Friedrich Steinle's account of exploratory experimentation, Smeaton's experiments differed in several important respects. Most importantly, Smeaton's use of parameter variation was not used as an alternative to hypothesis testing – instead, it was the means by which he tested hypotheses. This contrasts with parameter variation used as part of exploratory experimentation, which is not theory-driven, but focuses instead on finding regularities and forming concepts.[45] However, recent research on this topic has also highlighted the fact that scientific experimentation cannot be reduced to either theory testing or

---

44   Smeaton, "An Experimental Inquiry Concerning the Natural Powers of Water and Wind to Turn Mills, and Other Machines, Depending on a Circular Motion" 138.

45   Steinle, *Exploratory Experiments. Ampère, Faraday, and the Origins of Electrodynamics* 331.

JOHN SMEATON'S FIRE ENGINE TRIALS

exploratory research; instead, there are many different experimental strategies that incorporate various different degrees of theory-drivenness.[46] Smeaton's experimental method clearly falls into this category. The way he tests workmen's hypotheses and common practices is very different from the paradigm of the *experimentum crucis*, where a fully-fledged scientific theory can be put to the test with a single experiment specifically designed to only focus on the narrow topic of that theory. By contrast, parameter variation necessarily has a much broader focus, involving a more flexible experimental set-up and the production of larger quantities of data.

4 The Fire Engine

Smeaton was already well-known by the time he turned his attention to steam power, in the 1760s. The Newcomen engine had been invented by Thomas Newcomen (1664–1729) in 1712, and served as the main type of fire engine used during the eighteenth century, until Watt contributed his major improvements from 1769 onwards.[47]

Before looking at his research, it will be worthwhile to get a brief overview of the functioning of a Newcomen engine [Fig. 8.5]. A is the furnace in which water is heated by the fire below. When valve I is opened, the steam fills the cylinder C. Note that the steam does not push the piston (D) up – instead a counterweight G pulls the piston to the top of its stroke. Once there, valve J is opened and a jet of water from the reservoir L sprays into the cylinder, rapidly cooling the steam. This creates a vacuum in the cylinder, pulling the piston down (or rather the outside atmosphere *pushes* the piston down) and raising the pump rod H. Once the piston reaches the bottom of its stroke, valve I opens and the cycle begins again.

By the time Smeaton began working on the design of these engines, the mechanism had become more complicated, with the addition of a gear system to maintain the automatic opening and closing of the valves, as well as the addition of a device called a cataract to regulate piston rate. Also, a smaller pump was added which supplied water to the reservoir L, which had a small

---

46 See Schickore J., "'Exploratory Experimentation' as a Probe into the Relation between Historiography and Philosophy of Science", *Studies in History and Philosophy of Science* 55 (2016) 7 https://doi.org/10.1016/j.shpsa.2015.08.007 for a critical account of exploratory experimentation.

47 Cardwell, *The Fontana History of Technology* 121 and 161.

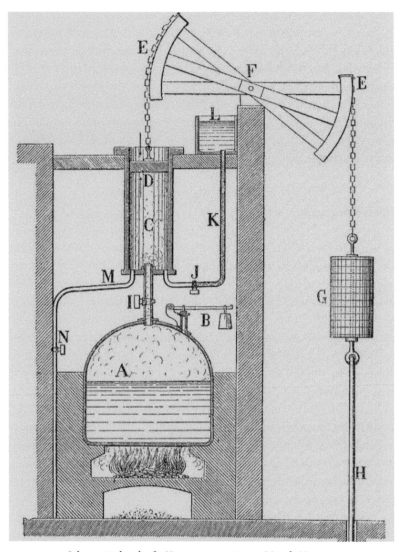

FIGURE 8.5 Schematic sketch of a Newcomen engine, in Max de Nansouty, *Chaudières et machines à vapeur* (Paris, Boivin et Cie: 1911) 61

pipe attached which allowed a small quantity of water to be poured onto the top of the piston D, which improved the seal around the piston.

Smeaton's first fire engine project was the New River engine, which he began working on in 1765. By this time, Smeaton was already a respected civil engineer, so the New River Company sought Smeaton's advice on how to maintain a steady uphill water supply at their site. Originally the water was raised over the 10-metre height difference using a windmill to work the pumps. This

JOHN SMEATON'S FIRE ENGINE TRIALS

system was replaced by a horse-mill, which works by horses walking in circles, turning an axle.[48] Smeaton's recommendation was to build a fire engine to replace the horse-mill. Smeaton had a wide experience with mill building, so he could calculate the potential performance of a waterwheel on the site with ease. Interestingly, he also appeared to possess a substantial understanding of fire engine performance, despite the fact that this was his first official fire engine project, although he did design a portable engine in 1765.[49]

Based on his calculations, a fire engine would be the most efficient way to raise the water. According to his preliminary report, Smeaton estimated that his fire engine would be roughly four times more efficient than the existing horse gin. As it turned out, the sources of Smeaton's fire engine know-how were not accurate – his engine performed less well than expected. Although he had carried out trials on a portable engine built in 1765, the results of these trials were not as rigorously established as hoped.[50]

Smeaton had hoped to improve the efficiency of the engine by slowing its action – this alteration would avoid wasting energy by continually stopping and starting the heavy pump mechanism.[51] Slowing the action of the engine was how he had improved the performance of overshot waterwheels in 1759, since a slow and measured action was better at transmitting power, compared to a jerky and erratic action. In the case of the waterwheel, this alteration avoided turbulence and loss of motive force in contact between the water and the waterwheel buckets.[52]

To slow the action of the pumps, Smeaton moved the fulcrum of the beam and increased the load on the pumps, in order to reduce the movement of the pump. However, this made the pumps increasingly heavy and difficult to move.[53] The cold water injection system was also insufficient, requiring the additional raising of water to increase the quantity of water injected into the cylinder.[54] As a consequence the cylinder stayed cool enough to place one's hand on it, meaning that energy was being lost in the repeated heating and cooling cycle. In short, the dimensions of the engine were far from optimal.

---

48   Kilburn Scott E., "Smeaton's Engine of 1767 at New River Head, London", *Transactions of the Newcomen Society* 19 (1938) 119, https://doi.org/10.1179/tns.1938.007.

49   Allen J.S., "Steam Engines", in *John Smeaton, FRS*, ed. A.W. Skempton (London: 1981) 180.

50   Rolt L. – Allen J.S., *The Steam Engine of Thomas Newcomen* (New York: 1977) 112.

51   Farey J., *A Treatise on the Steam Engine*, vol. 1 (London: 1827).

52   Morris, A.M.A., "John Smeaton and the *Vis Viva* Controversy: Measuring Waterwheel Efficiency and the Influence of Industry on Practical Mechanics in Britain 1759–1808", *History of Science* 56 (2018) 208, https://doi.org/10.1177/0073275317745455.

53   Farey, *A Treatise on the Steam Engine* I:158.

54   Ibidem.

Smeaton tried as far as possible to alter a number of variables in the engine, before discovering that, by once again reducing the load on pump, the engine ran much more smoothly. This result seemed to involve a degree of luck, thus convincing Smeaton to build his own engine in order to study the subject more systematically and vary parameters freely: 'I resolved, if possible, to make myself master of the subject, and immediately began to build a small fire-engine, which I could easily convert into different shapes for experiments'.[55]

5 Smeaton's Survey

Before embarking on his own experiments, Smeaton ordered a study of fire engines in the Newcastle area to be carried out, hoping to identify weaknesses and strengths of different engine designs. In order to evaluate these engines, Smeaton used the notions of 'great product', to measure the power output, and 'effect', to measure the efficiency of the engines.[56] These measures correspond to the slightly more modern notions of horsepower and duty. Horsepower is the weight lifted through a given height in feet, in the space of one minute, and it corresponds to the output of the engine. Eighteenth-century estimates varied for the value of one horsepower, between 22,000 and 33,000 lb.ft/min for horses working steadily throughout the day.[57] The effect, or duty, of the engine, measured how much input was needed to achieve a certain mechanical output, and was calculated as the weight lifted a certain height with one bushel of coal (where a bushel is a basket with a volume of roughly 36 litres – Smeaton equated this to 88 lbs of coal, or 40 kg). Some engines would have a good supply of coal, but would need to lift a large load, whereas other machines would not need to be as powerful, although it might be difficult or expensive to supply large quantities of coal. Fire engines used for lifting water out of mines, for example, had a plentiful supply of coal available, whereas the New River engine would have had to transport coal all the way to London.

In the survey of Newcastle fire engines, almost half (43 out of 100) no longer functioned properly.[58] Of the remaining 57, Smeaton selected 15 to study in greater detail. He provided tables relating their cylinder diameter, great product (horsepower) and effect (duty). Some of these engines were very economical indeed, being powered by the small pieces of coal that were left over after

---

55 Farey, *A Treatise on the Steam Engine* 1:158.
56 Allen, "Steam Engines" 181.
57 Smith, "Scientific Work" 54.
58 Farey, *A Treatise on the Steam Engine* 1:234.

JOHN SMEATON'S FIRE ENGINE TRIALS

mining was complete.[59] Some other engines required high quality coal to work sufficiently. As expected, the bigger engines were more powerful, although this increase in power was not proportional to the increase in size, since the load did not increase proportionally to the increase in cylinder size.[60] The pistons also pumped more slowly on larger engines. On the other hand, some smaller engines were able to compete with the larger engines when it came to efficiency, with the best duty being registered by one of the least powerful engines, which produced only 16 HP compared to the 37 HP produced by the bigger engines.[61] This is similar to what Smeaton discovered in the case of the waterwheel back in the 1750s: although a rapidly turning waterwheel might be more powerful when water is plentiful, in cases where water is in short supply a small, a slow-moving waterwheel is more efficient despite its smaller output.[62] In 1769, Smeaton carried out a similar survey of engines in Cornwall, focusing on eighteen large engines which were said to be of higher quality than the Newcastle engines.[63]

These surveys taught Smeaton that there was much to be improved about the functioning of fire engines. The principal sources of inefficiency were poorly made components. The cylinders were not made with enough precision, meaning that, during each stroke, the piston either rubbed unnecessarily against the cylinder, or gaps opened causing a loss of steam.[64] The same was the true for the pump work, which did not move water efficiently. The pump would ordinarily be a simple valve mechanism with upward moving flaps. As the pump rod descended into the water, the flaps would open, allowing water to collect above the piston. As the cylinder piston was pulled back down when the cold water was injected into the cylinder, the pump piston would be lifted and the weight of the water would hold the piston flaps shut, thus lifting the water. If this mechanism was poorly designed, water would leak out, lowering the performance of the engine.

One of the most significant technical achievements of the period following Smeaton's research was the improvement in manufacturing standards and precision.[65] The Carron company erected a new cylinder boring mill that could

---

59   Ibidem, 1:235.
60   Allen, "Steam Engines" 181.
61   Farey, *A Treatise on the Steam Engine* 1:234.
62   Morris, "John Smeaton and the *Vis Viva* Controversy" 202.
63   Rolt – Allen, *The Steam Engine of Thomas Newcomen* 109.
64   Farey, *A Treatise on the Steam Engine* 1:235.
65   Stewart R.J., "John Smeaton and the Fire Engine: 1765–1785", *The International Journal for the History of Engineering & Technology* 87 no. 2 (3 July 2017) 198, https://doi.org/10.1080 /17581206.2018.1463731.

produce round fire engine cylinders with much more precision.[66] And it was Smeaton himself who was employed to design this mill, which was powered, rather ironically, by waterwheels.[67]

Smeaton also noticed that the boilers were poorly designed.[68] They were generally too small and unable to maintain a steady supply of steam to the cylinder. The fire grate beneath the boiler was too low, and to remedy this the coals were piled high in the middle of the grate, meaning that the boiler was heated unevenly, with only the middle of the boiler receiving adequate heat.

Another problem noticed by Smeaton was the ineffective use of the length of the cylinder.[69] Since the piston did not move through its entire range of motion, there was what is called a dead space at the ends of the cylinder. From his work on air pumps, Smeaton knew that dead space was a significant flaw – the greater the dead space as a proportion of the cylinder volume, the lower the output of the engine.[70]

This issue combined with the uneven boring of the cylinder created a third problem: the accumulation of water on top of the piston.[71] Much like in eighteenth-century air pumps, a layer of liquid (water in this case, oil or water in the case of air pumps) was used to help seal the piston as it moved up and down the cylinder. If the piston did not fit the cylinder snugly, more liquid was needed to close the seal. However, if the piston did not rise right to the top of the cylinder on the up stroke, water accumulated on top of the piston, weighing the piston down and cooling the cylinder, which reduced the efficiency of the engine. A consequence of having this water above the piston was that it insulated the cylinder and kept it cool. This interfered with the efficient functioning of the engine, which required the cylinder to be heated and cooled as quickly as possible during each stroke.

A further weakness in these Newcastle engines was a tendency for substandard pipework.[72] The steam pipe and regulator did not allow enough steam into the cylinder with each stroke, meaning that the counter-weight did most of the work lifting the piston (which, as we have just seen, was already

---

66 Rolt and Allen, *The Steam Engine of Thomas Newcomen* 113.
67 Smith D., "Mills and Millwork", in *John Smeaton, FRS*, ed. A.W. Skempton (London: 1981) 65.
68 Farey, *A Treatise on the Steam Engine* I:235.
69 Farey, *A Treatise on the Steam Engine* I:235.
70 On air pumps, see Smeaton, John, "A Letter from Mr John Smeaton to Mr John Ellicott, F.R.S. Concerning Some Improvements Made by Himself in the Air-Pump", *Philosophical Transactions of the Royal Society* 47 (1752) 415–428, https://doi.org/10.1098/rstl.1751.0070.
71 Farey, *A Treatise on the Steam Engine* I:235–236.
72 Ibidem, I:236.

JOHN SMEATON'S FIRE ENGINE TRIALS

weighed down by the sealant water). Smeaton also noticed that the injection cistern, which held the water that was used to condense the steam and create a vacuum at the end of the up stroke, was generally not placed high enough to produce a powerful spray of water into the cylinder.

6   Austhorpe Engine Experiments

After building his first engine and carrying out enquiries on the engines in Newcastle, Smeaton decided to build a model engine on which he could carry out trials at his leisure. Already, with the New River head engine, Smeaton had begun to experiment by varying different aspects of the cylinder, furnace, pump etc. However, in order to carry out this type of experimentation in a systematic manner, a dedicated engine model was preferable: 'I resolved, if possible, to make myself master of the subject, and immediately began to build a small fire-engine, which I could easily convert into different shapes for experiments'.[73] Following this experimental research, Smeaton went on to build many fire engines, including the Chacewater engine portrayed in [Fig. 8.6], which was has been claimed to represent the pinnacle of this technology.[74] This engine was similar to the engine Smeaton built at home, and allows us to get a sense of the type of engine Smeaton was working with.

Smeaton carried out more than 130 experiments, over the course of four years, on the engine at his home in Austhorpe. By this point in his career, Smeaton was very experienced with carrying out trials of this kind.

Smeaton used a mercury barometer to measure the pressure of the steam.[75] He also measured the temperature. Smeaton was very familiar with the design and use of barometers, as he had invented his own – called the 'pear gauge' – as part of the air pump he designed in 1752.[76] By lowering the pressure and temperature in the furnace and cylinder, Smeaton could see if the engine would run on less fuel.

To vary the pressure and temperature, and also to determine the most effective set-up, Smeaton tried different techniques for burning coal, comparing a thin, clear, bright fire evenly spread over the grate, or a pile of coal left to burn

---

73   Farey, *A Treatise on the Steam Engine* 1:158.

74   Stewart, "John Smeaton and the Fire Engine", 206. Stewart's paper provides a detailed look at the different fire engines made by Smeaton in this period of his career.

75   Farey, *A Treatise on the Steam Engine* 1:168.

76   See Smeaton, "A Letter from Mr John Smeaton to Mr John Ellicott, F.R.S. Concerning Some Improvements Made by Himself in the Air-Pump".

FIGURE 8.6 John Smeaton, *Chacewater engine*, 1775, drawing, 63 × 44.5 cm, Smeaton Collection, vol. 3, fol. 111
ROYAL SOCIETY ARCHIVE

JOHN SMEATON'S FIRE ENGINE TRIALS

unattended in the middle of the grate. He found that the pile of coal only provided 5/6 of the mechanical power compared to the more thinly spread, well ventilated fire. Smeaton thus confirmed his suspicions about the inefficiency of the piles of burning coal that he had observed in the Newcastle engines, where coal was simply piled up in the middle of the grate instead of being evenly spread under the boiler.[77] On the other hand, Smeaton's recommendation would be more labour intensive, with the fire requiring regular supervision.

Smeaton also investigated the optimal height of the fire grate by testing different heights. He had observed that the grates were too low on the Newcastle engines, presumably to accommodate the large piles of coal that were ignited then left to burn unattended.[78]

He also varied the kind of fuel used in the furnace, comparing ordinary lumps of coal, smaller pieces of coal (commonly found in coal mines as a by-product of coal mining), chips of ash wood (a hard, dense wood) and coke (a smokeless, purified form of coal).[79] After determining that larger pieces of coal were the most efficient fuel, Smeaton carried out further trials of different kinds of coal, comparing Newcastle, Cannel, Hage-moor, Flockton, Middleton-wood, Welsh and Berwick-moor coals, determining with exactitude how much more efficient certain types of coal were, by comparing their performances against the common Yorkshire Halton coal. Cannel coal was found to be best, outperforming Halton coal by a ratio of 133 to 100.[80] Smeaton also verified whether the common practice of wetting the coals before burning them improved performance or not – unsurprisingly the answer was negative.[81]

The furnace powered the entire engine, which is why Smeaton wanted to cover some very basic points about fire management – where potentially large efficiency savings could be made – before getting into the technical details of fire engine design and dimensions.

Smeaton compared the engine's performance with and without the use of the cataract, which was a simple mechanism designed to slow the number of strokes per minute.[82] The cataract, also known as a "jack in the box," worked like a water clock: water slowly filled a trough which tipped over when full, activating a lever that triggered the injection valves to spray cold water into the cylinder, beginning the down stroke. Smeaton determined that using the

---

77   Farey, *A Treatise on the Steam Engine* 1:168.
78   Ibidem.
79   Ibidem, 1:172.
80   Ibidem.
81   Ibidem.
82   Ibidem, 1:169.

cataract wasted fuel, because the cataract paused the engine operation when the cylinder was filled with steam – the longer the wait, the more heat was lost through the walls of the cylinder compared to when the injection valve opened as soon as the piston arrived at the top of the up stroke.

To vary the load placed on the engine, Smeaton devised a simple mechanism to vary the height of the column of water raised by the pump, by adding or removing tubes of wood that extended or shortened the pump barrel, and therefore the quantity of water being lifted.[83] As well as determining the optimal load for the engine, Smeaton also established a relationship between the pump length and the piston cylinder length that would be widely applicable – he found that the best performance was achieved when the pump barrel was the same diameter as the piston cylinder, and fixed at 18 feet high.[84] He found that smaller loads were less efficient and output was also lower. Larger loads improved absolute output, but the efficiency of the engine (i.e., the duty rather than the horsepower) was lower.

As we saw in his survey of Newcastle engines, Smeaton was dissatisfied with the performance of the injection system. To remedy this, he placed the injection water tank as high as possible above the piston cylinder, in order to generate as much pressure as possible. Smeaton also equipped the injection hose with a variable opening that could be adjusted using a small micrometre screw.[85] Smeaton had designed his own micrometre, an astronomical instrument, which he had built, on and off, between the 1740s and 1770.[86] The injection hose's variable aperture was useful not only for determining the optimum aperture size, but also because the opening needed to be smaller when setting the engine into motion (usually different hose caps were used for this), and more water needed to be injected into the cylinder in summer, when it was warm.[87]

---

83   Farey, *A Treatise on the Steam Engine* 1:169.

84   Ibidem.

85   Ibidem, 1:168.

86   See Smeaton John, "Account of an Observation of the Right Ascension and Declination of Mercury out of the Meridian, near His Greatest Elongation, Sept. 1786, Made by Mr. John Smeaton, F. R. S. with an Equatorial Micrometer, of His Own Invention and Workmanship; Accompanied with an Investigation of a Method of Allowing for Refraction in Such Kind of Observations; Communicated to the Rev. Nevil Maskelyne, D. D. F. R. S. and Astronomer Royal, and by Him to the Royal Society", *Philosophical Transactions of the Royal Society* 77 (1787) 319, https://doi.org/10.1098/rstl.1787.0033.

87   Farey, *A Treatise on the Steam Engine* 1:194.

JOHN SMEATON'S FIRE ENGINE TRIALS

The resulting improvements included making the injection water pipe out of wood to insulate it from the surrounding steam. He also widened the supply pipe but narrowed the nozzle, and he raised the reservoir as high as possible, above the beam of the engine.[88]

During the course of these trials, Smeaton also noticed that the engine worked better when a small quantity of air was allowed into the cylinder, via a purpose-built valve. The advantage came from the insulating role played by the air, which prevented the hot steam from condensing against the walls of the cylinder by providing a buffer. Smeaton did not come up with this innovation on his own – it was already used in earlier Savery engines. However, as engines got older, their seals leaked, which made the air valves obsolete.[89]

As with his earlier technological experiments, Smeaton took contemporary know-how, and estimation of its limitations and weaknesses, as his point of departure.[90] In the case of steam power, Smeaton carried out a careful survey of existing engines to see which performed better, which performed worse, and he then hypothesised about the causes of this difference. Using parameter variation, Smeaton was able to test whether his proposed improvements did indeed improve the engine, or whether his hypotheses were falsified. For example, his early suggestion to slow the engine down did not turn out to be entirely correct – this he was able to determine by comparing the performance of the engine when running at full speed, and when it was slowed using the cataract, as well as comparing engines with longer or shorter stroke times. One important fact that his experiments uncovered was that bigger was not always better – whereas earlier fire engine designers had attempted "brute force" improvements by simply increasing the size of the cylinder, Smeaton's wide-ranging experiments showed that optimal performance and efficiency were not always achieved by simply making the cylinders larger.[91]

By varying many parameters, Smeaton was able to identify the best combination of features for an efficient engine [Fig. 8.7]. In this way, he used the experimental data he had gathered to make a table of recommended dimensions for new fire engines. This was in addition to more specific recommendations such as type of coal to use, height of the grate, size of the cooling jet nozzle, or which materials to use. Tables such as these were one of the major tools used in the eighteenth century to present research results in an accessible

---

88   Rolt and Allen, *The Steam Engine of Thomas Newcomen* 115.
89   Farey, *A Treatise on the Steam Engine* I:124.
90   This point is mentioned in Allen, "Steam Engines" 181.
91   Rolt and Allen, *The Steam Engine of Thomas Newcomen* 112.

Mr. Smeaton's Table for the Proportions of the parts of Newcomen's Engines, deduced from actual experiments, 1772.

Horse Power. 33 000 lbs. 1 foot per min.	Cylinder. Diameter.	Cylinder. Area.		Motion of the Piston. No. per minute.	Length of Stroke.	Motion per minute.	Boilers. Centre Boiler, Diameter.	Fire Surface.	Diameter of Steam-Pipe.	Injection Water. Square Aperture.	Height of the column.	Quantity. Per minute.	Per stroke.	Boiler's Feed per stroke.	Newcastle Coals per hour.	Power. Pumpage, or load on the piston.	Great Product per minute, or quantity of water raised 1 foot per min.	Effect per minute of 1 Bushel per hour.
	Inches.	Circ.Inc.	Sq. Inc.		Ft. In.	Feet.	Feet.	Sq. Ft.	Inches.	Inches.	Feet.	Cub. Ft.	Cy.I.Ft.	Cub. In.	Bushels.	Cyl. In. Ft.	Cyl. Inc. Ft.	Cyl. In. Ft.
1⅓	10	100	78	16¼	3 ,,10	63·3	5½	30	2·2	·28	14	1·2	13	9·3	·6	1800	113 940	200 000
1½	12	144	113	16¼	4 ,, 0	66·0	6	37½	2·5	·33	16	1·6	18	12·7	·7	2592	171 072	231 000
2½	14	196	154	16¾	4 ,, 2	67·7	6½	44	2·8	·37	18	2·0	23	15·9	·9	3528	238 881	263 000
3⅓	16	256	201	16	4 ,, 4	69·3	7	51	3·0	·42	20	2·4	28	19·7	1·1	4608	319 488	288 000
4¼	18	324	254	15¾	4 ,, 6	70·9	7½	58¼	3·3	·47	20	2·9	34	23·9	1·3	5832	413 372	311 000
5⅓	20	400	314	15½	4 ,, 8	72·3	8	66¼	3·6	·51	22	3·5	41	28·5	1·6	7200	520 800	334 000
6½	22	484	380	15¼	4 ,,10	73·7	8½	75¼	3·9	·56	22	4·0	48	33·6	1·8	8712	642 162	355 000
8	24	576	452	15	5 ,, 0	75·0	9	84¼	4·2	·60	22	4·6	56	39·3	2·1	10 368	777 600	374 000
9½	26	676	531	14¾	5 ,, 2	76·2	9½	94	4·5	·64	22	5·2	65	45·3	2·4	12 168	927 323	393 000
11⅓	28	784	616	14½	5 ,, 4	77·3	10	104	4·8	·69	24	5·9	74	52·0	2·7	14112	1 091 328	410 000
13	30	900	707	14¼	5 ,, 6	78·4	10½	115	5·1	·73	24	6·6	84	59·1	3·0	16 200	1 269 432	427 000
15	32	1024	804	14	5 ,, 8	79.3	11	126	5·4	·78	24	7·3	95	66·8	3·3	18 432	1 462 272	443 000
17½	34	1156	908	13¾	5 ,,10	80·2	11½	138	5·7	·82	24	8·0	107	75·0	3·6	20 808	1 669 010	459 000
19¾	36	1296	1018	13½	6 ,, 0	81·0	12	150	6·0	·87	26	8·8	120	84	4·0	23 328	1 889 568	472 000
22	38	1444	1134	13¼	6 ,, 2	81·7	12½	163	6·4	·91	26	9·7	134	93	4·4	25 992	2 123 806	486 000
24½	40	1600	1257	13	6 ,, 4	82·3	13	176	6·7	·95	26	10·5	148	104	4·8	28 800	2 371 200	498 000
27¾	42	1764	1385	12¾	6 ,, 6	82·9	13½	190	7·0	·99	26	11·4	164	115	5·2	31 752	2 631 606	510 000
30	44	1936	1520	12½	6 ,, 8	83·3	14	204	7·3	1·03	28	12·3	181	126	5·6	34 848	2 904 000	520 000
33	46	2116	1662	12¼	6 ,,10	83·7	14½	219	7·7	1·07	28	13·3	199	139	6·0	38 088	3 188 346	531 000
36	48	2304	1810	12	7 ,, 0	84·0	15	234	8·0	1·11	28	14·3	218	152	6·5	41 472	3 483 648	539 000
39	50	2500	1963	11¾	7 ,, 2	84·2	11 & 11	252	8·3	1·15	29	15·3	239	167	6·9	45 000	3 789 450	548 000
42½	52	2704	2124	11½	7 ,, 4	84·3	11¼ 11½	276	8·7	1·19	29	16·4	261	182	7·4	48 672	4 104 672	555 000
45¾	54	2916	2290	11¼	7 ,, 6	84·4	12 12	300	9·0	1·22	29	17·5	284	199	7·9	52 488	4 427 888	561 000
49½	56	3136	2463	11	7 ,, 8	84·3	12 13	326	9·3	1·26	30	18·6	310	216	8·4	56 448	4 760 448	567 000
52¾	58	3364	2642	10¾	7 ,,10	84·2	12 14	354	9·7	1·29	30	19·7	336	235	8·9	60 552	5 099 084	572 000
56¼	60	3600	2827	10½	8 ,, 0	84·0	12 15	384	10·0	1·33	31	20·9	365	255	9·5	64 800	5 443 200	575 000
60	62	3844	3019	10¼	8 ,, 2	83·7	12 & 2 of 11	402	10·3	1·36	31	22·2	396	277	10·0	69 192	5 792 062	579 000
63½	64	4096	3217	10	8 ,, 4	83·3	12 2 11½	426	10·7	1·39	32	23·4	429	300	10·6	73 728	6 144 000	581 000
67	66	4356	3421	9¾	8 ,, 6	82·9	12 2 12	450	11·0	1·42	33	24·7	464	324	11·2	78 408	6 498 455	581 000
70½	68	4624	3632	9½	8 ,, 8	82·3	12 2 12½	476	11·3	1·45	34	26·0	502	351	11·8	83 232	6 852 768	583 000
74½	70	4900	3848	9¼	8 ,,10	81·7	12 2 13	502	11·7	1·47	35	27·3	542	379	12·4	88 200	7 206 822	583 000
78	72	5184	4071	9	9 ,, 0	81·0	12 2 13½	530	12·0	1·50	36	28·8	586	409	13·0	93 312	7 558 272	581 000

Diam. of the outsideboilers.

The surface in the flue is taken at one-half.

FIGURE 8.7 Table of fire engine dimensions from John Farey Jr., *A Treatise on the Steam Engine*, vol. 1 (London, Longman: 1827) 183

way, and had been a feature of Smeaton's work since the 1750s.[92] Another general rule that he established was the relation between the volume of the piston cylinder and that of the pump cylinder.

Conclusion

Smeaton's research on fire engines was a crucial stepping-stone towards the highly efficient steam engines of the nineteenth century. This research can be understood as articulated around three themes, which are closely related to the three major themes of this book: knowledge production, technology transfer and economy.

Smeaton's fire engines were destined to serve in industry, and Smeaton was explicit about improving efficiency in order to save money. This was why

92 Mokyr, "Intellectual Origins of Modern Economic Growth" 305.

JOHN SMEATON'S FIRE ENGINE TRIALS 251

Smeaton came up with an early distinction between horsepower and duty (i.e. output and efficiency): the absolute amount of work done by the engine was of little value if the engine was too costly to run. This was why he also measured output per bushel of coal; improving this figure would involve studying the furnace itself in more detail, as well as taking into account the cost of different types of coal.

In trying to optimise his engine, Smeaton turned fire into an epistemic object. By studying fire as part of the fire engine, Smeaton was able to quantify it without having to deal with tricky methodological questions about what was being measured and how to measure it. The fire served to power the fire engine. If the engine performed poorly with a certain kind of fire, coal type or grate height, all other things being equal, then the weakness of the fire could be given a numerical value, irrespective of whether the fire was estimated to be hotter, brighter or larger than in other configurations. On the other hand, as we have seen, qualitative descriptions of the behaviour of the fire were still helpful for guiding the fire engine operator.

The study of fire as an epistemic object was only one part of Smeaton's broader approach, which has been the focus of this paper. We have seen how Smeaton developed a distinct, general method that was applicable to the improvement of almost any technology. Some commentators, such as Walter Vincenti or Donald Cardwell, have argued that Smeaton completed the development of a widely applicable engineering methodology that was distinct from strictly scientific methods.[93] Although this is partially true, recent research on scientific experimentation – in particular Friedrich Steinle's work on exploratory experimentation – suggests that Smeaton's methods did not constitute a distinct engineering methodology, but instead were closely related to the methods of experimental natural philosophy. This is particularly clear in the case of Smeaton's early electrical experimentation. A more detailed discussion of this issue is needed, but that is beyond the scope of the present paper.

Smeaton's experimental methods exemplified the new direction technological research took during the Industrial Enlightenment:

> The *Encyclopédie* and similar works of the eighteenth century symbolized the very different way of looking at technological knowledge: instead of intuition came systematic analysis, instead of tacit dexterity came an attempt to attain an understanding of the principles at work,

---

93　Vincenti W.G., *What Engineers Know and How They Know It. Analytical Studies from Aeronautical History* (Baltimore: 1990) 138; Cardwell, *The Fontana History of Technology* 195. See also Morris, "English Engineer John Smeaton's Experimental Method(s)".

instead of secrets learned from a master came an open and accessible system of training and learning. It also insisted on organizing knowledge in user-friendly compilations, arranged in an accessible way.[94]

It was in part thanks to Smeaton's drive to share the results of his research that we are able to appreciate his working methods so fully. So, although he likely didn't "invent" such a method of optimisation via hypothesis testing and parameter variation himself, the fact that as a practitioner, rather than a teacher (such as Desaguliers), Smeaton wrote on it so widely, was new.

Further, what takes Smeaton's research beyond that of publications like the *Encyclopédie*, is the fact that he was not content to merely *catalogue* artisanal techniques, he also *critically evaluated* them. In this respect he took a leading part in the transfer of these techniques from the realm of art to that of science.

Bibliography

Primary Sources

Bélidor Bernard Forest de, *Architecture hydraulique, ou L'art de conduire, d'élever et de ménager les eaux pour les différens besoins de la vie* vol. 1 (Paris, Charles-Antoine Jombert: 1737).

Desaguliers John Theophilus, *A Course of Experimental Philosophy Volume 1.* (London, John Senex: 1734).

Desaguliers John Theophilus, *A Dissertation Concerning Electricity* (London, W. Innys and T. Longman: 1742).

Farey J., *A Treatise on the Steam Engine* vol. I (London: 1827).

Smeaton John, "Experiments, with Some Queries on Electricity", *The Gentleman's Magazine* 17 (1747) 15–16.

Smeaton John, "A Letter from Mr John Smeaton to Mr John Ellicott, F.R.S. Concerning Some Improvements Made by Himself in the Air-Pump" *Philosophical Transactions of the Royal Society* 47 (1752) 415–428.

Smeaton John, "Account of an Observation of the Right Ascension and Declination of Mercury out of the Meridian, near His Greatest Elongation, Sept. 1786, Made by Mr. John Smeaton, F. R. S. with an Equatorial Micrometer, of His Own Invention and Workmanship; Accompanied with an Investigation of a Method of Allowing for Refraction in Such Kind of Observations; Communicated to the Rev. Nevil

---

94   Mokyr, "Intellectual Origins of Modern Economic Growth" 307.

Maskelyne, D. D. F. R. S. and Astronomer Royal, and by Him to the Royal Society". *Philosophical Transactions of the Royal Society* 77 (1787): 318–343.

Smeaton John, "An Experimental Inquiry Concerning the Natural Powers of Water and Wind to Turn Mills, and Other Machines, Depending on a Circular Motion". *Philosophical Transactions of the Royal Society* 51 (1759) 100–174. https://doi.org /10.1098/rstl.1759.0019.

Smeaton John, *A Narrative of the Building and a Description of the Construction of the Edystone Lighthouse with Stone* (London, H. Hughs: 1791).

Smeaton John, *John Smeaton's Diary of His Journey to the Low Countries 1755* (Leamington Spa: 1938).

Secondary Sources

Ackermann R., "The New Experimentalism". *The British Journal for the Philosophy of Science* 40, no. 2 (1989) 185–90. https://doi.org/10.1093/bjps/40.2.185.

Allen J.S., "Steam Engines", in *John Smeaton, FRS*, ed. A.W. Skempton, (London: 1981) 179–195.

Anderson J.D., *A History of Aerodynamics* (Cambridge: 1997).

Cardwell D., *The Fontana History of Technology* (London: 1994).

Channell D.F., "The Emergence of Engineering Science" in Channell D., *The Rise of Engineering Science* (Switzerland, Springer: 2019).

Gargiani R., *Concrete: From Archeology to Invention 1700–1769* (Lausanne: 2013).

Hacking I., *Representing and Intervening* (Cambridge: 1983).

Heilbron J.L., *Electricity in the 17th and 18th Centuries* (Berkeley: 1979).

Hilaire-Pérez L., "Une tendance récente: le cultural turn de l'histoire des techniques" in Barjot D. – Bellavitis A. – Haan B. – Feiertag O. (eds.) *Regards Croisés Sur l'historiographie Française Aujourd'hui* (Paris: 2020) 143–152.

Hilaire-Pérez L. – Négre V. – Spicq D. – Vermeir K. (eds.) *Le livre technique avant le XXᵉ siècle. A l'échelle du monde* (Paris: 2017).

Jones P.M., *Industrial Enlightenment. Science, Technology and Culture in Birmingham and the West Midlands 1760–1820* (Manchester: 2008).

Kilburn Scott E., "Smeaton's Engine of 1767 at New River Head, London" *Transactions of the Newcomen Society* 19 (1938) 119–126. https://doi.org/10.1179/tns.1938.007.

Kuist J.M., "A Collaboration in Learning: The *Gentleman's Magazine* and Its Ingenious Contributors" *Studies in Bibliography* 44 (1991) 302–317.

Lefèvre W., *Minerva Meets Vulcan: Scientific and Technological Literature – 1450–1750* (Cham: 2021), https://doi.org/10.1007/978-3-030-73085-7.

Mokyr J., "Intellectual Origins of Modern Economic Growth", *Journal of Economic History* 65 (2005) 285–351.

Mokyr J., *The Gifts of Athena. Historical Origins of the Knowledge Economy* (Princeton: 2002).

Morris A.M.A., "English Engineer John Smeaton's Experimental Method(s): Optimisation, Hypothesis Testing and Exploratory Experimentation", *Studies in History and Philosophy of Science Part* A 89 (2021) 283–294. https://doi.org/10.1016/j.shpsa.2021.07.004.

Morris A.M.A., "John Smeaton and the *Vis Viva* Controversy: Measuring Waterwheel Efficiency and the Influence of Industry on Practical Mechanics in Britain 1759–1808", *History of Science* 56 (2018) 196–223. https://doi.org/10.1177/0073275317745455.

Popper K., *The Logic of Scientific Discovery* (London: 1934).

Reynolds T.S., *Stronger than a Hundred Men. A History of the Vertical Waterwheel* (Baltimore: 1983).

Rolt L. – Allen J.S., *The Steam Engine of Thomas Newcomen* (New York: 1977).

Schickore J., "'Exploratory Experimentation' as a Probe into the Relation between Historiography and Philosophy of Science" *Studies in History and Philosophy of Science* 55 (2016) 7. https://doi.org/10.1016/j.shpsa.2015.08.007.

Smith D., "Mills and Millwork" in Skempton A.W., *John Smeaton, FRS* 59–83 (London: 1981).

Smith N., "Scientific Work" in Skempton A.W., *John Smeaton, FRS* 35–58 (London: 1981).

Steinle F., *Exploratory Experiments. Ampère, Faraday, and the Origins of Electrodynamics*, trans. A. Levine (Pittsburgh: 2016).

Stewart R.J., "John Smeaton and the Fire Engine: 1765–1785" *The International Journal for the History of Engineering & Technology* 87, no. 2 (3 July 2017) 190–225. https://doi.org/10.1080/17581206.2018.1463731.

Turner T. – Skempton A.W., "John Smeaton". in Skempton A.W., *John Smeaton, FRS* 7–34 (London: 1981).

Valleriani M. (ed.), *The Structures of Practical Knowledge* (Cham: 2017). https://doi.org/10.1007/978-3-319-45671-3.

Vincenti W.G., *What Engineers Know and How They Know It. Analytical Studies from Aeronautical History* (Baltimore: 1990).

Woolrich A.P., "John Farey and the Smeaton Manuscripts" *History of Technology* 10 (1985) 181–216.

PART 3

Fire in the Urban Space

∴

CHAPTER 9

The Outbreak of Fire: Inventions, Materials, and Combustion Knowledge in the Eighteenth Century

Marie Thébaud-Sorger

Abstract

The growth of urban density in the eighteenth century went in parallel with the development of artisanal and manufacturing activities, entailing a widespread use of fireplaces, boilers, but also new furnaces working with high temperatures – e.g., in glasswork – that increased the hazards caused by fire. The necessity of managing fire on a daily basis fostered the emergence of new methods and new technical processes. By focusing on the nature, the degrees of fire, and the types of materials, a great number of inventions were proposed to act on the outbreak of fire, drawing upon new knowledge of combustion and chemistry. In this paper, I will explore the inventors' attention towards this specific temporality of fire, namely the crucial moment when a fire started, which stimulated research on prevention strategies, fireproof materials, and fire suppression systems.

Keywords

urban conflagration – preventive methods – fireproof materials – public experiments – risks and insurances – Hartley – Mahon-Stanhope – Abbé Mann

In 1778, Abbé Théodore Mann published a memoir following a scientific journey he had just made to England where he draws a parallel between two competing inventions, those of David Hartley and Charles Mahon, both aimed to preserve buildings from fire.[1] Although their methods differed, the principle was similar. The idea was not to extinguish the fire, but to counter its progress by an initial treatment of the building that prevented air from enhancing the fire. Hartley's and Mahon's experimental devices consisted in setting fire to the

---

1 Mann Théodore Augustin, *Mémoire sur les diverses Méthodes inventées jusqu'à présent pour garantir les édifices d'incendie*, (Bruxelles, Imprimerie académique, 1778 1st edition).

ground floor of a purpose-built house: whereas some material damage was to be deplored – the windows melted, attesting to the power of fire – the flames did not manage to spread to either the upper floor or, *a fortiori*, to the outside: the outbreak of fire was thus contained, and the fire seemed to extinguish itself, without damaging the building. For the audience, including prestigious witnesses, the staging allowed them to experience the paradoxical process of a fire that did not burn.

Hartley's and Mahon's 'set-on-fire' but 'non-combustible' houses, as singular as they may seem, echoed similar experiments carried out from the 1760s onwards. By proposing to act in advance of any accident, these experiments highlighted the epistemological shift introduced by the search for fireproofing processes. They extended new approaches to fire management, combining precautionary and regulatory measures, that had emerged in various ways in European societies since the end of the seventeenth century.[2] Indeed, historiography has recently re-evaluated the traditional opposition between, on the one hand, medieval and modern cities, where the power of the unleashed element could reduce to ashes a dense, poorly organised urban space built mainly of combustible materials, and, on the other, the implementation in the nineteenth century of a rational management of risk through the control of fires (closed fireplaces, boilers, etc.), the professionalization of fire brigades, the multiplication of insurances, and, above all, the disappearance of wood from urban architecture. Far from surrendering to the force of circumstances, the increased sensitivity to accidents had favoured throughout the early modern period the collective implementation of actions which have made it possible, as David Garrioch has recently shown, not only to limit their impact but also to adapt to the changing nature of fires.[3]

In the wake of these historiographical revisions, this chapter is intended to explore the ways in which 'non-combustible methods', such as those of Hartley and Mahon, participated in the emergence of a different relationship with the economy and the risks of fire. Based on a better understanding of combustion processes, these methods promoted research into the treatment of materials and their degree of flammability. I will first discuss how the preventive approach deployed by daily fire administration fostered the collection of observations on the cause of fires and the mechanism of propagation. This

---

2　Bankoff G. – Lübken U. – Sand J. (eds.), *Flammable Cities. Urban conflagrations and the Making of the Modern World* (Madison, WI: 2012); Zwierlein C., *Der gezähmte Prometheus. Feuer und Sicherheit zwischen Früher Neuzeit und Moderne* (Göttingen: 2011).

3　Garrioch D., "Towards a Fire History of European Cities (Late Middle Ages to Late Nineteenth Century)", *Urban History* 46.2 (2019) 202–224.

new way of understanding fires led to a particular focus on the timing of fires. Hartley's and Mahon's methods are thus part of this perspective, articulating more radically the idea of prevention with that of 'resistance.' Secondly, I will discuss how the circulation of their methods contributed to a collective discussion on a European scale that never separated the material and technical dimensions from the social dimension of fires. Not limited to urban communities, thinking about materials as a means of fire control made it a broader issue of political economy.

1 Monitoring the Temporalities of Fire

1.1 *The Great Urban Disasters and the Daily Administration of Fire*
The great fire of London in 1666 may have seemed emblematic of the risks to which urban societies were subject. The spectre of this event is a common landmark of European culture, beyond Britain, echoed by the news here and there of major disasters.[4] However, by the end of the seventeenth century and throughout the eighteenth, there is a noticeable drop (*fire gap*) attesting to the rarity of these great conflagrations. This is all the more striking given that throughout the period, urban density increased sharply, with the appearance of real metropolises, notably Paris and London (with more than 550,000 and 800,000 inhabitants respectively by the end of the eighteenth century). The intensification of artisanal and economic activities, as has been shown for London for the last third of the eighteenth century, confronted the cities with new risks.[5] On the one hand, these risks were linked to an increased use of fire, for the needs of daily life (heating, cooking, lighting), new forms of consumption (assemblies, theatres, etc.), but also the resulting crafts and manufactories, especially food (bakeries, breweries, sugar refineries). On the other hand, this led to the accumulation of combustible materials and the proliferation of storage places and warehouses. The management of fire cannot be limited to urban tropism – I will discuss that later – but the effects of the propagation of fire in a city may have been dramatic in terms of material and human

---

4 Porter S., *The Great Fire of London* (Godalming: 1998); Garrioch D., "1666 and London's Fire History: A Re-evaluation", *Historical Journal* 59 (2016) 330; Nières C., *La reconstruction d'une ville au XVIIIe siècle: Rennes, 1720–1760* (Paris: 1972); Garrioch D., "Why Didn't Paris Burn in the Seventeenth and Eighteenth Centuries?", *French Historical Studies* 42.1 (2019) 35–65.

5 Hilaire-Pérez L. – Thébaud-Sorger M., "Urban Fire Risk and the '*Industrious Revolution*': The Case of London in the Last Third of the Eighteenth Century", *Le mouvement social* 249 (2014) 21–39.

damage.[6] It seems that urban societies, according to different chronologies and 'fire regimes' in different European areas, had been able to deploy relatively effective management to stem them. Not that their frequency was decreasing, but the major fires remained in most cases confined to large buildings such as hospitals, auditoriums, customs warehouses, etc. These latter are better documented than small scale fires, innumerable micro-events and daily snags, which rarely surface in the collective memory. The exploration of case studies from surviving police archives,[7] complaints or insurance policies – in countries where insurance developed – attests to the substantial part played by fire management in the web of ordinary practices in which collaboration between urban authorities and neighbourhoods was implemented, often at the street level.[8] While the proportion of buildings capable of withstanding fire was still very limited, the accumulation of preventive measures seems to have made it possible not only to reduce major disasters to a large extent, but also to adapt to the transformation due to the increase in activity and construction, the precariousness of buildings and the constancy of risk practices.[9]

1.2 *Adjusting the Means of Preventive Action to the Temporality of Fire*
In the range of preventive measures, technical solutions were also part of the relationship between knowledge of identified risks and firefighting. Thus, the municipalities – as well as the insurance companies in England – invested in the first fire-pump systems.[10] First developed in Germanic countries and the Netherlands – perfected by van der Heyden at the end of the seventeenth century – fire pumps were used to improve the traditional chain of rescue.[11]

---

6   Bankoff – Lübken – Sand (eds.), *Flammable Cities*; Garrioch D., "Towards a Fire History".

7   Archives municipales de Lyon, FF/21, Ordonnances et règlements particuliers de la ville (1536–1782), Ensemble de PV suite à des plaintes du voisinage (ou propriétaire ou locataire) enregistrées sur des risques d'incendies dans les logements, 1703–1777.

8   Denys C., *Police et sécurité au XVIIIe siècle dans les villes de la frontière franco-belge* (Paris: 2002); Berlière J., *Paris au siècle des Lumières: Les commissaires du quartier du Louvre dans la seconde moitié du XVIIIe siècle* (Paris: 2012); Dalmaz P., *Histoire des sapeurs-pompiers français* (Paris: 1996); Barke M., "The devouring element: the fire hazard in Newcastle-upon-Tyne, 1720–1870", *The Local Historian* 43 (2013) 2–13; Garrioch D., "Fires and Firefighting in 18th and 19th-Century Paris", *News and Papers from the George Rudé Seminar, French History and Civilization*, vol. 7 (2016).

9   Garrioch, "Towards a Fire History".

10   Cerise G., *Études sur l'ancienne France: La lutte contre l'incendie avant 1789* (Paris: 1893); Périer J., *Histoire de la lutte contre le feu et les autres calamités dans la ville de Lyon. Première époque de l'Antiquité à 1912* (Lyon: 2018); Hilaire-Pérez – Thébaud-Sorger, "Urban Fire Risk".

11   Donahue Kuretsky S., "Jan van der Heyden and the Origins of Modern Firefighting: Art and Technology in Seventeenth-Century Amsterdam", in Bankoff – Lübken – Sand (eds.), *Flammable Cities*, 23–42.

OUTBREAK OF FIRE: INVENTIONS, MATERIALS, COMBUSTION KNOWLEDGE   261

However, their use involved a prior organization on which their efficiency depended: logistics (location of depots), skills (to handle them), maintenance, better access to the buildings and water connections (pipes, fountains, filled tanks).[12] The generalized use of these devices therefore contributed to better manage the temporality of the fire in its development. Thus, despite – or thanks to – their obvious limitations, their inventive and commercial development favored a certain number of micro-improvements in order to act either as quickly as possible upstream with more manageable and portable systems, or by developing alternative means of action in places that the pumps could not reach, such as smothering the fire as soon as it started (alarms, extinguishing powders, 'fire bombs'), or downstream (emergency ladders).[13] Focusing on the tipping point between the onset of a fire and its taking hold increased vigilance on the conditions leading to its spread. This reinforced the implementation of preventive measures: the codification of precautions, the increase in the number of building construction rules, checking maintenance – in particular chimney sweeping – and the presence of combustible materials – construction material, furniture, warehouses.[14]

The repercussions of spectacular conflagrations (the fire at the Hôtel Dieu in 1772 or the Opera in 1781 in Paris, the Amsterdam Theatre in 1772, the Royal Store in Portsmouth in 1776, etc.) seem paradoxically all the more important as they tend to become rare. As constant reminders of the impossibility of eradicating fires, each of them led to new regulations but also encouraged improvements, mobilizing a multitude of actors: the artisans involved in the building trade, architects, administrators and town councilors, insurers, but also firefighters, engineers and entrepreneurs. Identifying the accident, determining its origin, pinpointing recurring elements form the basis of shared expertise on combustion processes, the firing point, the role of air, the way in which fire spreads according to the nature of the materials and their layout. Because beyond the spark, which is always possible, fire is above all the result of an environment that fosters its development.

---

12   According to Guillaume Cerise, the Parisian service included 39 pumps, 42 barrels, 12 water depots. See also Denys C., "Ce que la lutte contre l'incendie nous apprend de la police urbaine au XVIIIe siècle", *Orages. Littérature et culture 1760–1830* (2011) 17–36; Lyon-Caen N. – Morera R., *À vos poubelles citoyens! Environnement urbain, salubrité publique et investissement civique (Paris, XVIe–XVIIIe siècle)* (Ceyzérieu: 2020) 32.

13   For example, "Pompe nouvelle et portative pour les incendies et arrosements", *Journal de littérature, des sciences et des arts de M. l'abbé Grosier* (1780) 358–359.

14   Denys C., "La mort accidentelle à Lille et Douai au XVIIIe siècle: mesure du risque et apparition d'une politique de prevention", *Revue d'histoire urbaine* 2.2 (2002).

1.3 Acting Upstream: from Precautionary Measures to Material Treatment

Among the various causes recorded, apart from the small percentage due to natural phenomena, such as lightning, or hidden fires, produced by a slow and spontaneous ignition, human responsibility is particularly targeted. But the origin of fires cannot be limited to the topoi of human negligence or drunkenness. Listing the different conditions of its development also consists in recognising irreducible contingencies: the accident is an event whose irruption, if unexpected, is neither unforeseeable nor avoidable, but nevertheless occurs despite the control of practices. As immediate intervention is not always possible, attention was gradually focused on developing processes to make fire use safer, starting with the design of heating and cooking equipment, but also taking into account the immediate environment by integrating technical means into architectural structures.

As Valerie Nègre has shown, mortars, cement, various coatings, some of which were specifically designed to resist combustion, became a promising field of research because of their astonishing qualities of solidity and impermeability.[15] The Gazette du Commerce for example published several articles on this topic, such as the detail of a 'varnish' designed by a certain Glaeser in 1772, revealing the dynamism of German practitioners on that matter.[16] These methods of coating and protection constituted a body of knowledge, of which some known vernacular practices have been the subject of multiple improvements and which are sporadically and regularly mentioned in periodicals. In this advantageous context, entrepreneurs and inventors found a positive echo within institutions linked to architecture. In France, as early as the 1760s, the Academy of Architecture and the King's Buildings, sometimes concurrently, took up the examination of a number of proposals for mortars and fireproof structures.[17] In the last decade of the century, the newly formed club of 'Associated Architects' in London also set about making fire-fighting the core of its activity, against a background of land speculation and the securing of

---

15   Nègre V., "Surpasser la nature. Les ciments Loriot, La Faye et d'Étienne", *L'art et la matière. Les artisans, les architectes et la technique (1770–1830)* (Paris: 2016) 137–153.

16   Invention of a fireproof varnish communicated in *Gazette du commerce, de l'agriculture et des finances* on 26 September 1772, "Allemagne. De Suhl (dans le pays de Henneberg en Saxe), le 28 août". See Zwierlein, *Der gezähmte Prometheus* 188–192.

17   Nègre, "Surpassing nature"; Romano L., "Fighting Fires. Jean-Far Eustache de Saint-Far's Contribution to the Debate on Fireproof Constructions in France at the End of the 18th Century", in Campbell J. (ed.), *Iron, Steel and Buildings. Studies in the History of Construction. Proceedings of the Seventh Conference of the Construction History Society* (Cambridge: 2020) 379–386.

OUTBREAK OF FIRE: INVENTIONS, MATERIALS, COMBUSTION KNOWLEDGE   263

buildings in the wake of the development of manufacturing industry. Under the impulse of eminent architects such as Henry Holland and Robert Brettingham, a committee set up by the association compiled a list of all the fires and their damage in Britain (and the Empire) in 1791–92, month by month.[18] This collection work was undertaken in parallel with a series of fire prevention experiments conducted by the committee, which were re-examining the methods developed in the mid-1770s by David Hartley and Charles Mahon.[19]

The difficulty of dealing with fire and the multiplicity of causes led them to focus on the treatment of architectural structures. Rather than intervening on the regulation of customs and traditions, by inscribing the temporality of the register of preventive action inside the material arrangements, inventors tried to elaborate permanent technical solutions which would replace the lack of vigilance, reactivity and organization of rescue and assistance.

2   Incombustible Materials: Hartley and Mahon on Time Management

Throughout the eighteenth century, a variety of methods to contain fires were developed in England, including the development of an early insurance system: the companies issued fire regulations, funded fire brigades and the construction of fire pumps, and commissioned experts to assess the damage.[20] Following other disastrous episodes (notably the famous Cornhill fire of 1748 in London), the public authorities did not remain idle; parishes acquired pumps and a series of building regulations aimed at limiting the risk of fire, including the *Building Act* for London and the surrounding parishes, was enacted by Parliament in 1774.[21] The reiterated public demonstrations of David Hartley's and Charles Mahon's 'non-combustible' house models had thus a long echo,

---

18   London, Royal Institute of British Architects archives (RIBA), Henry Holland Archive. Papers relating to the Architects Club, 1791–1797: HoH/1/2–3, Notebook containing information on fires (accidental and deliberate) which had occurred in buildings in Great Britain, Ireland and Paris, France, in 1791–1792.

19   Gales, J., "Structural Fire Testing in the 18th Century", *Fire Safety Science News* 34 (2013) 32–33.

20   Pearson R., *Insuring the Industrial Revolution. Fire Insurance in Great Britain, 1700–1850* (Aldershot: 2004); Hilaire-Pérez – Thébaud-Sorger, "Urban Fire Risk".

21   *An Abstract of the Act of Parliament Made in the Fourteenth Year of the Reign of his Present Majesty, King George III. Entitled, an Act for the Further and Better Regulation of Buildings and Party-Walls* [...] *Calculated for the Use of Builders, and Workmen in General*, 2nd ed. (London, 1776) 6.

particularly in the context of the commemoration of the centenary of the Great Fire. The lack of combustion, which for Hartley 'eludes positive demonstration', is described by Mahon as a small 'paradox of physics'. These formulations emphasize that the unspectacular process challenges the audience's understanding vis-à-vis an unprecedented sensory experience. For, instead of a raging blaze that can be extinguished by the activation of a pump, it proposes the witnessing of a non-spectacle. By articulating the 'prevention' approach with that of the 'resistance' of structures, the inventors defined the contours of a radical action plan, that aimed to interrupt, or even reverse, the natural course of combustion in the open air, where the devouring flames would become powerless to propagate themselves.

2.1   David Hartley's Fire Plates

The principle of protective coatings was not new, but the inventive context contributed to its circulation, offering to some adventurous entrepreneurs new opportunities for trading in 'secret' compositions. This is how David Hartley (1732–1813) – the son of a notable physician, Fellow of the Royal Society and author of the famous essay *Observations of man* – was led to develop his fireproofing process: by observation, trial and substitution.[22] Educated at Oxford University, where he studied medicine at Merton College and was later elected as fellow, he became involved in various activities, mainly political and then diplomatic.[23] Elected to the House of Commons for the county of Hull, he led reforms in several areas, including engineering, but also economics, and took a more general interest in the fire-fighting issue. He was then (according to Mann's account) approached by a 'German' project maker, who tried to sell him a recipe for a fireproof varnish for £400 and left him some pieces of wood and canvas coated with this mysterious substance. The samples were scrutinized, tested and analysed. The Abbé Needham, founder of the newly established Royal Academy in Brussels, seems to have been in contact with Hartley at this time since, according to Mann, Hartley worked with him on the samples. Needham quickly determined the composition of the process, which consisted of plugging the pores of the wood to prevent air from activating combustion. However, Hartley preferred to use a mixture (using the properties of alum) and worked 'assiduously to discover something more effective for the same ends.

---

22   *Observations on Man, his Frame, his Duty, and his Expectations* (London, Samuel Richardson, 1749).

23   Guttridge G.-H., *David Hartley, MP, An advocate of conciliation 1774–1783* (Berkley: 1926) 231–340. Hartley was also a friend of Benjamin Franklin and an important contributor to the signing of the Paris agreements of 1782 (Treaty of Paris).

OUTBREAK OF FIRE: INVENTIONS, MATERIALS, COMBUSTION KNOWLEDGE   265

He imagined trying the effect of very thin plates of iron, nailed under a floor and had the 'satisfaction of succeeding beyond his expectations'.[24]

The principle consisted of a system of iron plates deployed within the buildings to secure the wooden floors by riveting 'fire-proof' plates under the floor, so that the fire could not reach the structure, mainly the beams and joists: 'In common houses, a floor on fire is a house on fire; but in a house secured by fire plates, as the floor will not take fire to make a body of heat, how is the house to be set on fire?'[25] Hartley therefore purposely built an experimental house at Putney Heath, on Wimbledon Common, where he carried out numerous tests. Notably on the King's birthday, June 4th, 1774, he held a large public experiment in the presence of the King, a number of London aristocrats and city officials. He also launched the production and sale of his fire plates after taking out a patent for his process of securing buildings and ships. But the production of the iron sheets turned out to be complex because, far from overseeing a manufacturing company, he made multiple contracts with different forges across England to manufacture and deliver the plates. Complex negotiations ensued between the forge masters and Hartley, in order to agree on a uniform quality and standard size – thinness, strength, adequate width – as well as an affordable price.[26] In spite of these difficulties, Hartley succeeded in getting his project onto the political agenda and the publication of his fireproof system attracted attention: he won exceptional support from his peers, who funded him up to 25,000 pounds sterling, several important contracts, notably for the Portsmouth docks, while his invention was celebrated by the City of London with the erection of an obelisk at Putney Heath.[27]

Hartley's method introduced a real novelty by using an incombustible material that could fit into existing architectural structures. However, the high cost, the rigidity of the plates and the number of plates that should have been used to make a wooden floor non-combustible sparked debates, leaving the field open to alternatives.

---

24   Mann, *Mémoire sur les diverses méthodes* 15.
25   Hartley David, *Account of a Method of Securing Buildings (and Ships) against Fire* (London: 1774) 11.
26   Berkshire Record office, Reading (BRO), Family Papers. Hartley Family. Hartley Business papers: microfilm B 4/1–5, Letters to David Hartley on fire plate experiments, orders for plates, letters from manufactures of iron and cooper plates at Pontypool, Bristol, Cleobury Mortimer in Worcestershire, Cobham (1773) and Iron mills in Yorkshire.
27   BRO, Hartley Business papers, B 4/3 1 bdl, correspondence on installing fire plates in Portsmouth dockyard.

2.2 *Count Mahon-Stanhope's Mortar*

The young aristocrat Charles Mahon (1753–1816) returned to England from Geneva, where he stayed from 1760 to 1774 and received a scientific education from Georges Louise Lesage, a physicist and chemist with a singular intellectual trajectory, and worked on his invention in the wake of Hartley, whom he admired.[28] From 1777 onwards, he undertook a series of experiments on his land in Chevening, Kent. Using the method of plastering, he developed a particular mortar which, when layered in a certain way, would achieve a similar effectiveness to that of Hartley's iron plates – and this for all-wood buildings, unlike the brick buildings envisaged by Hartley. The idea was to coat all the surfaces so that the pieces of wood that are then joined together are never in direct contact but protected by the mortar. Two 50-foot-long buildings with three floors in the middle were built on his property, in order to study the behavior of flames within the structure, but also to reproduce the possible propagation pattern in an urban context where buildings are close to each other.[29] Several tests were carried out from the summer of 1777, the success of which Mahon described to Lesage:

> I had a house built of lime wood and plaster made in the ordinary way but filled with chopped straw hay, and a little sand underneath the plates, which are all made of the most combustible pine wood [...] without the building undergoing the least real damage or those who were in the upper room feeling any heat, or any effect from the smoke. I have already had in one day several spectators who in truth did not understand too much how one could make a house incombustible by using only the most combustible materials. It is a small physical paradox which is however not difficult to solve. Besides the small merchants and the country people, there were about 70 ladies and gentlemen who live a few leagues from us.[30]

---

28 Chabot H., "Georges-Louis Lesage (1724–1803), un théoricien de la gravitation en quête de légitimité", *Archives internationales d'histoire des sciences* 53.150–151 (2003) 157–183; Bert J.-F., *Comment pense un savant? Un physicien des Lumières et ses cartes à jouer* (Paris: 2018).

29 Mahon Charles, "Description of a Most Effectual Method of Securing Buildings against Fire", *Philosophical Transactions of the Royal Society of London* 68 (1778) 884–894, esp. 893.

30 Bibliothèque Publique de Genève, Ms. suppl. 515, fol. 123: "Chevening July 6, 1777, Letter from Stanhope, Charles (comte, vicomte de Mahon) to Georges-Louis Lesage". Stanhope, who spoke French perfectly, corresponded with Lesage in this language.

Then, on 26 September, the Mayor and Aldermen of the City of London, the high aristocracy, learned foreign visitors and diplomats were invited, as well as some members of the *Royal Society*, of which he had become a member.

Hartley's and Mahon's processes were of course compared. But there was no real competition, and the two men, although belonging to different social spheres, followed parallel paths. Mahon, a high-ranking aristocrat, was then the brother-in-law of Prime Minister William Pitt by his recent marriage, and was elected to the House of Commons – until he entered the House of Lords in 1786, succeeding his father. This original character, with his progressive and radical views, invested himself ardently in various technical projects, developing numerous mechanical devices throughout his life.[31] In this case, he seemed however more inclined to consolidate his scholarly legitimacy than to develop an entrepreneurial project.[32] While waiting for a detailed printed explanation of his method (which was never published but of which his archives keep notes and drafts[33]) he published a brief description of his anti-fire process in 1778 in *Philosophical transactions*.[34] As he wrote to Lesage, what seems particularly surprising, artificial or counter-intuitive here is that, unlike Hartley's iron plate method, Mahon only used common, non-fire-resistant materials. He then developed a composition of a mixture of lime, sand and horsehair: it is the right proportion of the mixture and above all the way in which it is applied very simply with a brush, respecting the drying time according to the number of coats ('single' or 'double' reinforcement) that one wishes to apply, which ingeniously transformed this apparently vulnerable assemblage into a resistant structure. The joists and sleepers could be plastered separately, then once assembled one would interlock the wooden pieces in the plaster, making sure that there were no draughty gaps. Specifically designed for wooden houses, it also included the protection of stairs and certain vertical walls.

---

31 Musée des Arts et Métiers-CNAM, Inv. 43069, One-shot metal press by Lord C. Stanhope, ca. 1800.

32 Beatty F.M., "The Scientific Work of the Third Earl Stanhope", *Notes and Records of the Royal Society of London* 11.2 (1955) 202–221.

33 Maidstone, Kent Archive and Local History Service (Kent Archive), Stanhope Archives, U1590/C111: Papers concerning the methods of securing building. File 1 U1590/C111:1–5: U1590/C111/2: 'First Part etc.' – notes on methods; U1590/C111/3: 'Principle expounded'; U1590/C111/4: 'My method'; U1590/C111/5: 'Fire House Experiments'.

34 Mahon Charles, "Description of a Most Effectual Method of Securing Buildings against Fire".

2.3   Interrupt the Updraft, Compartmentalize the Fire

Regardless of the technical process, the ambition of Hartley's and Mahon's protective layers is the same: to make the structure resistant to fire, and in the first place the floors, by allowing the horizontal air draft responsible for the propagation of the fire to be cut off. All of their tests highlight the role of air and the vertical run of the fire: the structure is never strictly speaking 'fireproof' but the inventors' objective is to delay and confine the effect of the fire at its point of ignition in order to limit the damage.[35] Furthermore, as Hartley states, 'the principle is by no means the exclusion of air, but in the prevention of a draught of air, ascending vertically from below upwards'.[36] The management of air required therefore a certain subtlety, as it resulted from the layering of materials producing the necessary resistance by playing with structures and surfaces. The advantages and qualities of a method could thus be evaluated based on the effectiveness of such air management.

Thus, Hartley's system by making the floor completely airtight could encourage forms of 'dry rot' between the metal plates and the wood. The way in which the plates are riveted and overlapped at the joint is, according to him, essential to maintain a complex balance allowing to keep the necessary air filtration, while interrupting the action of air current. In this respect, coatings, which are more flexible in their application, can be used to adapt to particular shapes, whereas the plates are essentially reserved for flat surfaces. The qualities of the coatings, particularly of lime which strengthens as time goes by, are not to be proved further, but their ageing can end up making them porous and thus ineffective. However, the effect of the plates, unlike the coating, does not produce a slow combustion: according to Mahon, this could entail a significant smoke production which may prove to be very harmful.[37]

Their experiments reveal new understandings of the path of fire within buildings, the accumulation of heat at certain points in an enclosed space, the reaction of flames to air and to the contact with surfaces. The general discussion of Hartley and Mahon's procedures is not so much about the appropriateness of their methods – which are often considered equally suitable – as about their respective costs in relation to the contexts of use. Mahon's intention was precisely to find an accessible, simple and effective means; in his view, the

---

35   Gales, 'Structural Fire Testing'; Wermiel S., "The Development of Fireproof Construction in Great Britain and the United States in the Nineteenth Century", *Construction history* 9 (1993) 3–26.

36   Hartley, *Account of a Method*, 4.

37   Kent Archives, Stanhope Archives, U1590/C111/9: "Advantages".

fire plate method, whose invention he praises, should be reserved for specific uses: buildings storing valuable goods, warehouses for potentially hazardous materials, etc.[38] For Hartley, the identification of the high production cost of his plates as a negative aspect is incorrect, which a generalized adoption of the plates could easily solve.[39] This was the argument he still presented two decades later to the 'Associated Architects' club, which undertook a comparison of the two processes and some others (notably the rot-proof treatment of wood proposed by Henry Wood).[40] Their committee conducted numerous trials between June and December 1792, building experimental model houses for this purpose. Hartley, who had just published a pamphlet recalling the relevance of his method following another significant London fire,[41] received a distinction from the commission, which recognised the effectiveness of his plated structures.[42] Yet from a practical point of view, the debate remained open, and the use of plaster a viable option. In October 1796, following a fire that ravaged his castle, Mahon again asked the architects for an *in situ* assessment, and they reported that the part of the beams and joists treated with his mortar had withstood the damage better than the part that had not been protected.[43]

The cost of prevention through structural treatment was still high, but 'much more sure than any later remedy' in terms of damage.[44] The question prompts a comparison of the advantages and disadvantages of the two methods, not only in terms of their components, know-how and craftsmanship, but also in terms of social expectations and intended uses.

---

38   Ibidem.

39   Hartley David, *An Account of the Fire at Richmond House, on the 21st of December 1791, and of the Efficacy of the Fire-Plates, on that Occasion* (London, s.n: 1792).

40   Royal Institute of British Architects (RIBA), Henry Holland papers, HoH/2 Papers concerning fire-prevention experiments carried out at Nos. 5 & 10 Hans Place, London SW1, on behalf of the committee, 1792–1795 (13 items, mss), including some correspondence between Henry Holland and David Hartley, 1792, on Hartley's invention of iron fireplates and William Rothwell's recipe for a special plastering composition to prevent fire.

41   Hartley, *An Account of the Fire.*

42   *Resolutions of the Associated Architects, with the Report of a Committee by Them Appointed to Consider the Causes of the Frequent Fires, and the Best Means of Preventing the Like in Future* (London: 1793).

43   Stanhope Kent Archives, U1590/C111/folder 1: paper about fire at Chevening House (1796); RIBA, Henry Holland papers, HoH/1/4/1–2 and 1/5/1–2, Correspondence between Henry Holland, chairman of the committee, and Charles, 3rd Earl Stanhope, on Stanhope's fire-preventive invention, March 1792, and on a fire which occurred at his seat, Chevening Place, Kent, March 1797.

44   Hartley, *An Account of the Fire* 11.

3 Economy of Incombustibility and the Moral Economy of Risk

3.1 From Mann to Piroux: Synthesis of Knowledge on Fire and Preventive Action

The evaluation of fire prevention costs and techniques was not limited to debates between architects, experts and public authorities. The promoters of fireproofing materials and mortars belonged, to mention only a few examples, to circles other than those of the art of building: noble philanthropists, diplomatic politicians, physicians, and chemists. The publication and dissemination of Theodore Mann's memoir, which summarizes the comparative advantages of Hartley and Mahon's fireproofing techniques, is emblematic in this respect.

Mann's account is based on the first-hand experiments and explanations that the English inventors provided to him during his journey in November–December 1777, undertaken at the initiative of the Count of Starhemberg, Minister Plenipotentiary of the Austrian Netherlands, and piloted by the Brussels Academy of Sciences.[45] Mann, a naturalist, meteorologist, historian, demographer, mastered a wide range of knowledge fields – he even published a relatively unoriginal *Treatise on Elementary Fire*. In his account, after introducing the principles of combustion on which the processes are based, Mann successively describes the methods of Hartley and Mahon, and locates them in a broader reflection on fire prevention as an issue of social management.

The text, published in French, had a quick diffusion, and was republished in 1779 and enriched with corrections and additions. At first, Mann seemed to ignore the numerous experiments conducted in the German-speaking world, encouraged by the network of academies and economic societies. However, in the second edition, he included a well-informed passage on the experiments of the Suhl physician Johann Friedrich Glaser (1707–1783), who won the prize of the Academy of Göttingen in 1772.[46] Here, one finds a synthesis and summary of the various technical solutions that have been tried in the German area, and the history of their practical implementation. The English experiences gave rise to new trials on the continent while several academic prizes

---

45 Archives de l'Académie Royale des Sciences de Bruxelles (ARB), Registres de l'académie, Procès-verbaux, 1769–1794; ARB, dossier 48, séance de l'académie 11 novembre 1777; ARB, dossier 49, séance 13, January 1778: reading of Mann's paper and "choice to publish it separately and quickly".

46 Glaser Johann Friedrich, *Preisschrift, wie das Bauholz in den Gebäuden zu Abhaltung grosser Feuersbrünste zuzurichten*, (Hildburghausen: 1762); *Feuerlöschproben* (Marburg: 1786). Schilling R., "A Town-Physician in the 18th Century: Johann Friedrich Glaser", *Wurzbg Medizinhist Mitt* 30 (2011); Zwierlein, *Der gezähmte Prometheus* 188–192.

were open on the subject.[47] Thus, it is partly on Mann's work that Augustin Piroux's (1749–1805) *Moyens de préserver les édifices d'incendies et d'empêcher le progrès des flammes* (*Means of preserving buildings from fire and preventing the progress of flames*), awarded by the Royal Academy of Nancy in May 1781, is based.[48] This self-taught architect and lawyer from Lunéville, an avid reader of treatises and newspapers, organised his treatise through the prism of architecture.[49] Piroux classified the anti-fire procedures according to structural elements (stoves and fireplaces, floors, partitions, staircases, roofs, etc.), and listed practices, elements of vernacular knowledge and materials used in various regions of Europe, taking into account local particularities.

Mann's and Piroux's memoirs, accompanied by comparative tables, lexicons, and descriptive diagrams, aimed at popularizing fire-proofing methods. Their introductions contain a reminder of the general principles of combustion, the laws of ignition of bodies and propagation of flames, which constitute a theoretical framework for all further explanations: 'the ignition of combustible bodies' – as Piroux writes – 'is one of the greatest and most astonishing phenomena of nature; but at the same time it is one of the most difficult to conceive'.[50] Understanding the cause-effect relationship between the accumulation of heat and ignition, the expansive force of fire, and the process of air in combustion is the subject of many interpretations. Far from being restricted to physics, research on the 'nature' of fire was then boosted by the chemical approach, a cutting-edge field that was then in full reconfiguration. Mobilizing heterogeneous references (Hales, Boerhaave, Sage, even the works of Marat), they addressed the transformation of states of matter by emphasizing the role of *phlogiston*: an imponderable and invisible 'inflammable principle' (also called 'fire air') contained in the bodies which would be released precisely during combustion. Combustibility would therefore be relative to the degree of phlogiston contained in materials, as in the case of wood, which, depending on the species, is particularly rich in 'oily' phlogiston particles that react with the air, which serves as 'necessary food for the flame and materially contributes to

---

47 In Stockholm, Amiens, and Nancy.

48 The edition indicates booksellers in Avignon, Bern, Brussels, Hamburg, Lausanne, Liege, London, Lyon, Maastricht, Nancy, Neuchâtel, Paris, and Vienna.

49 Hottenger G.-A., *La vie, les aventures et les œuvres d'Augustin Piroux (1749–1805)* (Nancy: 1928); Bibliothèque municipal de Nancy, Meurthe-et-Moselle, Fonds Augustin Piroux, 1699–1805, Ms. Piroux 1- Ms. Piroux 175.

50 Piroux Augustin, *Moyens de préserver les édifices d'incendies et d'empêcher le progrès des flammes* (Strasbourg, les frères Gay: 1782).

its production'.[51] The more the fire burns in the open air and creates flames, the faster the combustion.

However, far from being limited to this theoretical framework, their works are also impregnated with the empirical knowledge developed in the context of the arts of fire, in cooking, or in practices using high temperatures (forge, pottery, glassware). The long domestication of fire in workshops highlights the precision of expert gestures of various craftsmen, blacksmiths, locksmiths, as they know how to make use of air or water ('which is reduced to steam, and forms an atmosphere that retains fire'[52]). Mann particularly refers to the skills of colliers, who can clearly distinguish between a combustion that produces flames and a slow combustion. For it is precisely the process of 'carbonization' of wood, which 'burns slowly, incompletely and above all without catching fire'[53] that is sought. Avoiding the flame being fed by the air, trying to 'make the wood charcoal,' this is what these anti-fire methods tend to do by making sure that the outside air 'can't get in between the constituent parts'.[54] The ambition to 'contain' the spread of fire, so that it either – in the best case scenario – smothers and dies by itself, or leaves time for an extinguishing intervention. In any case the damage would remain superficial. To be able to control the overflow of the fire by resisting it, implies to adjust material resources in a correct balance between the means (architectural structures, intrinsic qualities of materials, treatments of surfaces exposed to the fire) and the aims, that can differ from one context to another.

3.2 Territories and Collective Invention

For both Mann and Piroux, fire control is an issue of political economy and population government. They review all the conditions favorable to the appropriation of these anti-fire methods. The accessible resources, materials on a given territory (wood, bricks, stone, terracotta, or raw earth) and specific know-how (plastering, framing) are to be considered as much relevant as prevention policies and practices. Coatings are of particular interest because they allow the interior of existing structures to be treated at a cost that has no comparison with iron-reinforced structures (which Piroux estimates at 10 sol per square foot). Alongside traditional methods, such as stucco, clay, lime – one of the most widespread materials used in construction – whose methods and qualities have been compared, many other combinations are possible, such as

---

51   Ibidem, 7.
52   Ibidem, 135.
53   Mann, *Mémoire sur les diverses méthodes* 6–7.
54   Piroux, *Moyens de préserver les édifices* 4.

alum and glue, 'fatty earth and flour glue' etc.[55] Thus, for Piroux, 'the mysterious coating' of Mahon 'is none other than our white stuffing'.[56] The choice of the most suitable technique depends actually on its purposed uses and is a matter of political choices.

These anti-fire techniques thus attracted the interest of governments, and several enlightened monarchies supported the reproduction of experiments. The engineer Brequin de Demenge, at the request of Maria Theresa of Austria, conducted experiments in Vienna in August 1778, testing and comparing Mahon's techniques, but also Glaser's method, and carried out tests with clay.[57] An 'improved' coating 'after Mahon' was also put to the fire test in November 1779 in Saint Petersburg, by Kraft and Domaschnew. In France the experiment of the 'secret' of a private individual tending to preserve the houses from the activity of the fire held in place Louis XV in Paris with the agreement of the lieutenancy of police proved to be a failure.[58] But the policy of promoting inventions fostered other initiatives. While the Controller of Buildings investigated Hartley's method and ordered samples of his iron plates, local town councilors or intendants (king's representative in the provinces) supported experiments, such as those of the Lyon architect François Cointereaux.[59] The latter wished to promote incombustible houses in *pisé*, a technique of pressed earth 'improved by the Romans'.[60] Cointereaux's technique did not imply any coating, but he also built models to make fire tests in public, in order to develop a construction method accessible to all, once they learned the 'hand trick.' When the latter was installed in Amiens, at the request of the intendance in 1787, and could not find the necessary financial means to carry out his experiments, he sought the help of Mahon, whose commitment to these fireproof processes he knew.[61] Their objectives were quite similar: to look for a cheap, democratic process, for a use not restricted to urban areas,

---

55 Baud A. – Charpentier G. (eds.), *Chantiers et matériaux de construction: de l'Antiquité à la Révolution industrielle en Orient et en Occident* (Lyon: 2020).

56 Piroux, *Moyens de préserver les édifices* 68.

57 *Gazette of Vienna*, 22 August 1778.

58 Bibliothèque nationale de France, ms. 6682, fol. 422. Hardy Pierre-Siméon, *Mes loisirs, ou journal d'événements tels qu'ils parviennent à ma connaissance, mercredi 3 décembre 1777*.

59 Nicolas Henry Jardin is the architect commissioned by the King's buildings department, see Archives nationales de France, O/1/1294, Bâtiments, chemise 1, pièces 97–108: inventions 1757–1789, "procédé de M. Hartley pour garantir l'incendie".

60 Baridon L. – Garric J.-P. – Richaud G. (eds.), *Les leçons de la terre. François Cointereaux (1740–1830)* (Paris: 2015).

61 Leguay J.-L, "Apprendre et penser à 'se dépouiller de ses vieux usages, routines et préjugés de bâtir': promotion et réception des techniques de construction de François Cointereaux en Picardie", in Baridon – Garric – Richaud (eds.), *Les leçons de la terre* 139–152.

including also rural villages.[62] We do not know if the Earl responded favorably, but the reception of Cointereaux's works among English architects (notably Henry Holland) is well documented.

Oscillating between inventive strategies and political projects, the dissemination of these methods and prototypical models fueled a public debate, which brought together the social character of the materials and the civic dimension of the processes. The comparison of the various experiments and public contests opened by several learned societies also contributed to the construction of these anti-fire techniques as a field of collective knowledge. In this perspective, Piroux compares the recipes and the cost of the processes favoring their wide distribution, in contradiction to any protective and lucrative approach around the 'secret' of the compositions.[63] The choice of methods and materials used is part of an overall reflection which, for him, concerns the public good. In this respect, Hartley's approach of patenting his process had been criticized: these fire-fighting processes should above all be made available to everyone, without being the object of a lucrative trade. But in the configuration of the British market, since his method aimed primarily at wealthy owners or governmental acquisitions, it was based on another logic: while risk prevention is based on the development of an insurance system, Hartley's argument proposed to reduce the financial investment in insurance policies to the benefit of his fire-resistant system (a more long-lasting safeguard). The discussion on the resistance of buildings at the heart of the transformation of material life is thus strongly anchored to a debate on the mutualization of risks in society. The solutions depended on moral criteria as much as on economic ones, because the fire cannot be seized upon as damage of an essentially private nature, given its natural propensity to spread.

3.3   The Moral Economy of Fire

While the purchase and maintenance of fire pumps is the responsibility of a collective administration (whether public or private), to act in advance on buildings structures is to conceive the prevention technique as an action distributed among singular entities with heterogeneous status: dwelling house, public edifice, private warehouse. Between public and private action, the investment must also be measured relative to the cost estimated for the management of damage. Insurance, as it was conceived in eighteenth century England, refers also to a form of collective responsibility and civil mobilization,

---

62   Kent Archives, Stanhope Archives, U1590/C111/10, Notice of the *Mercury of France*, No. 34, 1785 and letter from Cointereaux to Stanhope.

63   Piroux, *Moyens de préserver les édifices* 77, 79 (chapter 2: 'Mortar').

OUTBREAK OF FIRE: INVENTIONS, MATERIALS, COMBUSTION KNOWLEDGE   275

but the efficiency of personal subscription did not meet the challenges posed by the expansive nature of fire, as the Abbé Mann emphasized (cited *verbatim* by Piroux):[64]

> Why, because I have insured my house and my belongings, caring little about fire, would I, without emotion, see the houses and belongings of my neighbors burned down if they had not taken the same precautions? Besides, when insurance against fire would be of such a widespread and general use [...] precautions against fire would not be more to be neglected.[65]

Very critical of the insurance system, Mann describes it as inefficient in terms of prevention, and socially unfair.[66] Moreover, according to him, the accountability for a fire starting cannot be attributed solely to individual misconduct, because accidents are mainly the result of 'ordinary negligence' which affects indifferently any 'sensible men'.[67] Every person must be protected. Consequently, the question arising for the community, beyond the method chosen, is that of the reparation cost of a danger that is collective by its very nature.

In terms of implementing fire protection methods, many authors proposed that the main focus be on large buildings in the urban space, such as hospitals, entertainment venues, large warehouses, docks and customs houses, which can be the source of devastating large-scale conflagrations. But focusing on sites of political and economic power and the properties of elites at the expense of more modest individual dwellings raises many questions. Especially since these dwellings, as well as barns, granaries and stables, are often more vulnerable to fire. The consequences of the destruction of property and the loss of human life must be assessed with a view to the general balance of society, taking into account all communities and territories in their disparity. Piroux, who emphasized that accidents are less frequent in cities than in villages, deployed a project of social reform based on fire control, centred on the protection of farmers and rural territories, which constituted the primary source of wealth

---

64   Clark G., *Betting on Lives: The Culture of Life Insurance in England, 1695–1775* (Manchester: 1999).

65   Mann, *Mémoire sur les diverses méthodes*, chapter "On the General Utility of Fire Precautions" 37.

66   Daston L., "The Domestication of Risk: Mathematical Probability and Insurance 1650–1830", in Krüger L. – Daston L. – Heidelberger M. (eds.), *The Probabilistic Revolution*, vol. 1 (Cambridge, Mass.: 1987) 237–260.

67   Mann, *Mémoire sur les diverses méthodes*, 37.

for kingdoms. For Mann as well, in the wake of physiocratic approaches, the calculation of the risk of fire damage cannot be limited to the value of more costly property losses. For, in addition to the loss of human life, failure to adopt effective means of prevention for ordinary people would mean having to bear the cost of the poverty of 'unfortunate families without a home after the devastation of the cities and towns': what would be the cost of prevention benefiting everyone with regard to the cost of repair? This political economy of fire considers men as a resource whose diminution or weakening would affect the proper functioning of societies.[68] Knowledge of materials and combustion was an integral part of reflection on the public good. Considering the fragility of buildings and of social conditions underlines the moral dimension of research on 'resistance,' that conceives the government of populations subject to the hazards of the elements as a matter of collective responsibility.

4 Conclusion

In the wake of the preventive approaches that developed around fire risk during the eighteenth century, the increased attention paid to the processes of combustion and fire propagation fostered the research for means of acting as far upstream as possible. The development of fireproof systems went hand in hand with the rise of a new architectural thinking: it initiated the revision of building standards, the search for new structural materials (steel) and coatings (cement, tar), the insertion of new devices (ovens, pipes, ventilation) encouraged by urban demand (assemblies, theatres), and industrial development.[69] The search for the answer in the structure establishes a new technical regime of fire because whatever the causes may be, the building seems to be able to respond.

The discussion of these methods and the numerous experiments they gave rise to throughout Europe from the 1770s onwards not only contributed to the transformation of material living conditions, but also to an evolution in the perception of risk. Paradoxically, prevention does not so much reveal a growing intolerance to fire risk but a certain acceptance of it. Rather than eradicating fire, it is more a question of developing means for its relative control and limiting the costs borne by the community (in terms of property and human

---

68 Rusnock A.A., *Vital Account Quantifying Health and Population in Eighteenth-Century England and France* (Cambridge: 2002) 179–208 (chapter 7: "Count, Measure, Compare: The Depopulation Debates").

69 Romano, "Fighting Fires"; Wermiel, "The Development of Fireproof Construction".

OUTBREAK OF FIRE: INVENTIONS, MATERIALS, COMBUSTION KNOWLEDGE 277

life), of normalizing it in a way. Maintaining life also means maintaining a society's capital and productive capacity, subject to the various vicissitudes of time and the vagaries of the elements. While conflagrations are an eminently sensitive subject involving the various strata of communities at several levels (administrations, inhabitants, builders and contractors, learned societies or experts), mastering the technical aspects of fire outbreaks embodies a social issue whose dimensions are moral as much as economic: materials are shaped by societies, but societies are also transformed by materials. In the (un)spectacular presentation of these fire-resistant houses, the wood and flammable materials, a metaphor for society, must be treated in such a way as to absorb the 'smouldering' fire without being destroyed by it. By leaving the structures standing and avoiding human casualties, the process suggested here a notion of embedded risk exemplified by these 'negative' demonstrations.

However, the principle had its limits. On 24 February 1809, the famous Drury Lane theatre built in 1794 by Henry Holland, implementing fire-resistant methods, caught fire in a spectacular fashion and was reduced to ashes in less than an hour. Alongside the performative discourse that emphasized the achievement of this resilient architecture, such dramatic conflagrations generated a spectacular aesthetic that exploited the fascination they engendered by being reminders of the limits of human action. Their sublime and fearsome dimension, staging the unleashing of an untameable element, unfolds as a counterpoint to the emergence of the industrial era – with industrial, domestic and collective investments – which aimed to radically transform the relations that are woven between human societies, materials and the environment.[70]

Bibliography

Primary Sources

"Invention d'un vernis à l'épreuve du feu" in *Gazette du commerce, de l'agriculture et des finances*, 26 September 1772.

"Pompe nouvelle et portative pour les incendies et arrosements", *Journal de littérature, des sciences et des arts de M. l'abbé Grosier* (1780).

An abstract of the Act of Parliament made in the fourteenth year of the reign of his present majesty, King George III. Entitled, an act for the further and better regulation of

---

70   Le Roux T. (ed.), *Les paris de l'industrie, 1750–1920. Paris au risque de l'industrie* (Grâne: 2013); Le Roux T. (ed.), *Risques industriels. Savoirs, régulations, politiques d'assistance, fin XVIIᵉ–début XXᵉ siècle* (Rennes: 2016).

buildings and party-walls [...] *Calculated for the use of builders, and workmen in general,* 2nd ed. (London, 1776).

Archives de l'Académie Royale des Sciences de Bruxelles (ARB), dossier 48, séance de l'académie 11 novembre 1777.

Archives de l'Académie Royale des Sciences de Bruxelles (ARB), dossier 49, séance 13, January 1778.

Archives de l'Académie Royale des Sciences de Bruxelles (ARB), Registres de l'académie, Procès-verbaux, 1769–1794.

Archives municipales de Lyon, FF/21, Ordonnances et règlements particuliers de la ville (1536–1782), Ensemble de PV suite à des plaintes du voisinage (ou propriétaire ou locataire) enregistrées sur des risques d'incendies dans les logements, 1703–1777.

Archives nationales de France, O/1/1294, Bâtiments, chemise 1, pièces 97–108: inventions 1757–1789, "procédé de M. Hartley pour garantir l'incendie".

Berkshire Record office, Reading (BRO), Family Papers. Hartley Family. Hartley Business papers: microfilm B 4/1–5, *Letters to David Hartley on fire plate experiments, orders for plates, letters from manufactures of iron and cooper plates at Pontypool,* Bristol, Cleobury Mortimer in Worcestershire, Cobham (1773).

Berkshire Record office, Reading (BRO), Hartley Business papers, B 4/3 1 bdl, correspondence on installing fire plates in Portsmouth dockyard.

Bibliothèque municipale de Nancy, Meurthe-et-Moselle, Fonds Augustin Piroux, 1699–1805, Ms. Piroux 1- Ms. Piroux 175.

Bibliothèque nationale de France, ms. 6682, fol. 422. Hardy Pierre-Siméon, *Mes loisirs, ou journal d'événements tels qu'ils parviennent à ma connaissance,* 3 December 1777.

Bibliothèque Publique de Genève, Ms. suppl. 515, fol.123: "Chevening July 6, 1777, Letter from Stanhope, Charles (comte, vicomte de Mahon) to Georges-Louis Lesage".

Gazette of Vienna, 22 August 1778.

Glaser Johann Friedrich, *Preisschrift, wie das Bauholz in den Gebäuden zu Abhaltung grosser Feuersbrünste zuzurichten* (Hildburghausen: 1762).

Hartley David, *Account of a Method of Securing Buildings (and Ships) against Fire* (London: 1774).

Hartley David, *An Account of the Fire at Richmond House, on the 21st of December 1791, and of the Efficacy of the Fire-Plates, on that Occasion* (London: 1792).

Kent Archive and Local History Service (Kent Archive), Stanhope Archives, U1590/C111: Papers concerning the methods of securing building. File 1 U1590/C111:1–5: U1590/C111/2: 'First Part etc.' – notes on methods; U1590/C111/3: 'Principle expounded'; U1590/C111/4: 'My method'; U1590/C111/5: 'Fire House Experiments'.

Kent Archives, Stanhope Archives, U1590/C111/folder 1: Paper about fire at Chevening House (1796).

Kent Archives, Stanhope Archives, U1590/C111/9: "Advantages".

OUTBREAK OF FIRE: INVENTIONS, MATERIALS, COMBUSTION KNOWLEDGE 279

Kent Archives, Stanhope Archives, U1590/C111/10, Notice of the Mercury of France, No. 34, 1785 and letter from Cointereaux to Stanhope.

Mahon Charles, "Description of a Most Effectual Method of Securing Buildings against Fire", *Philosophical Transactions of the Royal Society of London* 68 (1778) 884–894.

Mann Théodore Augustin, *Mémoire sur les diverses Méthodes inventées jusqu'à présent pour garantir les édifices d'incendie* (Bruxelles: 1778).

Musée des Arts et Métiers-CNAM, Inv. 43069, One-shot metal press by Lord C. Stanhope, ca. 1800.

Observations on Man, his Frame, his Duty, and his Expectations (London: 1749).

Piroux Augustin, *Moyens de préserver les édifices d'incendies et d'empêcher le progrès des flammes* (Strasbourg: 1782).

Resolutions of the Associated Architects, with the Report of a Committee by Them Appointed to Consider the Causes of the Frequent Fires, and the Best Means of Preventing the Like in Future (London: 1793).

Royal Institute of British Architects (RIBA), Henry Holland papers, Papers relating to the Architects Club, 1791–1797: HoH/1/2–3, *Notebook containing information on fires (accidental and deliberate) which had occurred in buildings in Great Britain, Ireland and Paris*, France, in 1791–1792.

Royal Institute of British Architects (RIBA), Henry Holland papers, HoH/2 Papers concerning fire-prevention experiments carried out at Nos. 5 & 10 Hans Place, London SW1, on behalf of the committee, 1792–1795 (13 items, mss).

Royal Institute of British Architects (RIBA), Henry Holland papers, HoH/1/4/1–2 and 1/5/1–2, Correspondence between Henry Holland, chairman of the committee, and Charles, 3rd Earl Stanhope, on Stanhope's fire-preventive invention, March 1792, and on a fire which occurred at his seat, Chevening Place, Kent, March 1797.

Secondary Sources

Bankoff G. – Lübken U. – Sand J. (eds.), *Flammable cities. Urban conflagrations and the Making of the Modern World* (Madison, WI: 2012).

Baridon L. – Garric J.-P. – Richaud G. (eds.), *Les leçons de la terre. François Cointereaux (1740–1830)* (Paris: 2015).

Barke M., "The devouring element: the fire hazard in Newcastle-upon-Tyne, 1720–1870", *The Local Historian* 43 (2013).

Baud, A. – Charpentier G. (eds.), *Chantiers et matériaux de construction: de l'Antiquité à la Révolution industrielle en Orient et en Occident* (Lyon: 2020).

Beatty F.M., "The Scientific Work of the Third Earl Stanhope", *Notes and Records of the Royal Society of London* 11.2 (1955).

Berlière J., *Paris au siècle des Lumières: Les commissaires du quartier du Louvre dans la seconde moitié du XVIIIᵉ siècle* (Paris: 2012).

Bert, J.-F. *Comment pense un savant? Un physicien des Lumières et ses cartes à jouer* (Paris: 2018).

Cerise G., *Études sur l'ancienne France: La lutte contre l'incendie avant 1789* (Paris: 1893).

Chabot H., "Georges-Louis Lesage (1724–1803), Un théoricien de la gravitation en quête de légitimité", *Archives internationales d'histoire des sciences 53* (2003).

Clark G., *Betting on Lives: The Culture of Life Insurance in England, 1695–1775* (Manchester: 1999).

Dalmaz P., *Histoire des sapeurs-pompiers français* (Paris: 1996).

Daston L., "The Domestication of Risk: Mathematical Probability and Insurance 1650–1830", in Krüger L. – Daston L. – Heidelberger M. (eds.), *The Probabilistic Revolution* (Cambridge: 1987) vol. 1.

Denys C., "Ce que la lutte contre l'incendie nous apprend de la police urbaine au XVIIIe siècle", *Orages. Littérature et culture 1760–1830* (2011).

Denys C., "La mort accidentelle à Lille et Douai au XVIIIe siècle: mesure du risque et apparition d'une politique de prévention", *Revue d'histoire urbaine 2.2* (2002).

Denys C., *Police et sécurité au XVIIIe siècle dans les villes de la frontière franco-belge* (Paris: 2002).

Gales, J., "Structural Fire Testing in the 18th Century", *Fire Safety Science News 34* (2013).

Garrioch D., "1666 and London's fire history: a re-evaluation", *Historical Journal 59* (2016).

Garrioch D., "Fires and Firefighting in 18th and 19th-century Paris", News and Papers from the George Rudé Seminar, *French History and Civilization,* vol. 7 (2016).

Garrioch D., "Towards a fire history of European cities (late Middle Ages to late nineteenth century)", *Urban History 46.2* (2019).

Garrioch D., "Why Didn't Paris Burn in the Seventeenth and Eighteenth Centuries?", *French Historical Studies 42.1* (2019).

Guttridge G.-H., *David Hartley, MP, An advocate of conciliation 1774–1783* (Berkley: 1926).

Hottenger G.-A, *La vie, les aventures et les œuvres d'Augustin Piroux (1749–1805)* (Nancy: 1928).

Hilaire-Pérez L. – Thébaud-Sorger M., "Urban fire risk and the 'industrious revolution': the case of London in the last third of the eighteenth century", *Le mouvement social* 249 (2014).

Lyon-Caen N. – Morera R., *À vos poubelles citoyens! Environnement urbain, salubrité publique et investissement civique (Paris, XVIe–XVIIIe siècle)* (Ceyzérieu: 2020).

Nègre V., *L'art et la matière. Les artisans, les architectes et la technique (1770–1830)* (Paris: 2016).

Nières C., *La reconstruction d'une ville au XVIIIe siècle: Rennes, 1720–1760* (Paris: 1972).

Pearson R., *Insuring the Industrial Revolution. Fire Insurance in Great Britain, 1700–1850* (Aldershot: 2004).

Périer J., *Histoire de la lutte contre le feu et les autres calamités dans la ville de Lyon* (Lyon: 2018).

Porter S., *The Great Fire of London* (Godalming: 1998).

Romano L., "Fighting Fires. Jean-Far Eustache de Saint-Far's Contribution to the Debate on Fireproof Constructions in France at the End of the 18th Century", in Campbell J. et al. (eds.), *Iron, Steel and Buildings. Studies in the History of Construction. Proceedings of the Seventh Conference of the Construction History Society* (Cambridge: 2020).

Rusnock A.A., *Vital Account Quantifying Health and Population in Eighteenth-Century England and France* (Cambridge: 2002).

Schilling R., "A Town-Physician in the 18th Century: Johann Friedrich Glaser", *Wurzbg Medizinhist Mitt* 30 (2011).

Wermiel S., "The Development of Fireproof Construction in Great Britain and the United States in the Nineteenth Century", *Construction history* 9 (1993).

Zwierlein C., *Der gezähmte Prometheus. Feuer und Sicherheit zwischen Früher Neuzeit und Moderne* (Göttingen: 2011).

CHAPTER 10

What Firefighting Tells Us about Eighteenth-Century Urban Police

Catherine Denys

Abstract

The regulations for the prevention and control of urban fires in the eighteenth century offer considerable insights into the transformations of early modern police. The repetitiveness of these texts conceals subtle shifts in police practice, and their analysis provides insights into the modes of writing regulations and the development of police know-how. Focusing on the case of Lille in French Flanders, this paper shows that regulations were part of a "patient pedagogy" that aimed not only to make the population aware of what could and could not be done, but also to remind officers of what behaviour was and was not allowed. Regulations thus invite us to think less in terms of the effectiveness of police action and more in terms of teaching the norm and building repertoires that have helped to turn police officers into a specialised professional category.

Keywords

firefighting – police ordinances – municipalities – circulation – professionalization

Fire* is at the heart of human history.[1] If today fires are rapidly dealt with in European cities, the destructive extension of contemporary forest fires, at all latitudes, reminds us of the helplessness of humans before this scourge. Early modern historians have long studied urban fires, and have often emphasized

---

* Text originally published in French under the title "Ce que la lutte contre l'incendie nous apprend de la police urbaine au XVIIIᵉ siècle", *Orages, Littérature et culture (1760–1830)* 10 (2011) 17–36. English translation by Marco Storni and Mark F. Rogers, revised by the author.

1 Dubois-Maury J., "Un risque permanent: l'incendie", *Annales de Géographie* 97.539 (1988) 84–95.

© KONINKLIJKE BRILL BV, LEIDEN, 2025 | DOI:10.1163/9789004521766_012

the extent of destruction and the repetition of disasters.[2] Specialists in urban planning have also highlighted the positive consequences of major fires, such as the reconstruction of London after 1666, or of Rennes after 1720.[3] It is to be noted, however, that cities did not always have the means to replace the smouldering building remains with modernized or beautified neighbourhoods. The fire at the Coudenberg Palace in Brussels in 1731 left the *Quartier de la Cour* in a state of ruin up until 1772, when the work on the *Place Royale* began.[4] Small cities had a hard time recovering from great fires, which often obliged the inhabitants to migrate or to live in precarious conditions in temporary shelters.[5] Firefighting therefore formed one of the most essential tasks of urban police in the Old Regime, along with the provision of supplies and the maintenance of the public space.

The "good police," source and guarantor of public safety, had to both prevent fires and organize rescues. The numerous regulations available at all municipal archives testify to this concern of the police. Urban historians have worked on such sources, and have already raised the question of the effectiveness of these dispositions: the relative rarity of serious fires in eighteenth-century European cities seems to suggest a positive answer, which rather goes against the widespread view of the inefficacy of police ordinances.[6] In this chapter, I wish to propose a reassessment of fire regulations promoted by early modern police, taking my cue from the well-documented example of the city of Lille, in French Flanders, to show how police archives provide much information – in fact, less about the object of dispositions than about the police itself, its

---

2 I do not intend to discuss here the risk of fires during wartime, which are neither prevented nor fought in the same way as fires occurring in time of peace. For an overview of the problem, see Nières C., "Le feu, la terre et les eaux", in Delumeau J. – Lequin Y. (eds.), *Les malheurs des temps. Histoire des fléaux et des calamités en France* (Paris: 1987) 367–382.

3 Reddaway T.F., *The Rebuilding of London after the Great Fire* (London: 1940); Nières C., *La reconstruction d'une ville au XVIIIe siècle: Rennes 1720–1760* (Paris: 1972).

4 Smolar-Meynart A. – Vanrie A. (eds.), *Le quartier royal* (Bruxelles: 1998).

5 Episodes of fires are well documented by local historians, whose publications are innumerable. To provide just a few examples: Andrey G. et al., *L'incendie de Bulle en 1805. Ville détruite, ville reconstruite* (Bulle: 2005); Vernus M., *L'incendie, histoire d'un fléau et des hommes du feu* (Yens sur Morges: 2006). See also Gresset M., "Quelques incendies de villes comtoises au XVIIIe siècle", and Lamarre C., 'Les incendies de villes dans les géographies du XVIIIe siècle', in Vion-Delphin F. – Lassus F., *"Les hommes et le feu de l'Antiquité à nos jours. Du feu mythique et bienfaiteur au feu dévastateur"* (Besançon: 2007) 275–281 and 283–291.

6 Part of the scholarship on fires has focused on the question of first aid for the victims: see Felicelli C., "Le feu, la ville et le roi: l'incendie de la ville de Bourges en 1252", *Histoire urbaine* 5.1 (2002) 105–134; Favier R., "Secourir les victimes. L'incendie au village dans les Alpes dauphinoises (seconde moitié du 18e siècle)", oral communication at the conference *Al Fuoco! Usi, rischi e rappresentazione dell'incendio dal Medievo al XX secolo* (Mendrisio: 2007).

practices, organization, and transformations.[7] My research is thus meant to contribute to the renewal of studies of the police in the Old Regime, by focusing on the discourses emerging from police writings, the concerns of historical actors, as well as the international influences which intersected and fed into the innovation of the police, in the framework of some great European debates of the eighteenth century.[8] While the analysis of preventive measures makes it possible to appraise the development of conceptions of the police in the Old Regime, the texts devoted to firefighting offer a global vision of the urban police apparatus, and reveal the existence of innovative currents that inspired police techniques in eighteenth-century Europe.

1 The "Good Police" of Fires: Classical Police and Global Prevention

Paolo Napoli's 2003 book *Naissance de la police moderne*, by integrating the approaches of Michel Foucault and Steven Kaplan, has provided a clear definition of the essence of the Old Regime "classical police."[9] Unlike "liberal" police, the classical police aimed less to punish crimes than to prevent them, by framing social activities in a network of tight rules, reinforced by the various normative institutions that characterized the Old Regime: the Church, the corporations, the municipalities, and the justice apparatus. While the post-revolutionary perspective highlighted the practices of espionage and political surveillance, misleading later commentators, the classical police perceived itself first as a regulatory body essential to social cohesion and to the well-being of the State, and especially of the urban organism.[10] Instead of the refusal of political or legal specialisation that would characterise the liberal state, the classical police was all-encompassing, in the sense that it could intervene in all areas of human activity, without any restriction.

---

7   In this paper, my goal is to expand on a few intuitions already expounded in a section of my PhD thesis. See the published version: Denys C., *Police et sécurité au XVIIIe siècle dans les villes de la frontière franco-belge* (Paris: 2002) 256–272.

8   Milliot V. (eds.), *Les mémoires policiers, 1750–1850. Écritures et pratiques policières du Siècle des Lumières au Second Empire* (Rennes: 2006). See also Berlière J.-M. – Denys C. – Kalifa D. – Milliot V. (eds.), *Métiers de police. Être policier en Europe, XVIIIe–XXe siècle* (Rennes: 2008); Denys C. – Marin B. – Milliot V. (eds.), *Réformer la police. Les mémoires policiers en Europe au XVIIIe siècle* (Rennes: 2009).

9   Napoli P., *Naissance de la police moderne: pouvoir, normes, société* (Paris: 2003).

10 Iseli A., *Bonne police. Frühneuzeitliche Verständnis von der guten Ordnung eines Staats in Frankreich* (Epfendorf – Neckar: 2003).

In the name of public safety – intended here less as a mode of imposed order than as a call for the collaboration of all citizens for collective security – the police authorities of the Old Regime produced a large number of ordinances intended to prevent accidents likely to start fires. The task was complex because the technical means of the time required the use of fire for manifold everyday activities. Lighting, heating, cooking, the production of foodstuffs or of metal objects: all these daily activities required the action of fire. The metal stoves containing a few embers, which one placed close to the bed to warm up children, could easily ignite the sheets, causing tragedies.[11] It has been pointed out that early modern cities were flammable because of the construction materials used, the narrowness of the streets and the accumulation of highly combustible materials in houses, cellars, attics, warehouses, and stables. All the great theatres of modern times burnt down at least once, libraries were often set on fire, and in military towns gunpowder depots added yet another source of danger.[12]

The regulations enacted by the municipalities, in collaboration with the royal intendants, aimed to classify the multiple occasions of the outbreak of fire.[13] The police found there the opportunity to deploy their technical knowledge of the city, and their control of the different layers of the urban space. Industrial activities made dangerous by the risk of fire must be kept away from urban centres, as much as the polluting, malodorous or violent activities (coal trade, tanneries, slaughterhouses), even if the necessities of daily supplies led to an exception for bakeries, scattered throughout the city. Prescriptions on construction practices are also well known: all cities struggled, in different ways

---

11 In Lille, the *écouages* (the removal of corpses) from 1713 to 1791 reveal a few dozen victims of tamed fires, out of 1,015 violent deaths: Denys C., "La mort accidentelle dans les villes du Nord de la France au XVIIIᵉ siècle: mesure du risque et apparition d'une politique de prévention", *Histoire urbaine* 2 (2000) 95–112.

12 In Valenciennes, in the night between 11 and 12 May 1722, a lightning bolt hit a tower of the Tournai city gate (*porte de Tournai*) that was used to store gunpowder. This started a fire, quickly tamed, although the inhabitants of the area, in panic, flew to the other side of town. See the Valenciennes Municipal Archives, AA 131/6, f°1. In Leiden, in 1807, a ship carrying gunpowder exploded, causing the destruction of the whole port area: see Reitsma H.J., "The explosion of a ship, loaded with black powder, in Leiden in 1807", *International Journal of impact engineering* 25 (2001) 507–514.

13 There is no opposition between the royal and the municipal police in firefighting. However, the division of competences has long attributed to municipalities a primary role (except for Paris, which is always exceptional). On the essential role of municipalities in firefighting in the nineteenth century, see Tanguy J.-F., "Les autorités municipales et la lutte contre l'incendie en Ille-et-Vilaine au XIXᵉ siècle", *Revue d'histoire du XIXᵉ siècle* 12 (1996) 31–51.

and at different rhythms, to eradicate thatched or timber roofs,[14] to replace cob and half-timbered walls with walls in brick or stone.[15] From the urban space as a whole to the disposition of the buildings, the police also extended their control inside houses and public buildings: the construction of chimneys was strictly regulated, as their size and position within the building must minimize the risks. Regular sweeping became essential and more or less controlled; in any case, its absence was sanctioned when it caused a chimney fire. After the frequent fires in old theatre halls, theatres started to be endowed with water tanks, equipped with numerous security exits and vast staircases to quickly evacuate the crowd. Finally, individual behaviour was also scrutinized by police regulations: amongst the prohibitions, there was that of entering a barn with a candle or any other unprotected flame, or of smoking a pipe therein, but also that of burning waste on the streets. The mistrust of the police with regard to popular entertainment is finally showcased by the ban on games involving gunpowder for reasons of safety: firecrackers, powder boxes, fireworks, bonfires, discharges of musketry, must disappear from the streets. The sale of gunpowder was also strictly controlled, if not completely moved out of town.

The multiplicity of these ordinances, their repetitive nature, are not signs of ineffectiveness. The police did not aim – back then as well as today – to completely eliminate dangerous or uncivil behaviour. Their goal was less to eradicate urban disorder than to contain it within limits acceptable to a majority of citizens, by adapting to the variations in sensibilities and the technical possibilities of the period. The police expounded, repeated, imposed norms, on people always ready to feign ignorance of such norms before the judges. The multiple editions of these ordinances thus aimed to instil awareness of what was forbidden, to transmit and demonstrate what could no longer be done, or what must be done. The fines – which were generally modest – that punished the violation of norms, did not seek to punish in the same way as criminal procedures, but to educate, even if that implied frequent repetitions. The regulation of the use of gunpowder in Lille illustrates this "patient pedagogy," whose aim was to raise awareness of the danger amongst a population that used gunpowder on festive occasions. The ordinance was reissued 21 times between 1667 and 1744. Aimed at all inhabitants of the city, it also systematically reminded police sergeants that they must prosecute offenders. Police pedagogy indeed began with police agents themselves, who could not, without

---

14   For instance, Berthet M., "Dans le Haut-Jura, toitures, transports, incendies" *Annales ESC* 8.2 (1953) 192–196.

15   Lavedan P., *Jeanne Hugueney et Philippe Henrat, L'urbanisme à l'époque moderne: XVIᵉ–XVIIIᵉ siècles* (Genève: 1982).

a targeted repetition, determine the priorities of their action amid an abundant and disordered body of regulations. The ordinances were therefore not only addressed to the inhabitants but also to police officers on duty, to remind them which behaviours had to be sanctioned. This facet of ordinances, which is "internal" to the institutional body of the police, is often neglected, as historians tend to read police regulations without considering the part played by the police actors of the time. The Lille Magistrate was well aware of the difficulty of transposing abstract regulations into the habits of the population, hence his insistence on reminding the city sergeants of their duty: the intention was to force people to adopt the rule, not by reading the ordinances, but by the fear of a fine. The preambles to the ordinances which deplore their non-application are, therefore, not an admission of powerlessness, but a rhetoric of motivation addressed to police officers.

Moreover, if the general outline of the text of the ordinances hardly changed, the repetitions could pertain to different aspects of that text. Eighteenth-century versions, for example, insisted on the childish practices of the use of gunpowder, in order to make parents responsible. Does this insistence reflect the abandonment of the festive uses of powder and its changing status as a children's game, or is it rather the sign of an increased concern for the safety of children? It is not possible to provide a definitive answer to this question. It is clear, however, that the phrasing, which is a mere detail in the general spirit of the ordinance, also points to the flexibility of police regulations, their adaptability – as theorised by Paolo Napoli – and their constructed, or even negotiated, character, that mediates between the urban will to formulate the rules of social order, and the common habits of people. With an ordinance enacted on 13 March 1747, which strictly organized the powder trade, the Magistrate of Douai took a more radical turn.[16] The object and the target of the text were always the same: to forbid children playing with gunpowder. But the responsibility of parents was this time coupled with that of retail traders. By formally prohibiting the latter from selling gunpowder to children, the aldermen enacted the traditional control that the municipality exercised over trades, in a city where the corporation police still belonged to the town hall. In any case, even in places more agitated by the liberal currents of the century, preventive policing always gave the right to local authorities to intervene in businesses that presented risks to health or public safety (apothecaries, surgeons, innkeepers, amongst others). In this sense, ordinances for firefighting, like all other municipal police ordinances, showcase – as Marco Cicchini has shown

---

16   Douai Municipal Archives, AA 104, fol. 25. Douai is a town very close to Lille, both geographically and culturally.

for the case of Geneva – a "factory of norms," which were more or less shared across the community.[17] Far from being archaic witnesses to the customary inefficiency of municipal powers, they express the strength and flexibility of police rules, and the encompassing capacities of the classical police.

2 The Fight against Fire: the Complexity of the Police Apparatus in the Old Regime

The same argument could be made for the ordinances which regulate the "rescue" from fires. The evolution of the Lille texts is in this sense, once again, exemplary. Between the ordinance of 1567 and the great regulation of 1773, nothing changed in the substance, but a lot changed in the methods and organization of firefighting.

As a matter of fact, a fire was an opportunity to deploy all the resources of the urban police, which were not limited to a few specialist agents, but concerned the entire community living in the city. The mobilisation of the bourgeoisie against fire was regarded as a civic duty, akin to the duty, typical of the previous century, for bourgeois military companies to take part in military activities. In the eighteenth century, in a garrison town like Lille, the bourgeois companies no longer assembled, and even the military guilds were dissolved.[18] The police were thus deprived of this structural organization of the inhabitants, but the Magistrate did not give up on the collective and voluntary nature of firefighting. However, as the century progressed, the tendencies towards the professionalization of police officers, the specialisation of administrative, judicial and police tasks, went against the attachment to the communal roots of firefighting.

The police must then learn to combine these two tendencies, namely the attachment to the collective – viz. amateur – nature of rescue, and a greater efficiency achieved through the professionalization of practices. The ordinances must therefore specify both the categories of people allowed to intervene in case of fires, and the terms of their intervention. Between the editions of the Lille regulations in 1721, 1742, 1755, 1756 and 1773,[19] the text of the decrees

---

17   Cicchini M., *La police de la République. Construire un ordre public à Genève au XVIIIe siècle* (Ph.D. dissertation, University of Geneva: 2009).

18   Denys, *Police et sécurité* 105–111.

19   Lille Municipal Archives, Registre aux Ordonnances du Magistrat 399, 4 January 1721; Affaires générales, box 563, dossier 13, 7 July 1742; Registre aux Ordonnances 405, 13 December 1755 and 18 December 1756.

WHAT FIREFIGHTING TELLS US ABOUT 18TH-CENTURY URBAN POLICE 289

lengthened, the articles became more numerous, at the risk of making it difficult to memorize them completely.[20] If the political philosophy that animates these texts was unchanged, and therefore certainly internalized for a long time by municipal officials and city dwellers, the extreme precision of the procedures, their complexity, the subtle sharing of skills formed a complex web, certainly difficult to master. The Lille municipal archives hold several copies of a small, printed booklet, which completes the 1773 ordinance and translates it into a sort of vade-mecum for the various police actors concerned. The booklet is 12mo, ninety-pages long, printed by Henry, the official printer of the city, and bears the date of 1784 for the second edition.[21] The small format suggests a different use from ordinances that were printed and plastered around town, or even from ordinances printed in the form of volumes. A booklet could be taken anywhere and eased learning; it could be used to verify the procedures even on the disaster scene. The text in question is not simply a copy of the 1773 ordinance; in fact, it overturns the order of the articles, in order to facilitate their use. The warning at the beginning of the booklet explains its concrete purpose:

The ordinance in execution of which this work was written recalls under different articles the policing which must be observed, and the rescuing strategies (*secours*) which must be employed in the case of fire or chimney fire.

Since the rescuing strategies had been described in a plan where the duties and obligations of groups and people are reported in different articles, sometimes remote from each other, we thought it necessary to ease their knowledge, and thereby accelerate the rescue, by bringing together under distinct paragraphs all the duties and obligations that concern the same groups or the same people, and which deal with the same issues.

This design in which we have absolutely departed from that of the ordinance, as we have divided the regulations without altering their meaning, has occasioned repetitions, which we have not sought to avoid, as they could be useful to those who need to read this work.[22]

The booklet was not only a professional guide for police forces charged with firefighting, but it was also part of an enlightened police literature in that it proposed what could further improve the organisation of firefighting:

---

20 Le 1721 ordinance features 26 articles, that of 1742, 40 articles, that of 1773, 55 articles.
21 Lille Municipal Archives, Affaires générales, box 563, dossier 15.
22 Lille Municipal Archives, Affaires générales, box 563, dossier 15.

We ended with an appraisal of the various aids that are used against fire, and of the zealous people on whom we might rely to elaborate the strategies that we think should be adopted.[23]

The booklet is anonymous, but it can only have been written by someone who exercised a function in the police. The official printer, the meticulousness and rigor of the presentation suggest that the author must be part of the police – indeed, a simple "project maker" would have dwelled on more general considerations. The author might perhaps be the provost of the city, an active man according to his superiors; but nothing can be proved with certainty.[24] Be that as it may, the transposition of the text of the ordinance into a booklet was directly aimed at the people who must take part in firefighting. The two texts therefore give a hint of all those who were required by the police to participate in the control of fire, making it possible to address the question of the police personnel active in the cities of the Old Regime, and that of the professionalization of the police.

In 1773, the firefighting service in Lille included over 300 people, divided into 17 groups:
- Fire commissioners (*commissaires au feu*)
- District commissioners (*commissaires de quartier*)
- Ammunition commissioners (*commissaires aux munitions*)
- Magistrates
- City angels (*angelots*)
- Cleaning sergeants (*sergens au nettoiement*)
- Deputy district commissioners
- The clerk of constructions (*clerc des ouvrages*)
- City sergeants and poor people sergeants (*sergens de ville et ceux des pauvres*)
- Lookouts (*guêteurs*)
- Water millers (*meuniers à l'eau*)
- Neighbours and other individuals
- Firefighters
- The patrols and the twenty-men (*les patrouilleurs et les vingt-hommes*)
- The city hall concierge
- Bailiffs and city valets (*huissiers à verge et valets de ville*)
- Brewers, innkeepers (*cabaretiers*), carters of the shores (*chartiers des rivages*), masons, carpenters, roofers, locksmiths, farriers, saddlers, wheelbarrow

---

23   Ibidem.

24   On the Lille provost and his relationship with the Magistrate, see Denys, *Police et sécurité* 37–46.

WHAT FIREFIGHTING TELLS US ABOUT 18TH-CENTURY URBAN POLICE 291

haulers (*brouetteurs aux poids*), wine-barrel carriers (*rouleurs de vin*), beer carriers (*porteurs de bières*), colliers, gaugers (*jaugeurs*), etc.

To this list, which is already long, one might add the monks who used to keep pumps and run to the rescue, as well as the soldiers of the garrison, who were also required, by their superiors, to assist the city in the event of a fire, by lending their equipment and their support.

The list can be divided into five categories of people rallied in the fight against fire. First of all, those who had a share in the municipal management. The fire commissioners and the ammunition commissioners, as well as most of the district commissioners were in fact members of the Magistracy in place, namely aldermen responsible for various commissions. The fire commissioners were the "rewart," a sort of second to the mayor for all roads and construction matters:[25] two of the six were works commissioners, responsible for inspecting buildings all year round, and two commissioners for ladders and hooks, responsible, as their name suggests, for ensuring the maintenance of firefighting equipment. All had to ensure, through regular inspections throughout the year, that the pumps were in good working order and that there were an adequate number of buckets at various depot locations. The ammunition commissioners simply had to be ready to provide lighting to the aldermen who got to the scene of the disaster and, if necessary, to have the neighbouring streets lit up by means of "tarred cokes (*tourteaux goudronnés*)."[26] The district commissioners, twenty overall, were not called all at once; the only one to be solicited was the one responsible for the district where the disaster was located. The Lille archives do not make it possible to specify the ordinary functions of the district commissioners, established in 1709, to whom was entrusted a general territorial supervision.[27] Each district commissioner must have been accompanied by his deputy, who oversaw on site the demand and proper use of the equipment stored in the neighbourhood. The role of the aldermen was limited to directing and organizing the emergency services on the spot, and to ensuring that everything was in good order.

Secondly, the regulations listed the duties of a small staff of municipal agents, whose diversity and the often-underestimated number are here made visible. The clerk of constructions inspected buildings, public or private, and supervised repair work. The eighteen city angels were construction workers,

---

25 Derode V., *Histoire de Lille et de la Flandre wallonne* (Paris – Lille: 1848), vol. 2, 418–419.

26 A "tarred coke" is a lighting device that consists of 'a ring formed of old lunt or of cords well beaten with a mallet' (*Scientific American Supplement* 586 [26 March 1887] 9361).

27 It would be a mistake to confuse these commissioners with the Châtelet commissioners of Paris or with the police commissioners created by the 1795 law.

employed by the city, who carried out minor repairs to public buildings or other *ad hoc* services, such as breaking the ice which obstructed the canals, the drinking troughs or the wells, during the severe winter frosts. The clerk called the building workers to the scene of the disaster, while the *angelots* brought the pumps and various materials from the depots of the city hall and the esplanade. The seven cleaning sergeants, usually responsible for ensuring that the inhabitants swept their thresholds, and that the cleaning contractors came to remove the piles of rubbish from all the streets each morning, were in these cases simply responsible for supporting a superintendent of the Magistracy, with a lantern to light his way, and to stay with him throughout the process to take his orders. The ten city sergeants, who usually represented the actual police in the city, and could thus give fines and arrest delinquents or criminals, had to monitor the lookouts and wake up the city hall concierge if necessary. Two sergeants had to stay at the city hall to guard the effects saved from the fire, the other eight had to go where the fire was, where they formed a barrier to protect the pump department. Their role was essential to organising the emergency services: they could let in only people useful to fight the fire and halt any intruders. This is an example of the function of maintaining order in a crowd, which was often the responsibility of city sergeants throughout the eighteenth century. The rather anachronistic permanence of the halberd in their equipment is no doubt explained by its use as a barrier, when held horizontally.[28] Their direct chief, the lieutenant of the provost, responsible for handling the police in town, but who was not part of the Magistracy, appears in the regulation only in the paragraph devoted to the duties of the city sergeants, who were directly under his orders. This surprising exclusion, while the lieutenant-provost was one of the main police chiefs of Lille, reflects perhaps the tensions between the provost, the lieutenant-provost and the aldermen. This absence does not imply the absence of the lieutenant-provost from the rescue operations, since it is certain that this active and zealous man participated in them.

In the same way, the regulations make only a brief allusion to the military support, while we know for a fact that the garrison was always very active in firefighting.[29] The zeal – sometimes untimely – of the soldiers, their brutal-

---

28 This can be seen in the painting *L'inauguration de l'écluse de Mardyck* by Jean-Baptiste Martin (best known as Martin des Batailles) preserved in the Musée des Beaux-Arts of Dunkerque.

29 Louis-Sébastien Mercier, in the *Tableau de Paris* attributes the disappearance of great fires to the purchase of pumps and to the assistance provided by the soldiers of the Paris

WHAT FIREFIGHTING TELLS US ABOUT 18TH-CENTURY URBAN POLICE

ity, also occasionally caused complaints from civilians. The military had their own equipment and could naturally dispose of significant forces. The grand ordinance about the *service des places et quartiers* (military service of places and districts) of 1 March 1768 also obliged them to help the cities and to maintain order at the rescue, as they did at public festivals. The barrier of city sergeants was therefore supplemented or doubled by that of the military, in a collaboration between urban police and army, very common in garrison towns. The nine poor people sergeants, usually in charge of the very unpopular task of capturing unauthorized beggars, were here required as simple support associated with the city sergeants. The city still employed lookouts, who stood guard at the top of the steeple of Saint-Étienne, the highest and most central church in Lille. When he spotted a fire, the lookout must sound the tocsin and warn of the district where the fire was occurring, during the day by a red flag in the direction of the district, at night by a bell code or by indications directly transmitted by a megaphone. The ringing of the tocsin had to be accentuated if the disaster increased. The wardens of the concerned parish and the neighbouring monasteries should also sound the tocsin.

Finally, there remain, amongst the municipal agents, twenty-four patrols and the twenty men, dubbed "of the shore" (*du rivage*), on which we have little information. Of the patrols, at most we know that they patrolled the streets of the city at night, but their status has not been debated in the great discussions in Lille on the night police. If these patrols really existed, the least that what we can infer from the judicial archives is that they hardly made any arrests, unlike the military patrols of the garrison. As regards firefighting, patrols were simply ordered to give the alarm in case they saw a fire, to notify the lookout, then to carry to the fire the axes, shovels, pitchforks, ropes, buckets and other materials kept at the city hall. The twenty men of the shore were in charge of ladders and hooks.

Thirdly, firefighting required the assistance of a whole series of well-identified professionals: the innkeepers must keep a certain number of buckets and transport them to the place of the disaster, otherwise their licence would be suspended; the water millers must stop their mills, closing the valves, in order to increase the volume of water available in the canals; the brewers must make their barrels available for the transport of water; the carters must prepare their carts. All trades related to transportation – wheelbarrow haulers,

Guard: 'This set up shows that it is possible to improve, the one after the other, all sections of the police; police organization was so defective twenty years ago, and yet now it has the admiration and gratitude of the citizens' ([Amsterdam: 1782] vol. 1, 210–211).

carriers of wine, oil, honey and syrup, beer carriers, carriers who used bags, colliers, gaugers of water, wood and fodder – should go to the fire as quickly as possible to provide help. The construction trades – masons, carpenters, roofers, locksmiths – were requisitioned, with their tools – ladders, hammers, axes, hooks – to demolish the houses wrecked by the fire. Finally, the farriers and the saddlers must be available to repair the leather casings of the pumps which risked being damaged during the fire. Of course, there was no question of summoning all of the transportation and construction professionals who worked in Lille. The Magistrate delegated the organization of the mobilisation to the guilds. Every year, on All Saints Day, the masters of trades (*maîtres des métiers*) must submit to the town hall a list of professionals and workers who could be rallied, with their names and addresses. A total of 93 professionals could be solicited, at the rate of 25 brewers, 12 carters from the shore, six masters (*maîtres*) and six workers amongst masons, roofers, carpenters, locksmiths and farriers; four *maîtres* saddlers with one worker each, and six representatives of the wheelbarrow haulers, carriers of wine, oil, honey and syrup, beer carriers, bearers using bags, colliers, and gaugers. In the event of a chimney fire, the city chimney sweep and one of his workers must be added to these 93 people. No one can leave town without making his replacement known to the Magistracy. This is naturally a maximum, not all the disposable tradespeople were called upon for a localized fire. But the involvement of guilds remained very strong.

Then come the firefighters. The regulation of 3 November 1773 specified their organization: there were then 70 firefighters divided into seven brigades of ten men for each of the seven large municipal pumps, deposited in various strategic places in the urban space, namely, the town hall, the furniture store on the esplanade, the monasteries of the Discalced Carmelites, the Recollets, the Capuchins and the Augustinians, and the general hospital. Each brigade was composed of a brigadier, who commanded the operations, a firefighter who was in charge of the fire hoses, and eight firefighters responsible for operating the arm of the pumps, to raise the water. The 1784 booklet provides the names and addresses of the 70 firefighters.

Finally, what about the participation of individuals and neighbors? The ordinance first encouraged residents of Lille to obtain small, efficient, and inexpensive hand pumps. Anyone who noticed an outbreak of fire must wake up all the inhabitants of the burning or threatened house. The neighbours must hang a lantern on their facade to light up the street and must place tanks full of water in front of their threshold. It seems that the neighbors were mainly expected to help the affected owners to shelter their furniture and personal effects. In any case, the city sergeants and poor people sergeants had the mission of pushing back far from the barriers 'the women, the children, and all other useless people'.[30]

The enumeration of the different actors involved in firefighting had therefore not really changed, in its inspiration, since the Middle Ages: the city relied on the mobilisation of its agents, encouraged the collective participation of the inhabitants, and imposed the mobilisation of professionals considered useful – all this under the supervision of the aldermen, who provided an order as implied by their role of "fathers of the people." Modernization here is not a matter of simple professionalization, which would have led to entrusting to specialist rescuers all the operations, to the detriment of citizens. All the people who could in fact be solicited reflect the ambiguities of police modernization at the end of the Old Regime. There is certainly a move towards specialisation, with the creation of fire brigades, exclusively responsible for the service of the pumps, or the insistence on the separation between the people who are authorized to participate in the rescue and the others. The physical and symbolic barrier that the city sergeants and the soldiers of the garrison must establish to control the passage of the help, and to keep out those considered "useless," was not really new, but had then reached such precision that all spontaneity disappeared from the population's voluntary participation in the rescue. Yet the barrier made by the sergeants did not simply divide professionals from amateurs. Indeed, apart from the city, poor people and cleaning sergeants, the lookouts and the concierge, all municipal agents gathered here were only semi-professionals: the city angels and the patrols only fulfilled these functions part-time as they had other jobs. Their status was basically not very different from that of the masters and workers of the trades solicited in the event of fire. As for the soldiers and the monks, their massive participation in firefighting, which stemmed from their dedication to the public service, obviously cannot be considered as the beginning of police professionalization. Need we remember, finally, that the aldermen themselves, who were the commissioners and members of a powerful Magistracy, were in no way professional administrators?

Even the firefighters who appeared for the first time in the form of an organized corps with brigadiers and territorial distribution were certainly not professionals. Just as current firefighters in small and medium-sized French towns, these people had a job and only participated in rescues when required by events. The ordinance of 3 March 1773 could certainly signal the creation, for the city of Lille, of a fire brigade, thereby joining the initiatives of other

---

30   Lille Municipal Archives, Affaires générales, box 563, dossier 15.

cities.[31] But the denomination of "brigadier" should not lead us to believe in the professionalization of rescues. In other places and throughout the nineteenth century, the firefighter would have been a sort of hybrid figure, embodying three contrasting models: the amateur rescuer, the policeman and the soldier.[32]

In the eighteenth century, if elements of professionalization were detectable in the evolution of certain police tasks, firefighting shows to what extent this trend towards professionalization was certainly not the only possibility envisaged by the urban authorities. The urban police systems of the Old Regime were complex. If one looks there for professional and specialist police officers such as those operating in contemporary institutions, one will neglect the considerable strengths that existed in public security structures, which came under the corporate links or microlocal neighbourhood notabilities, whose influence was probably declining less quickly than one might believe.[33]

3   Modernity and European Circulations

Does this mean that the prevention of and fight against fires did not change from the Middle Ages to the end of the Old Regime? Was this aspect of the "good police" of the city left out of the debates and exchanges that flourished in the second half of the eighteenth century? This is not so, of course. But modernity can lie in aspects different from (and less predictable than) the specialisation or professionalization of rescue. Historians have already noted the decrease in devastating fires in the large cities in the eighteenth century, which seems to imply real progress in fire prevention and control measures.[34] Occasions for the outbreak of fire were still not uncommon, but the fires of the eighteenth century seem to be contained more quickly than was previously done.

The reason lies primarily in the dissemination of technical advances that deeply transformed firefighting. In the event of fires on the roof or in the upper floors, the possibilities of fighting were very limited. In the sixteenth century, syringes were invented which made it possible to sprinkle the flames

---

31   Dalmaz P., *Histoire des sapeurs-pompiers français* (Paris: 1996) 14.

32   See, for instance, Python-Bernicot M., "Pompiers et gendarmes du XIXe siècle au service de la sûreté publique", in Luc J.-N. (eds.), *Gendarmerie, État et société au XIXe siècle* (Paris: 2002) 253–260.

33   On district police, see Denys C., "The Development of Police Forces in Urban Europe in the Eighteenth Century", *Journal of Urban History* 36 (2010) 332–344.

34   Delumeau J., *Rassurer et protéger. Le sentiment de sécurité dans l'Occident d'autrefois* (Paris: 1989) 534–539.

but could hardly raise the water very high. Progress here came at the beginning of the seventeenth century, with the "fire pumps," the first examples of which appeared in the Netherlands. These were simply hand pumps to send water from a tank under pressure into a leather pipe. Their distribution was quick from city to city: in the North of France, pumps were imported from Flemish and Dutch cities as of the 1670s. Paris took over in 1699, with François Dumouriez du Périer who sold machines on the Dutch model throughout the kingdom.[35] Pumps were continually improved throughout the eighteenth century, and the rise of a real market in European cities was solicited by inventors. Thus, Lille received a demonstration pump from Amsterdam in October 1750, a prospectus from the Parisian fireman Sieur Thillaye in 1781 and a Brussels inventor, Salomon Furst, who offered his machines in 1783.[36] The circulation of new pumps and their inventor-demonstrators joins exactly that of street lanterns in the 1780s. From town to town, officials exchanged information, experimented, sought the best management methods. The fight against fire therefore drove, like public lighting, a friendly competition between cities, to improve public safety and convenience. In this respect, the organization of fire rescue was part of the "Enlightenment police" movement (*Lumières policières*) of the eighteenth century.

The coincidence, or at least the temporal proximity, of the ordinances against fires reveals other possible circulations. If certain regulations were dictated by local considerations, such as the Lille regulation of 18 December 1756, which is somewhat redundant compared to that of 13 December 1755, which is explained by a fire at the town hall a few days earlier, others are probably a reflection of city-to-city text exchanges. It was indeed common for a city to draw inspiration from the police regulations of another city to produce a new ordinance. In the regional space of the former Netherlands, the circulation of police orders is widely attested for the eighteenth century, beyond the partition between France and the Habsburg Netherlands. The reputation of Paris also obliged more or less all urban officials to turn to the French capital before producing their own texts, as the reference to Paris had become commonplace in European police texts. Thus, in the Lille text of 1784, after recalling that the construction of furnaces and forges obeys precise rules and that the intervention of the clerk of constructions is essential, the writer added: 'Such is the practice of Lille, in accordance with that of Paris'. There were still circulations from big city to big city which did not always follow a very obvious geographical logic, but which were identified by striking similarities in the content. Thus,

---

35   Dalmaz, *Histoire des sapeurs-pompiers* 11–14.

36   Lille Municipal Archives, Affaires générales, box 564, dossier 2; box 561, dossiers 11 and 17.

although it is not possible to decide between pure coincidence or reciprocal inspiration, one might notice that the Lille ordinance of 1773, and a Regulation to be observed on the police in the event of fire enacted in Geneva on 4 June 1784 share many common traits.[37] This is only one particular example that a more systematic comparison of the texts in use in eighteenth-century Europe would certainly make it possible to generalize.

Amongst these common features, which show the international collective inspiration to improve the police service in the eighteenth century, a prominent one is an almost obsessive concern for ordering the mobilisation of people who had to run to the fire. The police resorted to techniques proven effective on other occasions: spatial identification of places and categorical identification of people by a visible sign. In Lille as well as in Geneva, the police themselves must form a barrier – as I have already pointed out – whose purpose was to sort out those who could be useful in the rescue and the intruders. The whole text from Lille breathes concern about the uncontrolled participation of the crowd, namely a refusal of spontaneity insofar as it creates disorder. The fire is a stage, similar to the theatre, where the police must illuminate the action, facilitate the acting of the actors, keep the spectators at a distance. The Enlightenment police thus sought to rationalize the rescue efforts to push back popular anarchy. The effectiveness of the control depended on the presence of the Magistrates, who provided order to the rescue, and who were the only depositories in the city of an uncontested authority over the inhabitants.[38] The possibility of control is also a responsibility of the masters of the trades, who were the relays of the aldermanic authority, hence the extreme detail of the contributions requested from the various professionals working in construction or transportation. Moreover, the police at the end of the eighteenth century intensified and systematized the principle of identifying their functions by their appearance, characterized by an official costume and a distinctive mark. The deputies of the Lille district commissioners must wear a badge with the city crest around their neck. The deputy of the district in which the fire broke out must be distinguished by a red ribbon; the city angels and the twenty men, as well as the solicited tradesmen, must wear the same ribbon at the buttonhole of their jackets. The various sergeants must carry their weapons and shoulder straps. Finally, as regards the firefighters, the text specifies as follows: 'in order that they can be easily recognized, all firefighters will have a leather shoulder strap, and a copper plate with the city crest with the number of the

---

37   Cicchini, *La police de la République* 391.
38   On the contrary, the brutality of soldiers and of non-commissioned officer is often reported.

WHAT FIREFIGHTING TELLS US ABOUT 18TH-CENTURY URBAN POLICE 299

pump of which they are charged; the brigadiers will also have a white cockade on their hat'.[39]

To help the immediate visual identification provided by the distinctive mark, the Lille police added a very advanced system of "attendance tokens." Their use in the police was not really a novelty, since already in the seventeenth century the attendance of sergeants on night shifts was guaranteed by the "maroons of patrol (*marrons de patrouille*)" that they must distribute throughout their patrol. But the regulations of 1784 reveal a whole series of marks or "seals" that the commissioners received from people who were ordered to help, or that they distributed to people involved. The day after the fire, the Magistrates gathered at the city hall and, on the basis of these marks, proceeded to the punishment of the absent and to the reward of the most zealous.

These measures of tracking were further complemented by spatial tracking, relayed by the display of multiple tables. The 1784 booklet gave the places of depots for the seven pumps of the city, the depots district by district for buckets, ladders and hooks, with their number. It then indicated the names, addresses and numbers of the 70 firefighters, classified by the pump to which they are assigned, then those of the people responsible for bringing the equipment to the fire. Construction and transportation professionals were not named, but the guilds must send the Magistrates the table of names and addresses of their representatives each year. Numbered medals were also provided to control their presence.

All this apparatus of numbers, tables, medals, badges and shoulder straps brought the fire police of the end of the eighteenth century well beyond the simple regulation of emergency services. What was left of the spontaneous mobilisation, local solidarity, the duty of mutual aid of the bourgeois in this superposition of finicky techniques of control? The nascent bureaucratic techniques, whose emergence is witnessed by other phenomena such as the control of the identity or movements of people, attest to a form of modernity in the Enlightenment police.[40]

4   Conclusion

The regulation of the prevention of and fight against urban fires in the eighteenth century opens up considerable insights into the transformations of the police. The banality and repetitiveness of ordinances hides the presence of

---

39   Lille Municipal Archives, Affaires générales, box 563, dossier 15.
40   Denis V., *Une histoire de l'identité France, 1715–1815* (Paris: 2008).

subtle inflections in police practices, in the development of writing and of police know-how. A fine-grained analysis of these texts invites us to think less in terms of the effectiveness of police measures than in terms of pedagogy of the norm and the constitution of repertoires that can first be used by agents, who were becoming professionals. Finally, this chapter suggests the opportunity of developing international comparative studies that shed light on the greater circulation of police Enlightenment practices across the European space.

Bibliography

Primary Sources

Douai Municipal Archives, AA 104, fol. 25.

Lille Municipal Archives, Affaires générales, box 563, dossier 15.

Lille Municipal Archives, Affaires générales, box 564, dossier 2; box 561, dossiers 11 and 17.

Lille Municipal Archives, Registre aux Ordonnances du Magistrat 399, 4 January 1721; Affaires générales, box 563, dossier 13, 7 July 1742; Registre aux Ordonnances 405, 13 December 1755 and 18 December 1756.

Valenciennes Municipal Archives, AA 131/6, fol. 1.

Secondary Sources

Andrey G. et al., *L'incendie de Bulle en 1805. Ville détruite, ville reconstruite* (Bulle: 2005).

Berlière J.-M. – Denys C. – Kalifa D. – Milliot V. (eds.), *Métiers de police. Être policier en Europe, XVIIIe–XXe siècle* (Rennes: 2008).

Berthet M., "Dans le Haut-Jura, toitures, transports, incendies", *Annales ESC* 8.2 (1953) 192–196.

Cicchini M., *La police de la République. Construire un ordre public à Genève au XVIIIe siècle* (Ph.D. dissertation, University of Geneva: 2009).

Dalmaz P., *Histoire des sapeurs-pompiers français* (Paris: 1996).

Delumeau J., *Rassurer et protéger. Le sentiment de sécurité dans l'Occident d'autrefois* (Paris: 1989).

Denys C. – Marin B. – Milliot V. (eds.), *Réformer la police. Les mémoires policiers en Europe au XVIIIe siècle* (Rennes: 2009).

Denys C., "The Development of Police Forces in Urban Europe in the Eighteenth Century", *Journal of Urban History* 36 (2010) 332–344.

Denys C., "La mort accidentelle dans les villes du Nord de la France au XVIIIe siècle: mesure du risque et apparition d'une politique de prevention", *Histoire urbaine* 2 (2000) 95–112.

WHAT FIREFIGHTING TELLS US ABOUT 18TH-CENTURY URBAN POLICE 301

Denys C., *Police et sécurité au XVIIIᵉ siècle dans les villes de la frontière franco-belge* (Paris: 2002).

Derode V., *Histoire de Lille et de la Flandre wallonne* (Paris – Lille: 1848), vol. 2.

Dubois-Maury J., "Un risque permanent: l'incendie", *Annales de Géographie* 97.539 (1988) 84–95.

Favier R., "Secourir les victimes. L'incendie au village dans les Alpes dauphinoises (seconde moitié du 18ᵉ siècle)", oral communication at the conference *Al Fuoco! Usi, rischi e rappresentazione dell'incendio dal Medievo al XX secolo* (Mendrisio: 2007).

Felicelli C., "Le feu, la ville et le roi: l'incendie de la ville de Bourges en 1252", *Histoire urbaine* 5.1 (2002) 105–134.

Gresset M., "Quelques incendies de villes comtoises au XVIIIᵉ siècle", in Vion-Delphin F. – Lassus F., *Les hommes et le feu de l'Antiquité à nos jours. Du feu mythique et bienfaiteur au feu dévastateur* (Besançon: 2007) 275–282.

Iseli A., *Bonne police. Frühneuzeitliche Verständnis von der guten Ordnung eines Staats in Frankreich* (Epfendorf – Neckar: 2003).

Lamarre C., "Les incendies de villes dans les géographies du XVIIIᵉ siècle", in Vion-Delphin F. – Lassus F., *Les hommes et le feu de l'Antiquité à nos jours. Du feu mythique et bienfaiteur au feu dévastateur* (Besançon: 2007) 283–292.

Lavedan P., *Jeanne Hugueney et Philippe Henrat, L'urbanisme à l'époque moderne: XVIᵉ–XVIIIᵉ siècles* (Genève: 1982).

Milliot V. (eds.), *Les mémoires policiers, 1750–1850. Écritures et pratiques policières du Siècle des Lumières au Second Empire* (Rennes: 2006).

Napoli P., *Naissance de la police moderne: pouvoir, normes, société* (Paris: 2003).

Nières C., "Le feu, la terre et les eaux", in Delumeau J. – Lequin Y. (eds.), *Les malheurs des temps. Histoire des fléaux et des calamités en France* (Paris: 1987) 367–382.

Nières C., *La reconstruction d'une ville au XVIIIᵉ siècle: Rennes 1720–1760* (Paris: 1972).

Python-Bernicot M., "Pompiers et gendarmes du XIXᵉ siècle au service de la sûreté publique", in Luc J.-N. (eds.), *Gendarmerie, État et société au XIXᵉ siècle* (Paris: 2002) 253–260.

Reddaway T.F., *The Rebuilding of London after the Great Fire* (London: 1940).

Reitsma H.J., "The explosion of a ship, loaded with black powder, in Leiden in 1807", *International Journal of impact engineering* 25 (2001) 507–514.

Smolar-Meynart A. – Vanrie A. (eds.), *Le quartier royal* (Bruxelles: 1998).

Tanguy J.-F., "Les autorités municipales et la lutte contre l'incendie en Ille-et-Vilaine au XIXᵉ siècle", *Revue d'histoire du XIXᵉ siècle* 12 (1996) 31–51.

Vernus M., *L'incendie, histoire d'un fléau et des hommes du feu* (Yens sur Morges: 2006).

CHAPTER 11

Organising the Chaos: Firefighting in Upper Lusatia in the Early Modern Period

Cornelia Müller

Abstract

In this chapter I examine how firefighting was organised in three early modern towns, Bautzen, Görlitz and Löbau, which were all located in the same legal area, Upper Lusatia (in eastern Germany). These towns had the right to make their own municipal ordinances, as they were relatively independent of the sovereign. Drawing on Henri Lefebvre's idea of space as a complex social construction, and on Susanne Rau's interpretation of Lefebvre's theory, I suggest that fire regulations should be read as "mental spaces" in which different aspects, both ideal and real, of the social, political and economic life of a city are represented. An analysis of the regulations shows that the urban space was not conceived as a unit, but was divided into zones with different functions and times of intervention in the event of fire, revealing specific social hierarchies. The distribution and visibility of firefighting equipment (buckets, ladders, syringes, etc.) also reflected social organisation. Interestingly, the regulations not only represented the mental plan of action elaborated by the city's magistrate, but also distilled past experience and used it to build future plans aimed at maintaining "good policing."

Keywords

urban history – Upper Lusatia – firefighting – fire regulations – spatial theory – Henri Lefebvre

1 Introduction

Urban fires occur in specific places: houses, streets, neighbourhoods. Fires have an elusive character, as they can occur, for instance, inside, outside or on both sides of the city walls. Their consequences are visible in the form of ruins, and reconstruction can take different forms. Yet to understand how fires can shape the urban space, one should not only consider the occurrence of fires, but also the strategies of fire prevention and firefighting. In this chapter, I will

© KONINKLIJKE BRILL BV, LEIDEN, 2025 | DOI:10.1163/9789004521766_013

discuss early modern examples of how the urban space could be organized on the basis of prevention measures, and how its configurations could change during fires.

To conduct my analysis, I will rely on Henri Lefebvre's idea of space as a complex social construction, and on Susanne Rau's application of it to historical research, notably urban history. Whereas Lefebvre identified different types of space – the 'imagined (*conçu*),' the 'perceived (*perçu*)' and the 'lived (*vécu*)' space – I will only focus on the most abstract one, namely the '*espace conçu*.' I will argue that a Lefebvrian analysis of the urban space as an imagined space can reveal several dichotomies, such as "open" versus "closed," or "public" versus "private." When it comes to firefighting, however, the most interesting dichotomies are other ones, namely "visible" versus "invisible," and "accessible" versus "inaccessible." As we shall see, approaching the history of early modern firefighting through the lens of a spatial analysis can contribute to renewing the history of firefighting, since one can go beyond the purely technical focus that still characterizes much historiography in this field.

I discuss the case study of the three early modern towns, namely Bautzen, Görlitz and Löbau, which were all located within the same legal area, Upper Lusatia. These towns had the right to adopt their own municipal ordinances. Such ordinances are a mirror of social and political power relations, as they were relatively independent vis-à-vis the sovereign. The latter could only issue general injunctions regarding firefighting, but could not decide the fire regulations for each town. Therefore, the question arises how these three towns regulated firefighting, and what differences, similarities and special features can be identified.

2   The Historiography of Fires and Firefighting

Fires have always been at the centre of public debate. Discussion about these events began immediately afterwards, in the early modern period as well as in modern times. For instance, fire sermons were often preached shortly afterwards represented a first attempt to interpret the event.[1] Together with

---

1   So-called fire sermons were often delivered immediately after a city fire or years later in memory of such an event. Especially in the eighteenth century, the sermons were printed and sometimes the proceeds from their sale were donated to the affected citizens. Marie Luisa Allemeyer has compared the sermons of various northern German cities in the early modern period, cf. Allemeyer M.L., *Fewersnoth und Flammenschwert. Stadtbrände in der Frühen Neuzeit* (Göttingen: 2007) and also Allemeyer M.L., "Wenn der liebe Gott einen Hauss Wirth mit Feuers Brunst heimsucht. Zur Deutung und Darstellung von Stadtbränden in religiösen, genossenschaftlichen, technischen und obrigkeitlichen Schriften", in Koppenleitner V.F. –

pictures and published reports, the sermons are elements of a culture of remembrance. The debate about the causes and consequences of fires could reach into scholarly circles. An example of this is the destruction that Lisbon suffered in November 1755. The earthquake, which was followed by numerous fires and a tsunami, shook the belief that humans could control nature.[2] Both the event in the city and the confrontation with it have been the subject of numerous publications until the present day.[3]

When it comes to serious and momentous fires, the following are often mentioned for the early modern and modern periods: the fire of London in 1666[4] or that of Moscow in 1813.[5] One of the best-known examples from the

---

Rößler H. – Thimann M. (eds.), *Urbs incensa: Ästhetische Transformationen der brennenden Stadt in der Frühen Neuzeit* (Berlin – Munich: 2011) 285–300. Thus, Johann Zeidler published his sermon under the title Tabeera Budissinae in 1634, after the severe fire of the town of Bautzen during the Thirty Years' War, cf. Zeidler Johann, *Tabeera Budissinae. Budissinische Brandstelle* (Dresden, Gimel Bergen: 1634). This sermon, supplemented by other sermons delivered at a later date by other pastors in memory of the destruction of 1634, was published again and expanded in 1660, 1686 and 1707. Similar prints are also known from Görlitz and Löbau.

2 See Molesky M., "The Great Fire of Lisbon, 1755", in Bankoff G. – Lübken U. – Sand J. (eds.), *Flammable cities. Urban conflagration and the making of the modern world* (Madison: 2012) 147.

3 See e.g. Walter F., *Geißel Gottes oder Plage der Natur. Vom Umgang der Menschen mit Katastrophen* (Stuttgart: 2016); Löffler U., *Lissabons Fall – Europas Schrecken. Die Deutung des Erdbebens von Lissabon im deutschsprachigen Protestantismus des 18. Jahrhunderts* (Berlin: 1999); Lauer G. (ed.), *Das Erdbeben von Lissabon und der Katastrophendiskurs im 18. Jahrhundert* (Göttingen: 2008); Georgi H., *Das Erdbeben von Lissabon. Wie die Natur die Welt ins Wanken brachte – von Religion, Kommerz und Optimismus, der Stimme Gottes und der sanften Empfindung des Daseins* (Wiesbaden: 2016); Günther H., *Das Erdbeben von Lissabon und die Erschütterung des aufgeklärten Europa* (Frankfurt: 2005).

4 For the causes of the fire see Garrioch D., "1666 and London's Fire History: a Re-Evaluation", *The Historical Journal* 59.2 (2016) 319–338; For (artistic) interpretations see, among others Heyl Ch., "'God's Terrible Voice in the City': Frühe Deutungen des Great Fire of London, 1666–1667", in Koppenleitner V.F. – Rößler H. – Thimann M. (eds.), *Urbs incensa: Ästhetische Transformationen der brennenden Stadt in der Frühen Neuzeit* (Berlin – Munich: 2011) 23–44; Koppenleitner V.F., "'Etiam periere Ruinae': Katastrophenereignis und Bildtradition in Darstellungen des Großen Brandes von London 1666", in Koppenleitner V.F. – Rößler H. – Thimann M. (eds.), *Urbs incensa: Ästhetische Transformationen der brennenden Stadt in der Frühen Neuzeit* (Berlin – Munich: 2011) 45–59; Another important contribution is: Keene D., "Fire in London: Destruction and Reconstruction, A.D. 982–1676", in Körner M. (ed.), *Destruction and Reconstruction of Towns*, vol. 1: *Destruction by Earthquakes, Fire and Water* (Bern – Stuttgart – Vienna: 1999) 187–211.

5 Scharf C. – Nordhof A.W. – Kessel J. (eds.) *Die Geschichte der Zerstörung Moskaus im Jahre 1812* (Munich: 2000).

German area is the fire of Hamburg in 1842.[6] But even the destruction of a single building could have far-reaching consequences. In response to the 1881 fire at the Vienna Ringtheater, which killed 384 people, fireproof curtains and the presence of a fireman during the performance became the norm.[7]

While especially large and severe fires have repeatedly been the subject of much scholarship for their consequences, firefighting itself has come into focus much less frequently. The fire brigades themselves have repeatedly addressed their history, especially from the perspective of the history of technology. Thus, since 1978, the Association for the Promotion of German Fire Protection has had a Department 11, which deals with the history of fire protection. In 2020, the Department of Fire Protection and Fire Service History was founded within the German Fire Service Association. The department is headed by Rolf Schamberger, director of the fire department museum in Fulda. But on that side the background of their establishment or the conditions before the foundation of the (voluntary) fire brigades in various towns and communities has hardly ever been questioned. This starting point is also reflected in the numerous fire brigade museums in Germany that focus their collections primarily on the twentieth century.

Almost all publications on fire brigades and their formation share two common features: first, they focus on the twentieth century, often with a technical emphasis;[8] second, while discussing the early phases of fire brigades, they mainly address political issues.[9] There is less discussion of how firefighting really functioned.

Surprisingly, the German historiography of policing has hardly paid any attention to firefighting, despite the large number of regulations that were issued over time. "Good policing"[10] – as the contemporary expression in early

---

6   See Klemm D., "Hamburg brennt: Zur bildlichen Darstellung der Jahrhundertkatastrophe", in Trempler J. – Bertsch M. (eds.), *Entfesselte Natur: Das Bild der Katastrophe seit 1600* (Petersberg: 2018) 80–89; Schubert D., "The Great Fire of Hamburg, 1842. From Catastrophe to Reform", in Bankoff G. – Lübken U. – Sand J. (eds.), *Flammable Cities. Urban Conflagration and the Making of the Modern World* (Madison: 2012) 212–234.

7   See Mikoletzky J., "Der Brand des Wiener Ringtheaters 1881 und seine Folge", *Ferrum: Nachrichten aus der Eisenbibliothek* 69 (1997) 59–68.

8   For Cologne see: Neuhoff St., *Feuer und Flamme. Die Geschichte des Brandschutzes in Köln* (Düsseldorf: 2014). For Augsburg and the surrounding area see: Pötzl W. (ed.), *Feurio! Es brennt. Zur Geschichte des Brandschutzes, der Brandbekämpfung und der Feuerwehren* (Augsburg: 2010).

9   The latest one is Briese O., *Für des Staates Sicherheit. Das Löschwesen im 19. Jahrhundert und die Gründung der ersten Berufsfeuerwehr Deutschlands in Berlin 1851* (Berlin: 2018).

10   The expression 'good policing' relates to good public order as a target for political decisions. With examples from different regions in Europe see e.g.: Stolleis M. (ed.), *Policey*

modern times goes – was about maintaining public order. Rulemaking, observance and control formed a triad. As Dirk Krüger has shown through the example of Dresden, fire protection measures were promoted and implemented in the city primarily by the 'traditional rule maker' as we can call, for example, the Electors in the Ancien Régime, and only occasionally on the initiative of the building owners themselves.[11] Other research on "policing" has focused on how the early modern state increasingly evolved through a variety of ordinances, mandates, and regulations. Achim Landwehr established a new point of view with his work on the town of Leonberg by not only looking at the

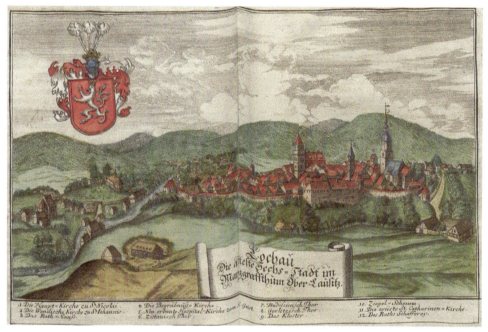

FIGURE 11.1 Hand drawn view of Löbau in Samuel Grosser, Lausitzische Merkwürdigkeiten Darinnen von beyden Marggrafthümern in fünf verschiedenen Theilen [...] (Leipzig – Budissin, David Richter – Immanuel Tietze: 1714) after page 83 in part III. Görlitz, Oberlausitzische Bibliothek der Wissenschaften
IMAGE © OBERLAUSITZISCHE BIBLIOTHEK DER WISSENSCHAFTEN GÖRLITZ

im Europa der Frühen Neuzeit (Frankfurt: 1996). See also: Iseli A., *Gute Policey: Öffentliche Ordnung in der Frühen Neuzeit* (Stuttgart: 2009).

11 Krüger D., *Die Entwicklung des baulichen Brandschutzes in Sachsen vom Beginn der Stadtgründungen bis 1945 im politischen, rechtlichen und wirtschaftlichen Kontext*, PhD dissertation (Technical University of Dresden: 2010).

FIGURE 11.2   Hand drawing of the 1634 fire in Bautzen in Samuel Grosser, Lausitzische Merkwürdigkeiten Darinnen von beyden Marggrafthümern in fünf verschiedenen Theilen [...] (Leipzig – Budissin, David Richter – Immanuel Tietze: 1714) between page 250 and 251. Görlitz, Oberlausitzische Bibliothek der Wissenschaften
IMAGE © OBERLAUSITZISCHE BIBLIOTHEK DER WISSENSCHAFTEN, GÖRLITZ

various orders as a means of rulership by the sovereign, but also by enquiring into the effects on the town's inhabitants.[12]

In this chapter, by contrast, firefighting will be examined from a different perspective. I will focus less on the actual implementation of regulations in everyday life, their application and control. Rather, the normative documents of the fire ordinances will be at the centre of the analysis. Despite their abstract character, norms were in fact created on the basis of concrete power relations, which they reflect. Differences in the fire regulations and thus the rules regarding firefighting can be found between the cities of Bautzen, Görlitz and Löbau, although they were close to each other. This should be understood as an invitation to examine not only the implementation or non-implementation of orders in the early modern period, but also the content of the documents themselves. They reflect a planned space with a specific social structure, and indicate aspects that are to be maintained or changed about it. This makes it possible to draw conclusions about the real conditions.

12   See Landwehr A., *Policey im Alltag: Die Implementation frühneuzeitlicher Policeyordnungen in Leonberg* (Frankfurt: 2000).

3   Firefighting in the *Espace Conçu*: the Fire Ordinances

The German city charters and bylaws of the sixteenth century already contain general provisions on fire prevention and firefighting. They state that under the command of the respective mayor, citizens, residents and especially certain craftsmen should fight the fire together.[13] From the seventeenth century onwards, these regulations were summarised in separate documents and printed. They were updated at irregular intervals. Repeatedly devastating fires showed that the previous approach to fire prevention and fighting was unsuccessful. For example, the Görlitz magistrate revised its fire regulations very extensively after the fire of 1691 which destroyed 200 houses.[14]

In addition to this revision 'from below,' an order of the sovereign could also be the cause of a revision. For example, the Bautzen fire ordinances of 1710 state at the beginning that a revision of the ordinances of 1671 had been deemed necessary and that there had also been a corresponding electoral order of 20 November 1709. Interestingly, the revisions in Görlitz (1709) and Löbau (1711) contain no reference to such an order. They merely refer to a basic duty to review and revise by the council and to the suffering caused by fires in their own town or in other towns. The establishment of fire ordinances in the standardisation and self-disciplining process of cities can be seen as an emancipation in terms of content. Instead of dealing with different subject areas in one document, individual ordinances were issued for different areas of daily life. There was a dress ordinance, a brewing ordinance, a fire ordinance, and so forth. Each of these contained rules that regulated precisely the daily life of the inhabitants, for example: who may wear what clothes and who may not; who may brew beer and when; when certain tradesmen are not allowed to work for reasons of fire safety, or what extinguishing agents everyone must keep on hand. Each ordinance represents its own theoretical space, or '*espace conçu*,' that is supposed to regulate the actions of citizens.

These social spaces include specific buildings, rules and apply only to individual inhabitants of the town. The ordinances define practices and hierarchies, designate points of reference such as buildings and places of action. They have a double function in relation to knowledge. On the one hand, they

---

13   Cf. the municipal ordinances of Bautzen (1596), Görlitz (1565) und Löbau (1616), published in Fröde T., *Privilegien und Statuten der Oberlausitzer Sechsstädte: Ein Streifzug durch die Organisation des städtischen Lebens in Zittau, Bautzen, Görlitz, Löbau, Kamenz und Lauban der frühen Neuzeit* (Spitzkunnersdorf: 2008).

14   Hartstock E., *Geschichte der Plagen der Oberlausitz: Von Feuersbrünsten, Epidemien und Dürrekatastrophen – Eine Chronik der Ereignisse von 1112 bis 1868* (Spitzkunnersdorf: 2017) 216. This fire is considered the most serious in the town.

are "vessels" that convey knowledge, for example about firefighting. On the other, they regulate how knowledge about competences and tasks is produced and practised. The orders are thus both producers and products of knowledge. The various orders were a means of ensuring urban order, namely the "good policing."

The ordinances – including those on fires and firefighting – described and defined different spheres of action for the various aspects of the economic, political and social life of the town. In Lefebvre's words, they are 'mental spaces' in the sense that they are 'the representations of power and ideology, of control and surveillance'.[15] The fire regulations give a good example of how strongly the town council tried to intervene in the everyday life of each household for reasons of fire prevention and to regulate firefighting.

The following overview shows when major fires occurred in the cities and when fire ordinances were revised. It is easy to see that smaller groupings of fire ordinances were formed in the last third of the seventeenth century, and the first decade and the last third of the eighteenth century. They thus remained valid for several decades. In the meantime, there were at most extensions or additions, as in Bautzen in 1720 and 1761. And in Görlitz, too, no new fire ordinances were issued in 1709, but an addendum to the regulations of 1692.

In the case of small towns in the Upper Lusatia, the responsibility for this lay with the respective patrimonial lord or with the sovereign. In the case of

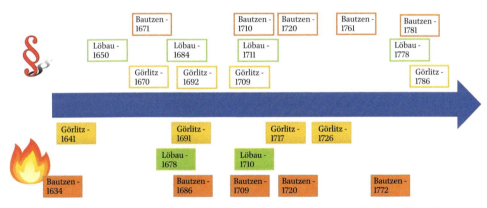

FIGURE 11.3   Fire ordinances of Bautzen, Görlitz and Löbau in the seventeenth and eighteenth century
OWN ILLUSTRATION

---

15   Soja E.W., "Die Trialektik der Räumlichkeit", in Stockammer R. (ed.), *TopoGraphien der Moderne: Medien zur Repräsentation und Konstruktion von Räumen* (Munich – Paderborn: 2005) 107; see also Soja E.W., *Thirdspace. Journeys to Los Angeles and other real-imagined places* (Malden: 1998) 67.

Fire ordinance of	Year	Number of paragraphs	Paragraphs on firefighting	In percentage terms
Bautzen	1671	47	27	57.4 %
	1710	58	28	48.3 %
	1781	87	33	37.9 %
Görlitz	1670	No counting	[pages 4-8]	
	1692	68	22	32.4 %
	1786	78	32	41.0 %
Löbau	1684	15	11	73.3 %
	1711	61	26	42.6 %
	1778	61	23	37.1 %

FIGURE 11.4   Scope of the fire ordinances with a share on the topic of firefighting
OWN ILLUSTRATION

Bautzen, Görlitz and Löbau, the magistrates themselves were responsible for setting the rules. The structure of the fire ordinances followed the same logic in most cases: before – during – after, i.e. preventive measures, active firefighting and aftercare.

Fig. 11.4 shows a breakdown of the number of active firefighting paragraphs in the fire ordinances. This shows that the number of paragraphs on active firefighting has – unsurprisingly – always been high. At the same time, a development can be observed. In Bautzen and Löbau the percentage is declining, while in Görlitz it is over a third. However, the total number and scope of paragraphs have increased significantly. This illustrates that especially the preventive, behaviour-related regulations are taking up more and more space.[16] At that time, the best way to fight a fire was to prevent it from breaking out. This changed only when the fire departments themselves became more and more professional from the second half of the nineteenth century and the technical means in the form of fire extinguishers or breathing masks, had improved greatly.

---

16   This analysis reflects only the numerical number of paragraphs, not their length or diverse content. Despite this limitation, it is clear that, notwithstanding the increasing volume of fire regulations, firefighting regulations had not increased to the same extent. The regulations on fire prevention concerned the handling of light and fire, building regulations for roofs and chimneys, regulations for potters and bakers concerning their working hours and the storage of flammable materials, for example. Since the eighteenth century, there were also smoking bans.

4   The Organization of Firefighting in Bautzen, Görlitz and Löbau

Let us imagine that the call 'Fire! Fire!' is shouted out. What happens? The urban space, and the entire rhythm of life and work change. All activities are interrupted. Each inhabitant has a task. The duties and powers of some, such as the mayor, were precisely described in the fire regulations. The duties of other residents, such as a housemaid, were less precisely described. But everyone was involved in fighting and monitoring the fire. Everyone acted on their knowledge of what should be done.

At the same time, the boundaries of the town changed: as soon as the bells warned about the outbreak of a fire, the gates were to be closed and the fortifications were to be manned with additional people. People were only allowed to enter the town with the express permission of the mayor. The town thus placed itself in a defensive position. It was threatened from within by an enemy, the fire, and in this weakened situation, when everyone was busy fighting the fire, it sealed itself off from the outside. Interestingly, this is where the first differences between Bautzen, Görlitz and Löbau appear. In contrast to the other two towns, in Löbau the gates remained open under heavy guard.[17] In this way, the inhabitants of the surrounding villages, who were subject to the landlordship of the town, could more quickly fulfil their duty to help Löbau. In Bautzen and Görlitz, the villages, which were also under obligation, first had to ask for admission at the gates.[18] This example shows how the dichotomy of "open-closed" can be used to concretely describe and analyse spaces or their changes. The fire codes solved the issue of gate closures in case of fire differently. As a result, the urban space in these three towns had different formations in case of fire.

5   Divisions of the Urban Space

The urban space was not a single unity, but was divided into different zones, as emerges from the fire ordinances. The cases of Bautzen, Görlitz and Löbau reveal different ways of dividing the urban space, and different temporalities.

---

17   See Rat der Stadt Löbau, *Revidirte und verbesserte Feuer-Ordnung 1711*, par. II.20; Rat der Stadt Budissin, *Verneuerte Feuer-Ordnung* (1710) par. 37 and Rat der Stadt Görlitz, *Verneuerte Feuer-Ordnung* (1692) par. II.14.

18   See Rat der Stadt Budissin, *Verneuerte Feuer-Ordnung* (1710) par. 26 and Rat der Stadt Görlitz, *Verneuerte Feuer-Ordnung* (1692) par. II.13.

In Görlitz, the various districts can be traced back to the tax books of the 1420s. However, they only found expression in the fire regulations at a later point in time. Although the responsibilities of the quartermasters have been mentioned since 1692, e.g. in the fire deputations, it was not until the fire regulations of 1786 that direct reference was made to the 4 inner-town quarters for active firefighting [Fig. 11.5]. Around 1700, the firefighting teams were primarily formed of members of the various guilds. Each was obliged to delegate a certain number of members to extinguish fires. In this way, 33 guilds provided a total of 107 people.[19] In addition, all the carpenters, bricklayers, chimney sweeps and brewers in the town were obliged to participate.[20] Given their expertise, they were needed above all for the targeted covering and tearing down of houses. Thus, firefighting in Görlitz was organised via the guilds.

With the fire regulations of 1711, Löbau divided the town into the so-called third quarters for the first time. This division was very unusual. The town was divided into three parts and not into four, as the word "quarter" would suggest. If one looks at the spatial division, it becomes clear that the existing conditions were taken into account in the decision. These are the Zittau, the Görlitz and the Bautzen districts. Looking at the dimensions as shown in Figure 11.6, it becomes clear that the town was thus divided into relatively equal districts. All of them contained buildings of political, economic or religious-social importance:[21] the cloth hall in the Zittau district, the St. Nicolas church and the St. John church in the Görlitz district and the town hall in the Bautzen district. In the event of a fire, the residents of the affected district were obliged to fight the fire first. The other residents were only supposed to help if the fire got out of control or spread to their own neighbourhood. Only then were the residents of the affected neighbourhood allowed to take care of their belongings.

In Bautzen, various special features must be taken into account. On the one hand, there were two special legal areas within the walls and thus in the core area of the town, over which the municipal authorities had no influence. These are Ortenburg Castle as the seat of the bailiff, to which some houses in the surrounding area belonged, and the village of Seidau below Ortenburg Castle. On the other hand, the cathedral chapter and the houses under it were excluded from the urban legal district. This urban area was also excluded from the fire ordinances. The town could not impose regulations on the inhabitants, concerning prevention or obligations to actively fight fires. The example of the

---

19 The guilds were to send between 1 and 30 men, depending on the number of their members. The clothiers sent the most.
20 See Rat der Stadt Görlitz, *Verneuerte Feuer-Ordnung* (1692) par. II.6.
21 Each highlighted in colour.

ORGANISING THE CHAOS: FIREFIGHTING IN UPPER LUSATIA 313

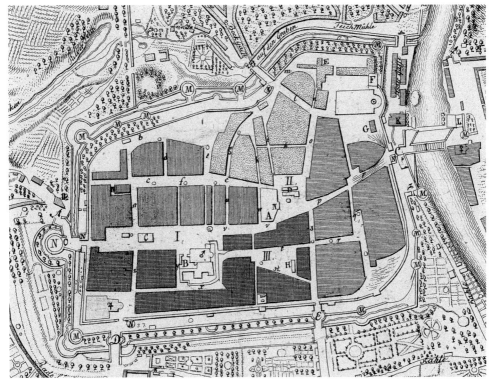

FIGURE 11.5   Section of the plan of the town of Görlitz from 1790. The draughtsman Liebsch has marked the various quarters within the town walls here: Dark the Frauenviertel, above it the Reichenbach quarter, to the right of the Frauenviertel the Neisse quarter and finally slightly above the Reichenbach and Neisse quarters the Nicolai quarter. See Liebsch "Plan der Churfürstlich Sächsischen Sechs Stadt Görliz", copper engraving from 1790, Görlitz, Oberlausitzische Bibliothek der Wissenschaften
© OBERLAUSITZISCHE BIBLIOTHEK DER WISSENSCHAFTEN GÖRLITZ

"purchase of extinguishing equipment" in the ordinances of 1710 makes this clear. Compared to the fire ordinances of 1671, more precise details were given in 1710, e.g. which extinguishing equipment should be kept ready. In 1671, it was still stated in very general terms that every citizen and resident with their own fireplace should hold the following extinguishing equipment: a hand-operated fire sprayer made of wood or metal, an axe, a fire hoe, a ladder and at least two leather buckets.[22] In the fire ordinances of 1710, a more complex system

22   Rat der Stadt Budissin, *Der Stadt Budissin Feuer-Ordnung: Von neuen Revidiret, verbessert / zu Mänigliches Wissenschafft publiciret und in offenen Druck gegeben im Jahre* MDCLXXI (Budissin, Christoph Baumann: 1671) par. 1.

FIGURE 11.6   Map of Löbau, 1843, Löbau municipal Archive; Own processing. Legend: *Blue*: Zittau Quarter, *Yellow*: Görlitz Quarter, *Green*: Budissin (aka Bautzen) quarter

was introduced. Now the owners and residents of certain streets were obliged to either go directly to the source of the fire with various fire extinguishers or protect certain public buildings in the town.

A closer look at the map [Fig. 11.7] shows very well at how small a scale the various obligations had been allocated. The entire town was divided along the streets. In the process, the obligations along the street and the two sides of the street alternated. This is a big difference with respect to the arrangements that one finds in Görlitz and Löbau. In Bautzen, the town area was very compartmentalized, which had a clear advantage: no matter where a fire broke out, inside or outside the town walls, different types of the usual extinguishing equipment were to be available in the immediate vicinity, thus enabling rapid intervention. Within a street, duties could change. No large concentration, e.g. per neighbourhood, was envisaged, but the various fire extinguishers were to be stocked throughout the town. The consequence of this regulation was

FIGURE 11.7　Section of the ground plan of Bautzen drawn by Johann Gottlob Krause, Grund-Riss der im Marggrafthum Oberlausitz gelegene Haupt Sechs-Stadt Budissin nebst der umher liegente Gegend, Handdrawing from 1781
Legend: *Blue*: Hand sprayer and fire bucket, *Orange*: Fire axes, *Green*: Fire hose and ladders, *Purple*: Upper and lower rifles in front of the town hall, *Light blue*: Syringes and buckets in front of the town hall, *Black*: Fire axes in front of the town hall, *Grey*: to St. Peter's Church, *Brown*: to the Michaelis Church, *Pink*: to the Protestant Town School, *Lavender*: to the Catechism Church, *Red*: to the Gewandhaus (Cloth traders' hall)
Own adaptation of the map: Sächsische Landesbibliothek – Staats- und Universitätsbibliothek Dresden (SLUB) / Deutsche Fotothek
© SLUB / DEUTSCHE FOTOTHEK

that rich residents, as in Reichengasse (Reichen alley) near the town hall, had the same duties as poor craftsmen in valley of the river Spree, in Gerbergasse (Tanner alley). Reserves of people and fire extinguishers were set up in front of the town hall in case of an emergency.[23]

However, this had one major disadvantage: with this change within the streets and street sections, it was probably not clear to every house owner or

---

23　The regulation concerned 'the inhabitants of the house to landlords and tenants', i.e. both the landlords and the other inhabitants of the house. See Rat der Stadt Budissin, *Verneuerte Feuer-Ordnung* (1710) par. 30.

resident which fire extinguishers to keep at hand. The town tried to compensate for this disadvantage by making sure that the fire regulations were regularly read out in public as part of the annual fire festival at the town hall.[24] Secondly, the guilds were to read the regulations to their members on a regular basis.[25] This latter practice was also common to the guilds in Görlitz and Löbau.[26]

6   Visibility and Availability of Firefighting Equipment

In addition to the formation of firefighting teams, it was crucial for successful firefighting that water resources and tools were available. As far as water was concerned, one could resort to the wells, ponds and rivers. Since the water supply in early modern cities is a broad and complex issue, I will not deal with it here.[27] Instead, I will exclusively focus on the existing means of firefighting in the form of buckets, ladders, hooks and syringes.

The key categories of visibility and accessibility play a crucial role when it comes to reflecting on the instruments of firefighting. To what extent were resources for firefighting visible in Bautzen, Görlitz and Löbau at the beginning of the eighteenth century? Which were directly accessible? From the fire regulations, one can deduce that several extinguishing agents were visible on the streets and squares at all times and available to everyone. In all three towns, filled water barrels and/or buckets were placed at the wells made of wood or stone.[28] There were also regulations that every homeowner had to place a filled

---

24   The Fire Festival was always held on the Wednesday after Misericordias Domini in memory of the destruction during the Thirty Years' War. On 2 May 1634, the town burned down completely except for a few houses. See Reymann R., *Geschichte der Stadt Bautzen* (Bautzen: 1902) 101f. and 822.

25   Rat der Stadt Budissin, *Verneuerte Feuer-Ordnung* (1710) 48.

26   Rat der Stadt Görlitz, *Verneuerte Feuer-Ordnung* (1692) 31f. and Rat der Stadt Löbau, *Revidirte und verbesserte Feuer-Ordnung* (1711) 38.

27   For water as an infrastructure see: van Laak D., *Alles im Fluss. Die Lebensadern unserer Gesellschaft* (Frankfurt am Main: 2018). For Bautzen see for example: Heinz M., "Wasser für Budissin. Bautzens Sorge um eine ausreichende Wasserbeschaffung war ein Problem von Jahrhunderten", *Bautzener Kulturschau* 27 (1977) 5–7; For Görlitz see for example: Anders I. – Roth E., "Das Röhrwasserleitungssystem in der Stadt Görlitz", *Görlitzer Magazin* 24 (2011) 43–56; Stiller W., *Von Brunnen, Zisternen und Rohrbütten zum Wasserwerk in Görlitz. 140 Jahre Wasserwerk in Leschwitz* (Görlitz: 2018).

28   Rat der Stadt Görlitz, *Verneuerte Feuer-Ordnung* (1692) par. 1.7; Rat der Stadt Budissin, *Verneuerte Feuer-Ordnung* (1710) par. 8, and Rat der Stadt Löbau, *Revidirte und verbesserte Feuer-Ordnung* (1711) par. 1.26.

water barrel before the front door. Bautzen stipulated that this had to be done at least between the beginning of April and the end of October and that the barrel had to be at least one tub in size. In Löbau it was stipulated that the barrels were to be placed on hot and dry summer days, when it was windy and during fairs.[29] This ensured that everyone had access to water from the public space at all times in order to be able to start fighting the fire as quickly as possible. The fire protection ordinances of Görlitz and Löbau also mention that ladders and fire hooks were hung on the walls of houses in the alleys.[30] These aids made it possible to quickly reach the upper areas of the affected buildings, e.g. to fight a possible flying fire or to rescue people.

Only for Bautzen is the existence of a fire engine house mentioned in a fire ordinance at the beginning of the eighteenth century. Between the town hall and the cathedral of St. Peter, there was 'a special building' in which a large metal fire engine with hose was kept. This building was therefore visible to everyone at all times, but not accessible. This fire engine, which was modern by the standards of the time, was only allowed to be operated by a member of a selected group of ten men who had skills in metal and leather works.[31] For Löbau, in 1711, the intention was also expressed to establish such a fire station at the Zittau Gate.[32]

Most extinguishing agents such as buckets, hooks and hand sprayers were stored indoors. They were therefore not generally visible from outside the buildings. Anyone moving within the rooms where these tools were stocked could see the extinguishers – in fact, a further distinction should be made as to whether the extinguishers were openly placed in the house or stored in a cupboard, for example. One might say that there existed different gradations in their accessibility.

Extinguishing agents that belonged to the town were kept in various public buildings – in the council cellar and in the weigh house, sometimes also in churches – from where, in case of emergency, they were made accessible. Access was therefore only temporary and tied to a specific emergency.

---

29 Rat der Stadt Görlitz, *Verneuerte Feuer-Ordnung* (1692) par. I.21, Rat der Stadt Budissin, *Verneuerte Feuer-Ordnung* (1710) par. 9, and Rat der Stadt Löbau, *Revidirte und verbesserte Feuer-Ordnung* (1711) par. I.26.

30 Rat der Stadt Görlitz, *Verneuerte Feuer-Ordnung* (1692) par. I.6, and Rat der Stadt Löbau, *Revidirte und verbesserte Feuer-Ordnung* (1711) par. I.25.

31 Rat der Stadt Budissin, *Verneuerte Feuer-Ordnung* (1710) par. 39.

32 Rat der Stadt Löbau, *Revidirte und verbesserte Feuer-Ordnung* (1711) par. I.25.

FIGURE 11.8
Willow fire bucket from the 19th century, Bautzen municipal Museum, Inventory no. L Opp. 333/1. These buckets had been made watertight on the inside with pitch
© MUNICIPAL MUSEUM OF BAUTZEN

Moreover, one could not simply enter the building and take out the required extinguishing agents, but their use often had to be supervised. In Bautzen, for example, the weighmaster, who opened the weigh house, where some of the buckets were stored.[33] Public buildings such as the town hall, the weigh house and the wine cellar were thus not only used for one purpose. In addition to their original and proper purpose, they were also storage spaces for fire extinguishing equipment. If we take the example of the weigh house, in the event of a fire its actual, economic purpose receded. Then the focus was primarily on the fact that buckets and extinguishing equipment were there available.

The extinguishing equipment, which the guilds were supposed to acquire and store, was also not visible to the public. They were accessible at all times to a limited group of people (the members of the guild) and were not stored in a

---

33   Rat der Stadt Budissin, *Verneuerte Feuer-Ordnung* (1710) par. 27.

FIGURE 11.9 Public buildings such as schools or churches should also keep extinguishers on site. This is an example from Angermünde with a leather fire extinguisher bucket from the 1830s, Angermünde municipal Museum, Inventory No. 271
© MUNICIPAL MUSEUM OF ANGERMÜNDE

publicly visible way. In an emergency, it was the task of the youngest masters to fetch the fire extinguishers.[34] The Löbau fire regulations give precise information about the number of fire extinguishers: the 13 guilds were to keep a total of 35 buckets, 31 tin hand sprayers and 15 axes.[35]

---

34   Rat der Stadt Löbau, *Revidirte und verbesserte Feuer-Ordnung* (1711) par. II.11.
35   Rat der Stadt Löbau, *Revidirte und verbesserte Feuer-Ordnung* (1711) par. I.23. For Görlitz it only says that 'a certain number of fire buckets and syringes' are to be stored in the guild houses. See Rat der Stadt Görlitz, *Verneuerte Feuer-Ordnung* (1692) par. I.34. The Bautzen guilds were not given such guidelines.

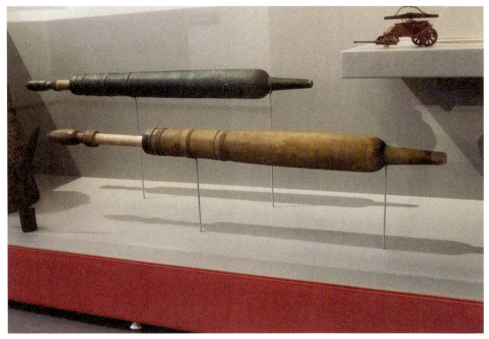

FIGURE 11.10   2 wooden hand sprayers from the first half of the 19th century, left in front the tip of a fire hook, municipal museum of Bautzen, Inventory no. 11196 a & b
© MUNICIPAL MUSEUM OF BAUTZEN

Most of the extinguishers were stored in the numerous breweries and private houses. These buckets, hand sprayers and axes were accessible to a limited group of people (the inhabitants of the houses) at all times, but did not necessarily have to be visible. They could be kept in a cupboard or placed in the attic. The three towns took slightly different paths in their commitments as to which fire extinguishing equipment should be purchased and in what quantity.

Consistently in all three cities, homeowners with brewing rights were the most heavily burdened. The other groups have different compositions: in Görlitz, craftsmen and 'other Wirth'[36] – tenants are not explicitly mentioned, it is rather assumed that in each case they are house owners. In Bautzen and Löbau, on the other hand, tenants, so-called *Hausgenossen*, are also included in the regulations. In Löbau, a small distinction is made as to whether the

---

36   A 'wirth' was usually the owner of the house or in rare cases the leaseholder of the house.

house is in the town or in the suburbs. In the suburbs, tenants must own a fire axe instead of a hatchet.

The fire ordinances of Bautzen and Löbau also stipulated that manual fire extinguishers in the workshops must be accessible at all times and visible to a certain group of people. In particular, certain craftsmen were to have extinguishing water available while working. In Bautzen, two cooling water barrels, leather buckets and hand-sprayer each were to be kept ready during malting, and three during brewing.[37] The confectioners and the smaller bakers as well had to have a barrel and a sprayer ready while they were working.[38] The bakers and carvers of Löbau even had to have two water barrels and 'some' hand sprayers with them for quick use.[39] It was added that all craftsmen who had to deal with fire should place a large water barrel and several hand sprayers in their workshops.[40] In this way, the outbreak of a fire in the workshops could be combatted quickly and effectively.

A special form of visibility of fire extinguishers, which was possible only at certain times, was practised in Löbau. Here the fire extinguishers were to be checked at least four times a year to ensure that they were in working order and their use was to be practised.[41] For this purpose, certain persons were appointed by the town administration. It is not possible to state with certainty whether, as in Bautzen, these were certain craftsmen who stood out because of their knowledge.

Another example of the visibility of prevention measures are various professions implied in prevention activities. First and foremost are the night watchman and chimney sweeps. Through their existence, they directly (chimney sweeps) or indirectly (night watchmen) remind people of the danger of fire.[42]

But other offices and areas of responsibility were also defined in the fire protection regulations. In Görlitz it was stipulated that the citizens entitled to brew had to have watchmen at their disposal, with water and various extinguishing devices.[43] When brewing in Löbau, the malt owner had to pay a watchman who, if necessary, sprayed the chimneys with water through the syringes kept ready. As additional supervision, two men were positioned by the fire

---

37 Rat der Stadt Budissin, *Verneuerte Feuer-Ordnung* (1710) par. 15.

38 Rat der Stadt Budissin, *Verneuerte Feuer-Ordnung* (1710) par. 19.

39 Rat der Stadt Löbau, *Revidirte und verbesserte Feuer-Ordnung* (1711) par. I.12.

40 Rat der Stadt Löbau, *Revidirte und verbesserte Feuer-Ordnung* (1711) par. I.13.

41 Rat der Stadt Löbau, *Revidirte und verbesserte Feuer-Ordnung* (1711) par. I.24.

42 The night watchman checked on the streets and in the inns not only because of the danger of fire, but also with regard to public order.

43 Rat der Stadt Görlitz, *Verneuerte Feuer-Ordnung* (1692) par. I.22.

deputation at the brewing pans.[44] In Löbau, the brewery owners each had to employ a watchman, who was engaged by the fire commission.[45] In Görlitz, fire watchmen were to patrol the alleys at night and knock on the affected houses when there was a smell of fire. Together with the alley masters, they checked the inns every evening.[46]

7 Control and Hierarchies

The fire ordinances do not only give a good insight into the respective working and living conditions, but they also reflect the existing power relations. Of particular interest is the question of who had what power of command in firefighting and who had the control and monitoring powers over the preventive measures and regulations. It was the task of the so-called fire deputation or fire commission to check whether the regulations were being observed.

At the end of the seventeenth century, in the fire regulations, deputations first appeared, which had to check the presence of extinguishing equipment. The first mention is found in the fire regulations of Löbau from 1684.[47] In Görlitz in 1692, four persons from the council were to be employed to watch over the implementation of the fire regulations alongside the quartermasters and the alley masters.[48] Together with the building scribes, a mason, a carpenter, a chimney sweep and a district master, they inspected the fireplaces three times a year and also had the fire equipment shown to them.[49] In Bautzen, too, the number of members remained unnamed; there as well, they were all members of the council. The term 'fire deputation' was used for the first time. It was chaired by the acting proconsul. The monthly meetings were attended by the lane masters, who reported on the districts entrusted to them. They had supervision over the houses and buildings and also inspected them.[50] In Löbau, the fire commission was composed of the acting proconsul and two elected

---

44 Rat der Stadt Budissin, *Verneuerte Feuer-Ordnung* (1710) par. 15.
45 Rat der Stadt Löbau, *Revidirte und verbesserte Feuer-Ordnung* (1711) par. I.13.
46 Rat der Stadt Görlitz, *Verneuerte Feuer-Ordnung* (1692) par. I.10 and I.36.
47 Rat der Stadt Löbau, Feuerordnung bey der Stadt Löbau [1684], in Weinart B.G. (ed.), *Rechte und Gewohnheiten der beyden Marggrafthümer Ober- und Niederlausitz*, vol. 4 (Leipzig, Jacobäer: 1798) 263, par. 2.
48 Rat der Stadt Görlitz, *Verneuerte Feuer-Ordnung* (1692) par. I.3.
49 Rat der Stadt Görlitz, *Verneuerte Feuer-Ordnung* (1692) par. I.4 and I.5.
50 Rat der Stadt Budissin, *Verneuerte Feuer-Ordnung* (1710) par. 1–3.

councillors.[51] As in Bautzen, the alley masters showed up at the monthly meetings and reported. However, the supervision here was much more detailed, as the alley masters were only responsible for individual streets. This allowed them to keep a very precise overview of an area. Here, too, their duties included checking the fireplaces of homes, as well as the storage of combustible material and the presence of the prescribed extinguishing equipment.[52] The three cities thus established control authorities comparatively late. In Cologne, the banner lords had already been controlling the fire facilities since 1583 and were obliged to report to the town council.[53]

The leading, controlling lordship role of the council is also evident in active firefighting. This can be seen in the command structure to be formed during this period. In all three cities, the acting mayor led the measures to contain the fire. In Görlitz and Bautzen, the lord mayors came to the scene of the fire on horseback, like a commander sending his troops into battle from a horse. In Löbau, the measures are directed from the town hall, where the mayor and the other council members assemble. The subordinate officials of the town administration have to report to them what is going on.[54] So decision making did not take place directly at the scene of the incident, but only after being transmitted by the subordinate agencies. On site, the members of the fire deputation had a directive function. They had to order and control the concrete measures. At the same time, an information chain was established to the town hall, whose instructions were to be obeyed.[55] While the members of the fire deputation were to govern the firefighting, they were nevertheless subject to the command structure of the mayor. They could not act independently, but had a good insight into the conditions in each house through regular inspections.

Insofar as every fire endangered public order, the "good policing" in the town, there was great interest on the part of the council and especially the incumbent mayor not to lose his leadership role in such situations. At the same time, citizens and residents expected the council to prevent unnecessary damage to the town. In doing so, they relied on the expert help of various craftsmen such as carpenters, bricklayers and chimney sweeps. For the tactic was mainly to try to contain the fire by unroofing or demolishing entire houses, as there were hardly any efficient fire hoses available. It would have been unthinkable

---

51 Rat der Stadt Löbau, *Revidirte und verbesserte Feuer-Ordnung* (1711) par. I.1 and I.2.
52 Rat der Stadt Löbau, *Revidirte und verbesserte Feuer-Ordnung* (1711) par. I.5 and I.6.
53 Neuhoff, *Feuer und Flamme* 30.
54 Rat der Stadt Löbau, *Revidirte und verbesserte Feuer-Ordnung* (1711) par. II.14.
55 Rat der Stadt Löbau, *Revidirte und verbesserte Feuer-Ordnung* (1711) par. II.16.

that, for example, the foremen of these guilds would have led the firefighting together, i.e. that professional competence would have had priority. At that moment, the existing hierarchy between the council and the guilds or the citizens would have been abolished, even reversed. The residents of the town still had to act within the framework of the existing power structures. Thus, even the plans for certain behaviours contained in the fire ordinance could only reflect the existing power structures. Even though it was an imaginary space in which tasks and roles were assigned to the various actors, no new power structures or hierarchies were formed there, but rather they built on and integrated into the existing structures.

8 Conclusion

Fires – whether caused by war, arson, lightning, or human error – meant a disturbance of public order, of "good policing." Based on the fire ordinances, it is possible to understand how the municipality wanted to prepare for the occurrence of such extreme events, namely what measures were laid down to ward them off. For there was a danger that a small fire in a backyard or in a dwelling would develop into a fire that would destroy entire neighbourhoods or the entire town.

In the seventeenth and especially in the eighteenth century, complex and detailed sets of rules were published on how to restore order to the ensuing fire chaos. In this paper, I have examined the fire ordinances adopted in Bautzen, Görlitz and Löbau at intervals of a few years. I have shown the role of different actors in firefighting, the divisions of the urban space as a useful means to organize rescues, the presence and visibility of the instruments of firefighting, and the hierarchies established to govern critical situations. Although the three cities taken into consideration – no more than 50 km apart – had comparable framework conditions in terms of the technical status of firefighting equipment, constitutional foundations, economic status, size and number of inhabitants, the organization of firefighting differed in some respects, which has been highlighted throughout the analysis.

I have suggested that Henri Lefebvre's theory of space, and Susanne Rau's interpretation of it, represent a useful framework for studying fire ordinances. Such documents lay down the *espace conçu*, or mental space, of firefighting. Not only do they represent the mental plan of action elaborated by the town's magistrate, but they also distilled past experience and use it to build future plans, aimed at maintaining 'good policing'. The study of these documents also reveals the power relations that existed within the cities and the hierarchies

ORGANISING THE CHAOS: FIREFIGHTING IN UPPER LUSATIA

among social groups. All this indicates that an analysis of the 'perceived' and 'lived' spaces (the two other types of space suggested by Lefebvre), which concern the level of individual existence and collective action, should always relate to the level of mental representations, as these provide the coordinates within which action unrolls.

Bibliography

Primary Sources

Grosser Samuel, *Lausitzische Merkwürdigkeiten, Darinnen von beyden Marggrafthümern in fünf verschiedenen Theilen* [...] (Leipzig – Budissin, David Richter – Immanuel Tietze: 1714).

Krause Johann Gottlob, Plan von Bautzen, 1791, SLUB / Deutsche Fotothek.

Liebsch, "Plan der Churfürstlich Sächsischen Sechs Stadt Goerliz" (1790), Görlitz, Oberlausitzische Bibliothek der Wissenschaften.

Municipal Archive Löbau, (without signature): Stadtplan von Löbau (1843).

Rat der Stadt Bautzen, *Feuer-Ordnung bey der Churfürstlich Sächßischen Hauot-Sechs-Stadt Budissin im Marggrafthum Ober-Lausitz* (Budissin, Johanna Eleonore verw. Scholzin: 1781).

Rat der Stadt Budissin, *Der Churfürstl. Sächß. des Markgrafthums Ober-Lausitz Haupt- und Sechs-Stadt Budißin Verneuerte Feuer-Ordnung, wie solche Nach wiederhohleter Revision, in einem und dem andern Stücke geändert, verbessert,* [...] (Budissin, Gottfried Gottlob Richter: 1710).

Rat der Stadt Budissin, *Der Stadt Budissin Feuer-Ordnung: Von neuen Revidiret, verbessert / zu Mänigliches Wissenschafft publiciret und in offenen Druck gegeben im Jahre MDCLXXI* (Budissin, Christoph Baumann: 1671).

Rat der Stadt Görlitz, *Feuer-Ordnung bey der Churfürstlich-Sächsischen Sechs-Stadt Görlitz* (Görlitz, Johann Friedrich Fickelscherer: 1786).

Rat der Stadt Görlitz, *Verneuerte Feuer-Ordnung* (Görlitz, without printer's name: 1692).

Rat der Stadt Löbau, *Anderweit Revidierte und verbesserte Feuer-Ordnung bei der Chur-Fürstl. Sächß. ältesten Sechs-Stadt Löbau im Markgrafthum Ober-Lausitz* (Löbau, Carl Friedrich Völkel: 1778).

Rat der Stadt Löbau, Feuerordnung bey der Stadt Löbau [1684], in Weinart B.G. (ed.), *Rechte und Gewohnheiten der beyden Marggrafthümer Ober- und Niederlausitz,* Volume 4 (Leipzig, Jacobäer: 1798) 262–270.

Rat der Stadt Löbau, *Revidirte und verbesserte Feuer-Ordnung bey der Churfürstl. Sächß. ältestten Sechs-Stadt Löbau im Marggrafthum Ober-Lausitz,* [...] (Zittau, Michael Hartmann: 1711).

Rat der Stadt Löbau, *Revidirte und verbesserte Feuer-Ordnung bey der Königl. Sächß. Stadt Löbau im Marggrafthum Ober-Lausitz* (Löbau, Johann Christian Schlenker: 1816).

Zeidler Johann, *Tabeera Budissinae. Budissinische Brandstelle* (Dresden, Gimel Bergen: 1634).

Secondary Sources

Allemeyer M.L., "'Wenn der liebe Gott einen Hauss Wirth mit Feuers Brunst heimsucht.' Zur Deutung und Darstellung von Stadtbränden in religiösen, genossenschaftlichen, technischen und obrigkeitlichen Schriften", in Koppenleitner V.F. – Rößler H. – Thimann M. (eds.), *Urbs incensa: Ästhetische Transformationen der brennenden Stadt in der Frühen Neuzeit* (Berlin – Munich: 2011) 285–300.

Allemeyer M.L., *Fewersnoth und Flammenschwert. Stadtbrände in der Frühen Neuzeit* (Göttingen: 2007).

Anders I. – Roth E., "Das Röhrwasserleitungssystem in der Stadt Görlitz", *Görlitzer Magazin* 24 (2011) 43–56.

Briese O., *Für des Staates Sicherheit. Das Löschwesen im 19. Jahrhundert und die Gründung der ersten Berufsfeuerwehr Deutschlands in Berlin 1851* (Berlin: 2018).

Fröde T., *Privilegien und Statuten der Oberlausitzer Sechsstädte: Ein Streifzug durch die Organisation des städtischen Lebens in Zittau, Bautzen, Görlitz, Löbau, Kamenz und Lauban der frühen Neuzeit* (Spitzkunnersdorf: 2008).

Garrioch D., "1666 and London's Fire History: a Re-Evaluation", *The Historical Journal* 59.2 (2016) 319–338.

Georgi H., *Das Erdbeben von Lissabon. Wie die Natur die Welt ins Wanken brachte – von Religion, Kommerz und Optimismus, der Stimme Gottes und der sanften Empfindung des Daseins* (Wiesbaden: 2016).

Günther H., *Das Erdbeben von Lissabon und die Erschütterung des aufgeklärten Europa* (Frankfurt: 2005).

Hartstock E., *Geschichte der Plagen der Oberlausitz: Von Feuersbrünsten, Epidemien und Dürrekatastrophen – Eine Chronik der Ereignisse von 1112 bis 1868* (Spitzkunnersdorf: 2017).

Heinz M., "Wasser für Budissin. Bautzens Sorge um eine ausreichende Wasserbeschaffung war ein Problem von Jahrhunderten", *Bautzener Kulturschau* 27 (1977) 5–7.

Heyl Ch., "'God's Terrible Voice in the City': Frühe Deutungen des Great Fire of London, 1666–1667", in Koppenleitner V.F. – Rößler H. – Thimann M. (eds.), *Urbs incensa: Ästhetische Transformationen der brennenden Stadt in der Frühen Neuzeit* (Berlin – Munich: 2011) 23–44.

Iseli A., *Gute Policey: Öffentliche Ordnung in der Frühen Neuzeit* (Stuttgart: 2009).

Keene D., "Fire in London: Destruction and Reconstruction, A.D. 982–1676", in Körner M. (ed.), *Destruction and Reconstruction of Towns: Volume 1: Destruction by Earthquakes, Fire and Water* (Bern – Stuttgart – Vienna: 1999) 187–211.

Klemm D., "Hamburg brennt: Zur bildlichen Darstellung der Jahrhundertkatastrophe", in Trempler J. – Bertsch M. (eds.), *Entfesselte Natur: Das Bild der Katastrophe seit 1600* (Petersberg: 2018) 80–89.

Koppenleitner V.F., "'Etiam periere Ruinae': Katastrophenereignis und Bildtradition in Darstellungen des Großen Brandes von London 1666", in Koppenleitner V.F. – Rößler H. – Thimann M. (eds.), *Urbs incensa: Ästhetische Transformationen der brennenden Stadt in der Frühen Neuzeit* (Berlin – Munich: 2011) 45–59.

Krüger D., *Die Entwicklung des baulichen Brandschutzes in Sachsen vom Beginn der Stadtgründungen bis 1945 im politischen, rechtlichen und wirtschaftlichen Kontext*, PhD dissertation (Technical University of Dresden: 2010).

Landwehr A., *Policey im Alltag: Die Implementation frühneuzeitlicher Policeyordnungen in Leonberg* (Frankfurt: 2000).

Lauer G. (ed.), *Das Erdbeben von Lissabon und der Katastrophendiskurs im 18. Jahrhundert* (Göttingen: 2008).

Lefebvre H., *The production of space* (Malden: 2011).

Löffler U., *Lissabons Fall – Europas Schrecken. Die Deutung des Erdbebens von Lissabon im deutschsprachigen Protestantismus des 18. Jahrhunderts* (Berlin: 1999).

Mikoletzky J., "Der Brand des Wiener Ringtheaters 1881 und sein Folgen", *Ferrum: Nachrichten aus der Eisenbibliothek* 69 (1997) 59–68.

Molesky M., "The Great Fire of Lisbon, 1755", in Bankoff G. – Lübken U. – Sand J. (eds.), *Flammable cities. Urban conflagration and the making of the modern world* (Madison: 2012) 147–169.

Neuhoff St., *Feuer und Flamme. Die Geschichte des Brandschutzes in Köln* (Düsseldorf: 2014).

Pötzl W. (ed.), *Feurio! Es brennt. Zur Geschichte des Brandschutzes, der Brandbekämpfung und der Feuerwehren* (Augsburg: 2010).

Rau S., *Räume: Konzepte, Wahrnehmungen, Nutzungen* (Frankfurt am Main: 2013).

Reymann R., *Geschichte der Stadt Bautzen* (Bautzen: 1902).

Scharf C. – Nordhof A.W. – Kessel J. (eds.), *Die Geschichte der Zerstörung Moskaus im Jahre 1812* (Munich: 2000).

Schubert D., "The Great Fire of Hamburg, 1842. From Catastrophe to Reform", in Bankoff G. – Lübken U. – Sand J. (eds.), *Flammable Cities. Urban Conflagration and the Making of the Modern World* (Madison: 2012) 212–234.

Soja E.W., "Die Trialektik der Räumlichkeit", in Stockammer R. (ed.), *TopoGraphien der Moderne: Medien zur Repräsentation und Konstruktion von Räumen* (Munich – Paderborn: 2005) 93–123.

Soja E.W., *Thirdspace. Journeys to Los Angeles and other real-imagined places* (Malden 1998).

Stiller W., *Von Brunnen, Zisternen und Rohrbütten zum Wasserwerk in Görlitz. 140 Jahre Wasserwerk in Leschwitz* (Görlitz: 2018).

Stolleis M. (ed.), *Policey im Europa der Frühen Neuzeit* (Frankfurt: 1996).

Van Laak D., *Alles im Fluss. Die Lebensadern unserer* Gesellschaft (Frankfurt am Main: 2018).

Walter F., *Geißel Gottes oder Plage der Natur. Vom Umgang der Menschen mit Katastrophen* (Stuttgart: 2016).

CHAPTER 12

Spectacle, Enthusiasm, Objectivity – Managing Fire as an Emoterial

Simon Werrett

Abstract

Recent attempts to better understand early modern material culture have proposed thinking about sociomaterials, stressing the bond between subjects and objects in early modern contexts. In this paper I use explorations of early modern ideas and practices around fire to propose a new category of the emoterial, a term that indicates the entanglement of emotions as much as sociability with materials in the early modern period. Fire was immensely evocative and transformative, delighting or destroying human and non-human bodies through everything from fireworks displays to domestic catastrophes. Human life and passions were said to reside in fiery economies inside the body, so managing social order depended on fire management. Natural philosophy claimed to discipline fiery spirits through knowledge, experiment, and self-discipline. Fire management therefore should be recognized as a critical feature of competing approaches to social, epistemic, economic and emotional order in the early modern period; and fire provides an excellent example of a more general phenomenon of emoteriality.

Keywords

fireworks – emotions – audiences – London – natural philosophy – art – physiology

An English print of 1749 shows a fireworks display held that year in London under the title *The Grand Whim for Posterity to Laugh at*. The firework, which took place in Green Park, was to be the grandest in Britain up to that time, marking the Peace of Aix La Chapelle and the end of the Seven Years War. The reason for the laughter was that this turned out to be a disastrous occasion. One of the wings of the ephemeral 'machine' or temple that had been built for the fireworks had caught fire and burned to the ground, prompting fights

© KONINKLIJKE BRILL BV, LEIDEN, 2025 | DOI:10.1163/9789004521766_014

FIGURE 12.1   Anon, *The Grand Whim For Posterity To Laugh At*, 1749, etching and letterpress printing, 469 × 351 mm, British Museum Department of Prints and Drawings
IMAGE © THE TRUSTEES OF THE BRITISH MUSEUM

MANAGING FIRE AS AN EMOTERIAL

among the artificers who were staging the display. In this unique image for the eighteenth century, the papers mocked the performance.[1]

Pyrotechnics represent a quintessential form of fire-management in the early modern period which will be the focus of this essay. Fire is a risky material to work with, providing part of the drama of a firework display but also a danger, as the 1749 artificers discovered. This is equally an emotional story. This essay explores the connection between the emotions and the material in fire management and insists that both need to be understood as intimately connected. The verses beneath the 'Grand Whim' talk about it being a 'fantastic idol show' put on in a time after a war when there was no money to look after veterans and pay the nation's expenses. Fireworks were an insulting waste of money, a 'wretched' 'foolish' 'devilish' 'folly'. Observing the fireworks was a deeply passionate experience, evocative of delight, no doubt, in audiences, but also anger, frustration or disdain. Another print made at the time entitled 'The Contrast' showed an Englishman standing in front of the Green Park temple biting his nails in a sweat because he had run out of money after spending it all on fireworks. A jolly and portly merchant stands next to him above the banner 'Money with Commerce'.

Fire is also deeply emotional in the present. At a time of environmental crisis we are confronted with desperate scenes of destructive fire so that understanding the interrelations between human emotions and fire management must be important historical work. This essay is a brief contribution to that process, and seeks to use fireworks as a way to think about the relations of materials and emotions more generally. It begins with some general comments about material culture and the common habit of discussing it in terms of 'objects' distinct from human 'subjects', the one inert, passive matter and the other living, emotive being. Historians of science have long criticised this division for being more ephemeral than it seems, appearing only in the early nineteenth century and changing its meaning over time.[2] Many alternative perspectives on materials and emotions are available from anthropology, history of art, and other disciplines. Many cultures divide people and things very differently to Europeans, and within Europe ideas of active and sentient matter have been common at various times. Pamela Smith, for example, has discussed an 'artisanal epistemology' in which renaissance artisans viewed

---

1   On the 'Grand Whim', see Simon Werrett, "From the Grand Whim to the Gasworks: Philosophical Fireworks in Georgian England", in Dear P. – Roberts L. – Schaffer S. (eds.), *The Mindful Hand: Inquiry and Invention from the Late Renaissance to Early Industrialisation* (Amsterdam and Chicago: 2007) 325–348.
2   Daston L. – Galison P., *Objectivity* (New York: 2007).

material things as living active matter with intentions and characters they needed to negotiate in order to produce art.[3] In *Thrifty Science* (2020), I have argued that early moderns did not distinguish objects from subjects but rather spoke of 'bodies' all with living characteristics.[4] One might then say that prior to the nineteenth century there was no such thing as 'material culture' because materials in the sense of passive objects did not exist. There is then a whole process by which the object-subject division came to be which needs a history. Bruno Latour's Actor Network Theory recognizes both human and non-human agencies, a distinction increasingly blurred in his thinking.[5] Werrett and Roberts have urged thinking in terms of the 'sociomaterial' to bridge this divide.[6] Drawing on the sociology of the workplace, this concept insists on a permanent and interactive combination of the social and material. People and objects are always interacting with one another and transforming their identities in the process. However, none of these approaches incorporates the emotional dimension of bodies or things. As Jacques Derrida wrote, 'tears and not sight are the essence of the eye'.[7] The profoundly emotional aspect of bodies needs to be included in their histories.

Scholars have explored this avenue, for example in Oliver Harris and Tim Flohr Sørenson's *Rethinking Emotion and Material Culture* (2010).[8] Harris and Sørenson propose three categories for linking emotions and materials: an 'affective field' of dynamic relations between 'people, places and things', changing over time; 'attunement', or 'how people notice, observe, perceive and recognize moods and emotions in themselves and others'; and 'atmosphere', the way collectives of people generate emotional environments that effect the persons involved. All three categories will be useful for exploring the emotional dimensions of fire management in pyrotechnics.

The novelist Svetlana Aleksievich is another inspiration in this regard. Alekseevich has sought to write emotional histories of events in recent Soviet and Russian history such as the Chernobyl disaster by interviewing participants and their families, often mothers and daughters, to capture not

---

3   Smith P.H., *The Body of the Artisan: Art and Experience in the Scientific Revolution* (Chicago: 2004).

4   Werrett S., *Thrifty Science: Making the Most of Materials in the History of Experiment* (Chicago: 2019) 98–102.

5   Latour B., *After Lockdown*, trans. J. Rose (Cambridge: 2021).

6   Roberts L. – Werrett S. (eds.), *Compound Histories: Materials, Production, and Governance, 1760–1840* (Leiden: 2017).

7   Dick M.D., *The Derrida Wordbook* (Edinburgh: 2013) 85.

8   Harris O. – Sørenson T.F., "Rethinking Emotion and Material Culture", *Archaeological Dialogues* 17 (2010) 145–163.

the 'factual' unfolding of events but the personal and emotional experience of them.[9] Aleksievich might be accused of adhering to a gender stereotype of equating emotion with women's experience and facticity with men, but her extraordinary writing re-presents these events in ways that capture their intense and fundamental emotions. This is what a history of fire management might also do. Here I propose the category of 'emoterials' as a way to achieve this. An emoterial is a process rather than a thing and it refers to the unfolding emotional relationship that bodies have with one another when they interact together, being co-constitutive of identities, feelings, experiences and physicality. Rather than talk about objects and subjects one might speak of emoterial bodies whose various interactions generate respective identities, experiences, properties and feelings. At the same time, it is necessary to think about the processes of reification that make divisions of object and subject seem natural and normal. If early modern bodies were not divided in this way, what happened to make it seem so? My suggestion will be that part of the answer lies in another area of fire management, since early moderns conceptualized emotions in quite pyrotechnic terms. The body was literally a source of 'enflamed' passions, and when these passions became overwhelming and dangerous, they needed to be tamed. The resulting goal of more tempered emotions and bodies marked a moment in the rise of objectivity as a desirable basis for civil and epistemic conduct. It also indicates that the reification of materials is a political process, a form of power. Managing fire is a form of managing people. Fire management is then the move from the emoterial to the merely material.

1 Pyrotechnic Performances

In early modern Europe, fireworks were already an ancient technology. Spreading from China and India in the Middle Ages, fireworks were by 1600 being shown around the world. In Europe they were typically performed as courtly and civic spectacles such as royal weddings, coronations, and births, celebrations of martial triumph or holidays and festivals such as the New Year.[10] Displays were grand, expensive and were typically performed by expert artisans or 'artificers' or the local artillery. Styles of performance varied across

---

9   Aleksievich S., *The Unwomanly Face of War*, trans. R. Pevear – L. Volokhonsky (London: 2017); Alekievich S., *Chernobyl Prayer: Voices from Chernobyl*, trans. A. Tait – A. Gunin (London: 2016).

10   On the history of fireworks, see Werrett S., *Fireworks: Pyrotechnic Arts and Sciences in European History* (Chicago: 2010).

Europe but usually centered on a piece of ephemeral architecture known as a 'machine', often in the form of a castle or classical temple.[11] Out of this, artificers fired an array of pyrotechnics similar to those still in use today, including rockets that flew into the air to burst with stars, wheels that span in circles of sparks, and fountains that gushed fire into the night sky. Other pyrotechnics were distinct to the period, such as 'aquatic' fireworks that disappeared below the surface of a river then erupted back up again to spout fire and sparks. Performances were typically staged amidst allegorical decorations that gave the display a narrative, like a play, telling of princely magnificence or martial might. The evidence that remains of these displays is sparce. To commemorate them, large, elaborate engraved prints were often issued depicting the decorations and their significance. Printed descriptions might accompany these, and audiences sometimes recorded their experience on seeing a show.[12]

Turning to the emotional dimension of these fireworks displays, Harris and Sørensen's category of 'affective field' prompts consideration of the emotional states that artificers sought to invoke in audiences through pyrotechnic effects and the nature of those states induced by observing and experiencing fireworks. Artificers, we may infer, like musicians, were keenly aware of their audiences and sought to manage their reactions to sights and sounds in a manner that would give the audience pleasure and entertainment. Literature on the production and staging of fireworks in the early modern period is quite scarce and was usually limited to recipes and explanations of how to make particular pyrotechnic pieces, and discussion of this emotional orchestration are very rare. Prints, however, occasionally tried to capture the process of a display, depicting how it unfolded to generate certain responses in audiences. In 1637, for example, Claude Lorrain made a series of etchings to record fireworks staged in Rome for the election of the new Holy Roman Emperor Ferdinand III. Kevin Salatino has noted how Claude's series mimics the way the layers of a pyrotechnic castle, perhaps of his own design, gradually enflamed and were consumed to reveal a new decoration below. A contemporary account said this display 'ravished the eyes and made them sleepy with wonder and delight'.[13]

Few authors theorized how pyrotechnic effects would induce different states in audiences. However, in the *Encyclopédie* of Diderot and d'Alembert,

---

11   Bonnemaison S. – Macy C. (eds.), *Festival Architecture* (London: 2007).
12   Salatino K., *Incendiary Art: The Representation of Fireworks in Early Modern Europe* (Los Angeles: 1997).
13   Manzini Luigi, *Applausi festivi fatti in Roma per l'elezzione di Ferdinando III* (Rome: 1637), quoted in Salatino, *Incendiary Art* 54.

MANAGING FIRE AS AN EMOTERIAL   335

Jean-Louis de Cahuzac attempted to prescribe such action.[14] For Cahuzac, fire-works should be a coherent spectacle, not just the setting off of one pyrotech-nic after another. 'In all the Arts it is necessary to paint. In the one that we call *Spectacle*, it is necessary to paint with actions'. To instantiate this claim, Cahuzac described a firework staged in Paris in 1730 celebrating the birth of a new dauphin. The motions and contrasts of fire induced effects in the audi-ence. By illuminating the river Seine, the scene of the display, with lamps using 'the greatest speed ... the surface of the river suddenly offered an enchanting spectacle'. Above this rose two artificial mountains decorated with plants, waterfalls, sirens and sea monsters, so that 'One saw a pleasant variety on these mountains, where nature was imitated with a great deal of art'. Garden terraces and rocks were illuminated with so many lamps that 'the eyes were dazzled, and the darkness of the night entirely dispelled. The movement of the lights, which by confusing them gave them even more brilliance, produced such an effect at a certain distance that one fancied seeing sheets and cascades of fire'. The fireworks set off from this machine were a 'pleasant spectacle' which built through several acts to a climax in several 'girandes' or bursts of stars from rockets above the whole scene. The grand size of the display, its speed of per-formance, its location on the river, the mix of fire and water, and the diversity of attractions building to a crescendo all contributed to a sense of pleasure and magnificence. Evidently a *sublime* experience, the artificers evoked in their audience a dazzling pleasure and enchantment, achieved through the unfold-ing manipulation of speeds of ignition and the growing scale of fiery effects.[15]

Harris and Sørensen's category of 'attunement' refers to the interactivity of a community in shaping and expressing emotional states. Fireworks brought together diverse audiences of spectators and communities of performers, whose organization in space, location, movements and relations generated and followed emotions. Fireworks prints indicate a diverse range of audience distributions, from strictly ordered positions for the court to a pell mell of spec-tators jostling to see the pyrotechnics. No doubt different arrangements were constitutive of different experiences and emotions. In the 1749 Green Park display ticketed seats were offered for different classes of people, indicating a social regimentation that would have correlated with different perspectives

---

14   Cahuzac Jean-Louis de, "Feu d'artifice", in Diderot Denis – d'Alembert Jean Le Rond (eds.), *Encyclopédie, ou dictionnaire raisonné des sciences, des arts et des métiers*, 17 vols. (Paris: 1751–65), vol. 6 (Paris: 1756) 639–640.

15   Salatino takes the sublime as a fundamental category for understanding fireworks. See *Incendiary Art* 47–98.

on, and emotional reactions to, the display and other audience members.[16] Alternatively, other images show spontaneity as people make merry around bonfires and decorations. In an article on fireworks audiences, I have argued that fireworks were occasions for social distinction, as aristocratic audiences made observations of more 'vulgar' audiences that served to distinguish them.[17] A literature on fireworks emerged in the sixteenth and seventeenth centuries aimed at an audience of 'virtuoso' gentlemen, who attained enough knowledge to know how a variety of technical arts were accomplished but without the particular expertise that marked the artisans who pursued such arts. Gentle audiences expressed this knowledge when writing of fireworks displays by saying that whereas they understood the artifices that created pyrotechnic effects, the 'vulgar' did not, and therefore found it hard to distinguish art from reality. Crucially these contrasting awarenesses were manifested in different emotional reactions to fireworks. The comprehending gentlefolk claimed to be unperturbed by pyrotechnics that they knew to be artificial, but spoke of vulgar audiences as fearful and distressed by fireworks, whose mechanisms they did not understand. No doubt the true picture was quite different – after all, the 'vulgar' included artisans who knew perfectly well how fireworks worked. But to express constraint and nonchalance before sublime pyrotechnic explosions and effects was considered a mark of gentility. In 1629 fireworks erupted over the Seine around decorations depicting Louis XIII as Perseus, rescuing Andromeda from a seamonster, an allusion to the French capture of La Rochelle from Protestant forces. Canova-Green has noted how the official record of the display described a contrast between noble and vulgar audiences watching the display. '[N]aïve bystanders along the shore … were stunned and frightened and … thrown about pell-mell […] The Court alone remained still and undaunted, being well acquainted with the vanity and ostentation of these diversions'.[18] Another representation of French fireworks, staged in Meudon in 1735, depicted the courtly audience engaged in nonchalant conversation in

---

16   An example of a 'Ticket for St. James's Park, April 27th 1749, for the Royal Fire Works' may be found in the Bodleian Libraries, University of Oxford, digital collections, see https://digital.bodleian.ox.ac.uk/objects/bf9e5c2a-cba5-4862-86a4-0338f01ce2f9/ (accessed on 18 December 2023).

17   Werrett S., "Watching the Fireworks: Early Modern Observation of Natural and Artificial Spectacles", *Science in Context* (special issue: *Lay Participation in the History of Scientific Observation*, ed. Vetter J.) 24 (2011) 167–182.

18   Quoted and translated by Canova-Green M.-C., "Fireworks and Bonfires in Paris and La Rochelle", in Mulryne J.-R. – Watanabe-O'Kelly H. – Shewring M. (eds.), *Europa Triumphans: Court and Civic Festivals in Early Modern Europe* (Farnham: 2004) 145–153, on 150.

MANAGING FIRE AS AN EMOTERIAL   337

front of an explosive scene of pyrotechnics, again suggesting a capacity not to
be outwardly moved by the display. Elsewhere, I have noted the similarity to
the frontispiece of Bernard de Fontenelle's cosmological primer *Conversations
on the Plurality of the Worlds* in which an astronomer dispassionately instructs
a lady before a sublime fiery apparition of the heavens in the skies above
them.[19] A scientific nonchalance thus reflected the courtly control of emotions
that fireworks demanded.

The third category used by Harris and Sørensen to discuss emotions and
materials is 'atmosphere' – how do collectives of bodies in space create an emotional environment that effects those within it? Many fireworks began with a
prelude in the form of lighting illuminations, whereby thousands of lamps were
ignited to create a glowing picture of scenery or architecture in outline before
the display began. Spectators recalling displays invariably expressed wonder
at the atmosphere illuminations produced. The most famous illumination was
that of the Castel Sant'Angelo in Rome, which was illuminated before fireworks
celebrated the election of a new Pope.[20] According to Vannoccio Biringuccio,
who gave the first detailed account of the display in his *Pyrotechnia* of 1540, all
the embrasures and merlons of the castle were studded with 'two small lanterns made out of a sheet of white paper over a mound of clay in which a tallow
candle is put. When they are lit at night it is a very beautiful thing to see that
shining and transparent whiteness in many rows as far as the eye can reach'.[21]
Following its completion in 1615, the nearby Basilica of St. Peter's was illuminated for the festivals of Holy Week and the feasts of Saints Peter and Paul.
Enduring into the nineteenth century, this display began with the marking out
of the dome in the light of large paper *lanternoni*, before, as an observer in the
1820s recalled:

> instantly ten thousand globes and stars of vivid fire seemed to roll spon-
> taneously along the building, as if by magic; and self-kindled, it blazed in
> a moment into one dazzling flood of glory [...] In the first instance, the
> illuminations had appeared to be complete, and one could not dream
> that thousands and tens of thousands of lamps were still to be illumined.

---

19   Werrett, *Fireworks* 139–141.
20   Salatino, *Incendiary Art* 78–83.
21   Biringuccio Vannoccio, *The Pirotechnia* [1540] *of Vannoccio Biringuccio: The Classic
      Sixteenth-Century Treatise on Metals and Metallurgy*, trans. & ed. C. Stanley Smith –
      M. Teach Gnudi (New York: 1990) 442.

Their vivid blaze harmonized beautifully with the softer, milder light of the lanternoni.[22]

Illuminations thus created an extraordinary atmosphere that dazzled the spectator with beauty, scale, and harmony. Johann Wolfgang von Goethe saw the illuminations in Rome in 1787, recording how, 'a spectacle has to be really grand before I can enjoy it [...] To see the colonnade, the church and, above all, the dome, first outlined in fire and, after an hour, become one glowing mass, is a unique and glorious experience'.[23] Such experiences were examples of the sublime. In his essay on the sublime of the 1750s, Edmund Burke proposed that, 'A quick transition from light to darkness, or from darkness to light' would generate a great effect.[24] Audience-members frequently noted this aspect of illumination displays, 'by the darkness of the night, the splendour of the effort to convert night into the brightness of day was the more conspicuous'.[25]

Emotions and material circumstances in fireworks displays were thus intimately connected. The interplay of human and other bodies, of audiences, artificers, pyrotechnics, the environment, performance techniques and atmospheres always correlated with various emotional reactions and experiences ranging from pleasure, wonder and delight to fear and confusion. Various agents sought to manage these experiences by managing fire, with more or less success. Things could go wrong, as the 'Grand Whim' testifies, and accidents were common in displays. But the risk of 'playing with fire' added to the pleasure of beholding them and countless testimonies record these pleasures. Not everyone agreed these displays were pleasing. Critics expressed anger and frustration at the expense or lamented fireworks as evidence of moral decline. In the *Expedition of Humphry Clinker* (1771) Tobias Smollett lambasted the commercial pleasure gardens as corrupting spaces:

> The diversions of the times are not ill suited to the genius of this incongruous monster, called *the public*. Give it noise, confusion, glare, and glitter; it has no idea of elegance and propriety [...] Vauxhall [pleasure garden] is a composition of baubles, overcharged with paltry ornaments, ill conceived, and poorly executed; without any unity of design, or propriety

---

22   Hone W., *The Every-Day Book* (London: 1826) 885–888.

23   Goethe J. W, *Italian Journey (1786–1788)*, trans. W.H. Auden – E. Mayer (San Francisco: 1982) 344.

24   Burke Edmond, *A Philosophical Enquiry into the Origin of Our Ideas of the Sublime and Beautiful* (London, printed for R. and J. Dodsley: 1757) 62.

25   Anon., "The Proclamations of Peace", *Gentleman's Magazine* 72 (May 1802) 456–462.

of disposition. It is an unnatural assembly of objects, fantastically illuminated in broken masses; seemingly contrived to dazzle the eyes and divert the imagination of the vulgar.[26]

Fireworks were also political, in the sense that they constituted a social order both physically and through the representations made of the event. This becomes more evident if we focus more closely on early modern ideas of emotional states.

2   Pyrotechnic Physiologies

Early moderns did not divide 'object' from 'subject' but spoke of bodies of various kinds and capacities. For many, the soul differentiated human from other bodies, but similar vocabularies and ideas might be applied to diverse kinds of bodies. The Polish artillerist Casimir Simienowicz noted that 'Saltpeter may be properly called the very Soul' of gunpowder.[27] Pyrotechnic language and ideas also penetrated the human body, so that the internal management of fires within the body was as much a part of pyrotechnic culture as the external fire management involved in staging displays. This section explores these pyrotechnic physiologies.

Werrett has argued that the human body was long understood as an economy of fire.[28] The renaissance saw a revival of the ancient notion of the *flamma vitalis*, the idea that some vital flame enlivened the body. Aristotle had said the heart was where 'the soul is set aglow with fire' and renaissance courtiers such as Baldassarre Castiglione urged the necessity of managing one's fiery passions and burning desires in order to survive at court.[29] This management of internal fires grew into a complex pyrotechnic physiology in the seventeenth and early eighteenth centuries, entangled with religious and political ideas. The English fellow of the Royal Society John Mayow proposed that some micro-mechanism of gunpowder explosions inside the body might be responsible for life. Mayow argued that air contained 'nitrous particles' like the saltpetre in gunpowder, which were needed for respiration, and which gave off heat

---

26   Smollett Tobias George, *The Expedition of Humphry Clinker* (London, W. Johnson and B. Collins: 1771) 187.

27   Simienowicz Casimir, *The Great Art of Artillery*, trans. G. Shelvocke (London, J. Tonson: 1729) 87.

28   Werrett S., "Sparks of Life", *Cabinet* 32 (2009) 91–98.

29   Werrett, "Sparks of Life" 92.

when they burned. When air was exhausted of this nitro-aerial spirit things ceased to combust, while animals deprived of it died.[30] The Oxford physician Thomas Willis expanded on these ideas and proposed muscular action as the product of micro-explosions of such nitrous and sulphurous particles inside the muscles. The nerves passing from the brain served as 'the fiery inkindling or the match [...] [to] blow up the Muscle'.[31]

An important site, in England at least, where this 'pyrotechnic physiology' becomes apparent is in seventeenth and eighteenth-century discussions of enthusiasm, the religious and political fanaticism that early modern English felt to be responsible for the Civil War of the mid-century and constant threats and plots after the Restoration of the monarchy in 1660. This tendency to extremist and unbendable convictions and erroneous beliefs was put down to an excess of fiery spirits igniting inside the bodies of enthusiasts, turning them into 'firebrands' with an overheated disposition who were liable to use fire to destroy England. Worst of all were the Catholic Jesuits, lambasted in *Pyrotechnica Loyolana, Ignatian Fire-works. Or, the Fiery Jesuits Temper and Behaviour* (1667), for devilish attempts to burn London,

> That, the Jesuits are ambitious, their Founders name signifies a fire-brand, *quasi ab igne natus*; and that his disposition was Fiery, and his profession Military; whereupon they affirm he came to send Fire. Hence *de jure* they profess the Art of making and casting about Fire-balls and Wild-fire to burn Houses and Cities: to promote which, they have two Colledges, one at Madrid, another at Thonon [in France] to advance the study of Artificial Fire-works, and to subdue Protestants by fraud and Arms: they keep stores of powder in their Colledges.[32]

Incendiary plotting and internal pyrotechnics were thus linked together as equally in need of control. Historians of science are familiar with the account of Steven Shapin and Simon Schaffer who argued in *Leviathan and the Air-Pump* that the new science of the seventeenth century was in part an attempt to use natural knowledge-making as an antidote to enthusiasm.[33] Against the

---

30 Werrett, "Sparks of Life" 93–94; Guerlac H., "John Mayow and the Aerial Nitre – Studies on the Chemistry of John Mayow I", *Actes du Septième Congrès Intenational d'Histoire des Sciences, Jérusalem 4–12 Août, 1953* (Paris: 1954) 332–349.

31 Quoted in Werrett, "Sparks of Life" 93.

32 Anon., *Pyrotechnica Loyolana, Ignatian Fire-works. Or, the Fiery Jesuits Temper and Behaviour* (London, Printed for G.E.C.T: 1667) 124; See also Werrett, *Fireworks* 77–79.

33 Shapin S., Schaffer S., *Leviathan and the Air Pump: Hobbes, Boyle and the Experimental Life* (Princeton, NJ: 1985).

MANAGING FIRE AS AN EMOTERIAL 341

personal inspiration of fanatics, fellows of the Royal Society in Restoration England argued that knowledge should be made communally and in a way that avoided as much as possible the heated and fiery exchanges typical of enthusiastic forms of life. Focusing on the pyrotechnic language used in the period manifests the intense emotions at play in these debates. Just as enthusiasts were over-heated and incapable of controlling the fires raging within them, so experimental philosophers were supposed to be 'cool' and temperate, working against fiery passions, not least by explaining and domesticating them through natural knowledge of pyrotechnic physiology. Apologists for the Royal Society and experimental philosophy celebrated it as a means to bring sobriety and therefore national security against 'violent and fiery' tempers that could ignite another civil war. Thomas Sprat praised experimentalists for having 'onely the discreet, and sober flame, and not the wild lightning of the others Brains'.[34] In the early eighteenth century, John Harris attacked a foe of the experimenters as an enthusiast, 'the Heat and Fire of your Spirit and Temper makes you, as it doth all Men, ignorant and precipitant in your Judgments of things: your fiery Zeal is without Knowledge'.[35] Only those who managed these spiritual fires could be reliable makers of knowledge and hence worthy citizens.

These debates were fundamental in establishing a division between science and non-science that has remained strong ever since, equating control of the emotions with rational knowledge and sensible behaviour and a more fiery disposition with a lack of control and more inspired, and so subjective, knowledge. It is the emotional economy allied with the gradual separation of 'objects' from 'subjects' that took place in the eighteenth and nineteenth centuries, which may thus be seen as in part a process of fire management. Crucially, 'materials', the sense that things are fundamentally different things to persons, objects to subjects, were the rarefied outcome of these efforts of distinction in the name of social order. Exactly how this process unfolded awaits a history but there were many scientific enterprises of fire management that belong to the process of subduing fiery spirits in the name of reason. In another seventeenth-century example, the Royal Society experimented with blood transfusion in the hope that the blood of a lamb, representative of Christ, would cool and calm the fiery spirits in the blood of a madman named

---

34 Sprat Thomas, *History of the Royal Society of London, for the Improving of Natural Knowledge* (London, 1667) 38.

35 Harris, quoted in Werrett, *Fireworks* 96.

Arthur Coga.[36] The experiments failed but their logic of heated and cooling spirits makes apparent the pyrotechnic economy of passions behind them.

3　Conclusion

Pyrotechnic fire is only one kind of fire, and similar stories might be told in relation to, for example the 'electric fire' in the eighteenth century, which gradually replaced pyrotechnic fires in explanations of a range of natural phenomena, including internal physiological processes. Here, however, I have sought to draw attention to two interacting economies of fire management, in the organization of fireworks displays and in the comprehension of internal physiology in pyrotechnic terms. The integration of these economies suggests the need for a sociomaterial approach, that recognized the constant connection of human and material lives, and which avoids taking for granted the object-subject division that these economies produced over time. I have argued that additionally, such sociomaterial histories need to be emoterial, recognizing the permanent emotional dimension of such episodes. Emotions may be mapped, as in the approach of Harris and Sørensen to describing affective fields, attunements and atmospheres, but it also needs to be recognized that comprehensions of emotional states were intrinsic to the whole form of life that generated these experiences in the first place. Natural and physiological knowledge set and reflected the social, political, and religious stakes that communal events like fireworks displays were seeking to manifest.

Bibliography

Primary Sources

Anon., "The Proclamations of Peace", *Gentleman's Magazine* 72 (May 1802).

Anon., *Pyrotechnica Loyolana, Ignatian Fire-works. Or, the Fiery Jesuits Temper and Behaviour* (London: 1667).

Burke Edmond, *A philosophical enquiry into the origin of our ideas of the sublime and beautiful* (London, printed for R. and J. Dodsley: 1757).

36　On Coga, see Schaffer S., "Regeneration: The Body of Natural Philosophers in Restoration England", in Lawrence C. – Shapin S. (eds.), *Science Incarnate: Historical Embodiments of Natural Knowledge* (Chicago: 1998) 83–120, on 100–105.

Cahuzac Jean-Louis de, "Feu d'artifice", in D. Diderot and J. d'Alembert (eds.), *Encyclopédie, ou dictionnaire raisonné des sciences, des arts et des métiers*, 17 vols. (Paris: 1751–65), vol. 6 (Paris: 1756).

Hone William, *The Every-Day Book* (Londo, Plurabelle Books Ltd: 1826).

Manzini Luigi, *Applausi festivi fatti in Roma per l'elezzione di Ferdinando III* (Rome, Facciotti: 1637).

Simienowicz Casimir, *The Great Art of Artillery*, trans. G. Shelvocke (London, J. Tonson: 1729).

Smollett Tobias George, *The Expedition of Humphry Clinker* (London, W. Johnson and B. Collins: 1771).

Sprat Thomas, *History of the Royal Society of London, for the Improving of Natural Knowledge* (London, T. R.: 1667).

Secondary Sources

Alekievich S., *Chernobyl Prayer: Voices From Chernobyl*, trans. A. Tait – A. Gunin (London: 2016).

Aleksievich S., *The Unwomanly Face of War*, trans. R. Pevear – L. Volokhonsky (London: 2017).

Biringuccio V., *The Pirotechnia of Vannoccio Biringuccio: The Classic Sixteenth-Century Treatise on Metals and Metallurgy*, trans. & ed. C. Stanley Smith and M. Teach Gnudi (originally Venice, 1540; New York: 1990).

Bonnemaison S. – Macy C. (eds.), *Festival Architecture* (London: 2007).

Canova-Green M.-C., "Fireworks and Bonfires in Paris and La Rochelle", in Mulryne J.R. – Watanabe-O'Kelly H. – Shewring M. (eds.), *Europa Triumphans: Court and Civic Festivals in Early Modern Europe* (Farnham: 2004).

Daston L. – Galison P., *Objectivity* (New York: 2007).

Dick M.D., *The Derrida Wordbook* (Edinburgh: 2013).

Goethe J.W. von, *Italian Journey (1786–1788)* trans. W.H. Auden and E. Mayer (San Francisco: 1982).

Guerlac H., "John Mayow and the Aerial Nitre – Studies on the Chemistry of John Mayow 1", *Actes du Septième Congrès Intenational d'Histoire des Sciences, Jérusalem 4–12 Août, 1953* (Paris: 1954).

Harris O. – Sørenson T.F., "Rethinking Emotion and Material Culture", *Archaeological Dialogues* 17 (2010).

Latour B., *After Lockdown*, trans. J. Rose (Cambridge: 2021).

Salatino K., *Incendiary Art: the Representation of Fireworks in Early Modern Europe* (Los Angeles: 1997).

Schaffer S., "Regeneration: The Body of Natural Philosophers in Restoration England" in Lawrence C. – Shapin S. (eds.), *Science Incarnate: Historical Embodiments of Natural Knowledge*, (Chicago: 1998).

Shapin S. – Schaffer S., *Leviathan and the Air Pump: Hobbes, Boyle and the Experimental Life* (Princeton, NJ: 1985).

Smith P.H., *The Body of the Artisan: Art and Experience in the Scientific Revolution* (Chicago: 2004).

Werrett S., "From the Grand Whim to the Gasworks: Philosophical Fireworks in Georgian England", in Dear P. – Roberts L. – Schaffer S. (eds.), *The Mindful Hand: Inquiry and Invention from the Late Renaissance to Early Industrialisation* (Amsterdam and Chicago: 2007).

Werrett S., "Sparks of Life", *Cabinet* 32 (2009).

Werrett S., "Watching the Fireworks: Early Modern Observation of Natural and Artificial Spectacles", *Science in Context: Special Issue: Lay Participation in the History of Scientific Observation*, ed. J. Vetter, 24 (2011).

Werrett S., *Fireworks: Pyrotechnic Arts and Sciences in European History* (Chicago: 2010).

Werrett S., *Thrifty Science: Making the Most of Materials in the History of Experiment* (Chicago: 2019).

Conclusion: the Technicity of Fire in Modern Europe – a Historiographical Crossroads

Liliane Hilaire-Pérez

Abstract

The conclusion provides a critical discussion of the research results presented in the contributions to the volume, and reflects on the open perspectives for future research on early modern fire.

Keywords

fire – technique – technology – knowledge – practices

The texts in this book express a highly original historical approach to fire, guided resolutely by the centrality of techniques.[1] Whether we are talking about fire as a social object, as a productive and energetic agent, as an object of knowledge at the heart of modern science, or as an emotional springboard and a spectacular device, the history of techniques constitutes a new and promising angle of attack for its study. From the contributions gathered here, it emerges that this approach puts back on the table several issues, usually dealt with by the human and social sciences, first and foremost ethnology, anthropology and philosophy, from Claude Lévi-Strauss and Gaston Bachelard to Stephen Pyne. In fact, what can the history of techniques add to the study of the human relationship to fire in seventeenth- and eighteenth-century European societies?

As Gianenrico Bernasconi and Olivier Jandot have shown,[2] one of the key elements in the study of this period, which saw major transformations in the relationship with fire, lies in the articulation of technical and symbolic aspects.

---

1 In translating this text into English, the word "*technique*" and related lemmas ("*technicisation*," "*technicité*," etc.) have been transliterated, to preserve the difference that exists in French between "*technique*" and "*technologie*." The word "technology" has been adopted only to translate the French "*technologie*."

2 Bernasconi G., "Cuisine et cultures du feu. De l'âtre au 'feu enveloppé' (XVIIIe–début du XIXe siècle)", *Food and History* 20.2 (2022) 109–128; Jandot O., *Les délices du feu. L'homme, le chaud et le froid à l'époque moderne* (Ceyzérieu: 2018).

© KONINKLIJKE BRILL BV, LEIDEN, 2025 | DOI:10.1163/9789004521766_015

This statement echoes that of Daniel Roche, who, in *Les choses banales*, called for an analysis of 'the juxtaposition of lifestyles and symbolic values'.[3] This link between techniques and symbolism is not self-evident. In the presentation of the proceedings of another conference on 'technique and religion,' Guillaume Carnino, Sébastien Pautet and I pointed to the distinction 'between the primacy that emerging sociology accorded to the religious fact as a basic social phenomenon, and instrumental practices, henceforth circumscribed to the infra-social domain of individual organic subsistence'.[4] We noted, however, that 'the materiality, indeed the technicality, of religion and spirituality is giving rise to renewed research': in this research the Neuchâtel team is taking an active part, for example through the past collaboration on an international conference on repair. Repair is an invisible if not trivial act, yet charged with multiple symbolic dimensions – from the shields of ancient Greece to today's repair cafés – that update the mediating function of technical objects, namely their role as mediums.[5] Such approaches are consistent with current research linking technology, emotions, affects and symbolization – a new page in the history of techniques, and a powerful tool for questioning the omnipresence of techniques in the collective unconscious. This inter-relationship between technique and symbolism is one of the paths traced by this book on fire, and one of its most innovative contributions.

The development of a technical history of fire raises another set of questions, widely addressed in the volume. This history brings to the fore the practices, the gestures, the body, as well as the artisans' and labourers' know-hows: the role of fire in the sphere of work is indeed one of the most fascinating themes in the volume, be it a matter of apothecary furnaces (Bérengère Pinaud), brick kilns (Cyril Lacheze) or atmospheric machine boilers (Andrew M.A. Morris). The technical study of fire reveals the breadth, diversity and sophistication of a body of knowledge that owes nothing to the intervention of the scholarly world, nor even to the regulations of the trades. In short, it is a body of specific knowledge that cuts across multiple activities – what Marco Storni calls 'technical knowledge of heat', designating as such 'the intermediary dimension between sensuous technology and philosophical knowledge'. The acknowledgment, through the history of fire, of the existence of a specific technical

---

3   Roche D., *Histoire des choses banales. Naissance de la consommation dans les sociétés traditionnelles (XVIIᵉ–XIXᵉ siècle)* (Paris: 1997).

4   Carnino G. – Hilaire-Pérez L. – Pautet S., "Avant-propos", in Carnino G. – Hilaire-Pérez L. – Pautet S. (eds.), *Technique et religion. Cultures techniques, croyances, circulations de l'Antiquité à nos jours* (forthcoming).

5   Bernasconi G. – Carnino G. – Hilaire-Perez L. – Raveux O. (eds.), *Les réparations dans l'histoire. Cultures techniques et savoir-faire dans la longue durée* (Paris: 2021).

CONCLUSION: THE TECHNICITY OF FIRE IN MODERN EUROPE

culture, irreducible to science and to the trades – namely of the concept of "technique," even though the word did not yet exist as such[6] – is crucial. One cannot help but think of the 'coal fuel technology', the technical culture of coal in England, highlighted by John R. Harris a generation ago:[7] this practical knowledge, essential to industrialization, owed nothing to applied science, but to the populations' centuries-long mastery of coal combustion in a wide array of domestic and productive activities. This volume takes a step further. On the one hand, it contributes to current debates on artisanal techniques as seen through the prism of the Enlightenment,[8] and on the other, it opens up a reflection on the role of fire as an operational agent, and thus on the technical schemes underpinning the very constitution of modern science.

These two aspects – the symbolic dimension of fire techniques, and the interferences between the technical culture of fire and the scholarly worlds – are the main thrust of this book, and open up promising avenues of research. They directly echo the Bachelardian epistemology, which placed the symbolic power of fire at the very heart of modern science, but also at the foundation of the intimate, concrete, sensitive experience of intellection and reverie. It is clear that the history of techniques, as it is currently being redefined as connected to the dimensions of emotions, symbols, sensations, and personal commitment,[9] owes much to the philosophy of science and technology in France – a point of which the history of fire is a fine illustration.

1 Techno-Symbolism of Fire

In the present volume, Simon Werrett, while discussing pyrotechnics, notes the habit of historians of material culture to focus on the study of objects,

---

6 Camolezi M., "Technique, *Technics, Technik*, substantifs: les mots à travers les dictionnaires du XIXe siècle", *Artefact* 15 (2021) 61–106.

7 Harris J.R., "Skills, Coal and British Industry in the Eighteenth Century", *History* 61.202 (1976) 167–182.

8 Lanoë C., *Les ateliers de la parure. Savoirs et pratiques des artisans en France XVIIe–XVIIIe siècles* (Ceyzérieu: 2024); Hilaire-Pérez L. – Lanoë C., "Pour une relecture de l'histoire des métiers: les savoirs des artisans en France au XVIIIe siècle", in Milliot V. – Minard P. – Porret M. (eds.), *La grande chevauchée. Faire de l'histoire avec Daniel Roche* (Genève: 2011) 334–358. These authors, by stressing the autonomy of artisanal knowledge, diverge from Bertucci P., *Artisanal Enlightenment: Science and the Mechanical Arts in Old Regime France* (New Haven: 2017).

9 Traversier M., *L'harmonica de verre et miss Davies. Essai sur la mécanique du succès au siècle des Lumières* (Paris: 2021); Montègre G., *Voyager en Europe au temps des Lumières. Les émotions de la liberté* (Paris: 2024).

rather than subjects, the one passive matter, the other living beings endowed with sensibility and power of symbolization ('the common habit of discussing it in terms of "objects" distinct from human "subjects," the one inert, passive matter and the other living, emotive being'). Werrett, by contrast, invites the reader to consider the emotional dimension of fireworks as inseparable from a material, corporeal and technical experience, echoing a major current in cultural history. Several recent works, such as those by Marie Thébaud-Sorger, Mélanie Traversier and Gilles Montègre, have shown that techniques in the early modern era are embedded in a range of sensory and emotional devices that multiply the symbolic resonances of objects, materials and mechanisms.[10] Many inventions of the eighteenth century took on the hybrid status of "boundary objects": they were at the crossroads of scientific concerns, the evolving tastes and sensibilities, the skills of craftsmen, and the multiple mediations offered by markets, institutions, social circles and audiences. An entire historiographical trend has highlighted the originality of this technical culture, inseparable from the meaning given to the "arts," a unifying and synthetic concept. Technicity itself was perceived as a source of aesthetic emotion,[11] and was invested with a symbolic power that industrial modernity never ceased to amplify. If the history of science, as Werrett points out, has shown the inanity of a caesura between objectivity and subjectivity in the modern era, the history of techniques, and in particular the history of the relationship with fire, is crisscrossed by symbolism – "halos," in Simondonian terms – whose temporalities intertwine, provoking an interplay of superimpositions that constantly nourishes the mythification of techniques. The inherited symbolism of the hearth and the candle, the embers and the pyre, is superimposed by the wonder and dread caused by energetic intensification, as in the glowing pictures of English forges by Joseph Wright and Philippe-Jacques de Loutherbourg,[12] at the very moment when volcanic eruptions fascinated scientists, painters and entertainment entrepreneurs, providing metaphors for representing the new coal-fuel metallurgy but also political passions.[13] In this regard, Gilles Montègre has recalled the interest shown by factory inspector (*inspecteur des manufactures*)

---

10 Traversier, *L'harmonica de verre*; Montègre, *L'expérience des Lumières*; Thébaud-Sorger M., *Une histoire des ballons. Invention, culture matérielle et imaginaire, 1783–1909* (Rennes: 2009).

11 Guffroy Y., *Représenter l'invention: étude de l'évolution du dessin d'objet technique en Angleterre (ca. 1750–1850)* (Ph.D. dissertation, École polytechnique fédérale de Lausanne: 2023); Hilaire-Pérez L., "'Techno-esthétique' de l'économie smithienne. Valeur et fonctionnalité des objets dans l'Angleterre des Lumières", *Revue de synthèse* 135.4 (2012) 495–524.

12 Fox C., *The Arts of Industry in the Age of Enlightenment* (New Haven – London: 2010).

13 Brewer J., *Volcanic: Vesuvius in the Age of Revolutions* (New Haven – London: 2023).

CONCLUSION: THE TECHNICITY OF FIRE IN MODERN EUROPE

François de Paule Latapie, on a mission to Italy, in 'the astonishing "Vesuvian machine" of the English ambassador to Naples, William Hamilton', 'a picture of a Vesuvian eruption painted in "transparent colour"'.[14]

One of the book's stakes is to consider, in all its facets, the great hiatus between the world of open fire – the hearth, the brazier, the torch, and more generally the diffused and shimmering fire, that fosters socialization – and a world of hidden and controlled fire – that of ovens, furnaces, boilers, namely a multi-functional fire, specialist, technologized, intensified, and made dangerous by the unbridled quest for caloric, as many of the authors show. Following in the footsteps of Daniel Roche, the aim here is to identify, characterize and question this great transformation. The chapters reveal the complexity of this transition, the ambiguities, interferences and coexistence between these two worlds. The history of techniques, which has become the place *par excellence* where temporal pluralities are considered, underlines the fact that these two worlds do not succeed one another, but interweave, suggesting that early modern people have lived through several systems of reference, which correspond to different technical cultures.

In *Le peuple de Paris*, in the chapter entitled *Savoir consommer* (*Knowing how to consume*), Daniel Roche masterfully paved the way for a historical anthropology of fire, subsequently updated in *Les choses banales*. Daniel Roche there set an analytical framework based on social sciences and philosophy. While discussing the stove – relying on Jean Baudrillard (*Le système des objets*, 1968) – Roche stated: 'Now hidden, abstract as it were, mastered and functional, [fire] no longer contributes in the same way to the atmosphere of the dwelling'.[15] Quoting Gaston Bachelard (*La flamme d'une chandelle*, 1964), Roche described the end of a period when man 'kept an action of Prometheus', 'because he could see the moment when it was necessary to help the wood to burn and place the extra logs in time'[16] – an age when the fireplace concentrated life in the single rooms of popular dwellings, where 'wood and fire are inseparably mixed'.[17] The closed world of fire is that of technicality, of the technical mediation of all activities, that creates a world of networks, of interferences, a reticular and abstract world, which gradually evacuates any concrete and emotional relationship with fire. Roche evoked 'a symbolic mutation', 'the substitution of a relationship to elements and things', the 'transformation of the sense of intimacy', the loss of the reflections of flame that expressed 'the concreteness

---

14   Montègre, *L'expérience des Lumières* 414.
15   Roche D., *Le peuple de Paris: essai sur la culture populaire au XVIIIe siècle* (Paris: 1981) 141.
16   Ibidem.
17   Ibidem, 139.

of values'.[18] The stove and its utensils, and even its functionality, signaled the advent of an abstract, homogeneous world, controllable – even in the ambient air – pushing reverie back into an internalized, metaphorical fire, or a fire sublimated by the contemplation of images, as Marie Thébaud-Sorger suggests.

But the dominant symbolic and technical universe is still that of the open fire, accurately portrayed by the volume's authors. The fireplace is at the heart of several descriptions. Using the probate inventories of Parisian apothecaries (and following a methodology initiated by Daniel Roche), Bérengère Pinaud highlights the role of the kitchen and its hearth as a space for the manufacture of remedies (beyond the apothecary shop) and, at the same time, as the core of domestic life, thus depicting an over-invested, multi-functional space. Gianenrico Bernasconi shows that cooking over an open hearth was the most common form of cooking (while the technicality of the oven or the spit is reserved to specific culinary professions). Bernasconi recalls the multiple constraints posed by the management of fire in the fireplace, its supervision, the incessant work to feed it, the hardship of cooks, the system of objects, through the range of utensils, but also the centrality of soup, a key element in the sociability of the open fire.

This world is by no means static, and this represents an innovative line of research. Olivier Jandot examines the role of masons and architects in the construction and, above all, the improvement of fireplaces, at a time when stoves were being developed and manufactured by new trades, such as stove makers and heating engineers (*fumistes*). Marie Thébaud-Sorger also shows that architecture played a key role in the prevention of urban fire that accompanied the improvements to fireplaces. In this way, the book's authors open up to a system of interferences that is one of the main thrusts of their collective thinking, identifying the complexity of the relationship with fire, and the ambiguities involved in the transformations at work. The book takes as its starting point the coexistence of technical generations and the plurality of temporal rhythms – "heterochrony," to use a fundamental concept in the current literature.[19] The tone is set right from the introduction. Gianenrico Bernasconi and Marco Storni emphasize the importance of persistence: while coal has been introduced for a variety of uses, the use of wood as the main

---

18   Ibidem.

19   Van der Straeten J. – Weber H., "Technology and its Temporalities, a Global Perspective", in Carnino G. – Hilaire-Pérez L. – Lamy J., *Global History of Techniques (19th–21st Centuries)* (Turnhout: 2024) 261–280; Herr-Laporte C., *Être à l'heure: Coordonner les mobilités en France au XVIIIᵉ siècle* (forthcoming).

CONCLUSION: THE TECHNICITY OF FIRE IN MODERN EUROPE

fuel endured, particularly in France – as Renan Viguié has shown, this was still the case for the countryside in the early twentieth century.[20] For the case of industry, Serge Benoit has highlighted the enduring variety of energies, thanks to innovations that made older equipment profitable, and ensured the long life of well-mastered techniques.[21]

The observation that traditional techniques are not immobile also means to highlight the abundance of inventiveness in all areas of fire use. Beyond the mix of energy sources and technical cultures, what was at stake in the eighteenth century, as Daniel Roche suggested, was the technical intensification and technicization of fire, whether in terms of new devices "enclosing" fire, or traditional open-fire equipment. Specialization, multi-functionality, technical mediation are revealed in all acts of life, right down to portable "counter heaters" for apothecaries or their apprentices, in the shop or in the kitchen (Bérengère Pinaud). Invention belongs to ordinary techniques, which are never trivial though. Not only might foot warmers (*chaufferettes*) be seen as variations on Hermann Boerhaave's portable stoves, but their technical simplicity is in itself an innovation, attesting to a conception of progress as a pledge of sharing through simplification and accessibility – so much so that they can still be reproduced today.[22] These light, adaptable and flexible techniques, which characterize fire management in the eighteenth century, were the credo of philanthropists, societies of arts, inventors and project promoters. The stakes were social and political: "improvement" and the public good. The portable stoves studied by Gianenrico Bernasconi are a pledge of autonomy in the kitchen, that lighten tasks, allow the wealthier to reduce domestic staff, and to simplify (and make safer) cooking techniques in households that can afford a stove. This inventive dynamic aroused the interest of a variety of audiences, via all kinds of mediation, above all the press and manuals, but also exhibitions of industrial products, for example. In several chapters (Gianenrico Bernasconi, Olivier Jandot, Marie Thébaud-Sorger), we see the rise of a polymorphous technical literature that accompanies the spread of this management of fire, and constructs the social effectiveness of techniques. The knowledge of fire, whether open or closed, belongs to the world of practice, and outlines a technological dynamic in the strongest sense of the term, operative, transductive, nourished by multiple interferences and reticulations between objects, actors,

20   Viguié R., *Bien au chaud. Histoire du chauffage au XXe siècle* (Paris: 2023).
21   Benoit S., *D'eau et de feu: forges et énergie hydraulique. XVIIIe–XXe siècle. Une histoire singulière de l'industrialisation française* (Rennes: 2020).
22   Hendriksen M.M. – Verwaal R., "Boerhaave's Furnace: Exploring Early Modern Chemistry through Working Models", *Berichte zur Wissenschaftsgeschichte* 43.3 (2020) 385–411.

trades, spaces of activity and even sites of knowledge production. Fire management epitomizes operative patterns that are at the heart of the development of modern science.

2 Fire Technologies: Operational Knowledge and Practical Resources

The technicization of fire goes in parallel with a thermal intensification, supported by multiple technical devices that mobilize all kinds of actors – craftsmen, industrialists, engineers, local and central administrators, users, and amateurs – whose interactions fostered knowledge circulation and fruitful networking, which are sources of innovation. One of the key areas is firefighting, a real public policy issue, as detailed by Marie Thébaud-Sorger, Catherine Denys and Cornelia Müller. The closed hearth, but also the fireplace and its eighteenth-century refinements as shown by Olivier Jandot, and the multiple heating devices used in workshops and factories, are the supports of a new energy intensity. Reducing heat loss and increasing heat were required in all areas of use, for a variety of reasons, as the authors point out: wood shortages, the search for economy in popular consumption, changes in sensitivity to cold and to the effort required to maintain an open fire, to which we can add the growing need for a lively, sustained fire for cooking (as the consumption of fresh meat increased), in industry, but also for institutional and collective demand, as well as public places, in the same way as lighting and ventilation.[23]

Technical engagement in thermal intensification was systematic, and generated a multiplicity of networked devices to control thermal output. Fire prevention and firefighting took centre stage, and as such are extensively analysed in this volume. Marie Thébaud-Sorger and Cornelia Müller highlight the use of increasingly targeted techniques to deal with the force of fires, which were rare as large-scale fires (although the setting ablaze of public places and the explosion of boilers, exceptional as they were, marked a turning point) but recurrent as ordinary fires, as the economic boom led to the multiplication of furnaces in urban spaces and the storage of flammable goods, such as

---

23   Thébaud-Sorger M., "Capturing the Invisible: Heat, Steam and Gases at the Crossroads of Inventive Practices, Place and Social Milieu in France and Great Britain in the Second Half of the Eighteenth Century", in Roberts L.L. – Werrett S. (eds.), *Compound Histories: Materials, Governance and Production, 1760–1840* (Leiden: 2017) 85–105; Bothereau B., "Illuminated Publics: Representations of Street Lamps in Revolutionary France", *Technology and Culture* 61.4 (2020) 1045–1075.

CONCLUSION: THE TECHNICITY OF FIRE IN MODERN EUROPE   353

textiles.[24] The development of firefighting techniques, against a backdrop of an intensified use of heat in industry and interiors, has been highlighted by analysts of the Anthropocene such as Stephen Pyne.[25] The disappearance of controlled, regenerative surface fire and slash-and-burn agriculture (echoing the rituals of walking on braziers[26]) would have given way to the enclosure of fire, thermal intensification and the burning of fossil fuels, a source of warming and gigantic fires. The deployment of fire-extinguishing and fire-prevention techniques, a hyper-technicalized field of action, coupled with a political logic of surveillance, is in line with the desire to stifle fire, evoked by Bachelard, nostalgia for the open fire, for fire as a "social being," a figure of parental prohibition, of the desire for transgression and self-confrontation with risk (recalled by rituals). The fear of the intensity of enclosed fire (extinguishers, firewalls, fire alarms, fireproof constructions) is part of the inventive proliferation of the Age of Enlightenment, described in detail by Marie Thébaud-Sorger; this dread is linked to a sacralization of technical solutions, at a time of "political technology."[27] Vigilance and urban surveillance techniques, mobilizing the full range of local sociabilities and municipal administrations, created a veritable techno-political regime (Cornelia Müller), involving a host of technical devices and objects, from lighting to street cleaning.[28]

While there are obvious points of contact between philosophical analysis and historical research, the latter is distinguished by its close study of the social processes underlying thermal intensification and its corollary, firefighting. To the symbolic register investigated by philosophers, historians add considerations of the 'continuity' of the 'technical environment', to borrow the historian Catherine Lanoë's use of a concept introduced by André Leroi-Gourhan, thus placing interrelations at the heart of the analysis, over and above 'professional

---

24 Hilaire-Pérez L. – Thébaud-Sorger M., "Risque d'incendie en milieu urbain et *industrious revolution*: le cas de Londres dans le dernier tiers du XVIIIe siècle", *Le Mouvement Social* 249 (2014) 21–39.

25 Pyne S., *Fire in America: A Cultural History of Wildland and Rural Fire* (Seattle, WA: 1982); Pyne S., *Fire. A Brief History* (Seattle, WA: 2001); Pyne S., *The Pyrocene. How We Created an Age of Fire, and What Happens Next* (Berkeley: 2021).

26 Franchina L., "'Bien marcher sur le feu' à l'île de La Réunion", *Techniques & Culture* 78 (2022) 50–65.

27 Kaplan S.L., *Les ventres de Paris. Pouvoir et approvisionnement dans la France d'Ancien Régime* (Paris: 1984).

28 Bothereau, "Illuminated Publics"; Lyon-Caen N. – Morera R., *À vos poubelles citoyens! Environnement urbain, salubrité publique et investissement civique (Paris, XVIe–XVIIIe siècle)* (Ceyzérieu: 2020).

and/or institutional differences'.[29] The pursuit of fire control is based on clusters of interactions, transpositions and technical convergences between trades, activities, registers of use and circles of interest. Bérengère Pinaud notes, with reference to the apothecaries' stoves, that knowledge of fire is based on networks of specialized craftsmen, and not on a particular trade. Beyond collectives, fire management is emblematic of the reticular and transductive nature of practical knowledge,[30] which creates a 'technical space', as Hélène Vérin has called it,[31] irreducible to the institutional space of trades and academic science – 'a technical knowledge of heat' as Marco Storni puts it.

The technicality of fire management, as it unfolds and even becomes autonomous, goes hand-in-hand with the production of a rich technical literature – of a technology – that many of the authors analyse. Beyond the public knowledge of fire and its uses, open to multiple mediations and redaction/reduction processes as highlighted by Hélène Vérin, it is a relational technical intelligence based on analogical, imitative, operative understanding – in short, on synthetic thinking – that defines this "technical environment." The proliferation of heating appliances described by Olivier Jandot shows the polysemy of the furnace, that has various typologies: built in bricks or metal, fixed, mobile, adapted to a diversity of uses. This variation around a basic principle, a form or a function, and the addition of elements and functionalities, is a major feature of inventive transpositions, particularly in vogue in the eighteenth century, at the same time that markets for private and public equipment developed.[32] Another manifestation lies in the hybridization of fireplaces and stoves, also described by Olivier Jandot, or in substitutions, such as those of fuels (peat). Fire management techniques are a perfect illustration of a world of transpositions and combinatories, governed by mixedness, heterogeneity and synthesis. Some inventions are based on assemblies and the complexification of tools and instruments (or even places, such as the interiors of the apothecaries' homes

---

29   Lanoë C., "Réflexions à propos du 'secret' de fabriquer 'les perruques au métier' (XVIIe–XVIIIe siècles)", in Depretto L. – Renoux C. – Speroni C. – Vickermann-Ribémont G. (eds.), *Cultures du secret à l'époque moderne. Raisons, espaces, paradoxes, fabriques* (Paris: 2023) 309–326; Lanoë, *Les ateliers de la parure*.

30   Lanoë C., "'Mettre les gants en couleurs'. La construction des savoirs des gantiers-parfumeurs, XVIe–XVIIIe siècles", *Zilsel* 2 (2021) 219–236; Thébaud-Sorger M., "Changing Scale to Master Nature: Promoting Small-scale Inventions in Eighteenth-Century France and Britain", *Technology and Culture* 61.4 (2020) 1076–1107.

31   Vérin H., *La gloire des ingénieurs. L'intelligence technique du XVIe au XVIIIe siècle* (Paris: 1993).

32   Hilaire-Pérez L. – Thébaud-Sorger M., "Les techniques dans l'espace public. Publicités des inventions et littérature d'usage en France et en Angleterre au XVIIIe siècle", *Revue de synthèse* 2 (2006) 393–428.

and shops, which Bérengère Pinaud describes as 'hybrid and unstable') – a "deconcretization," in Simondonian terms; in other cases, on the contrary, functions are integrated, parts are adapted, their interplay, their adjustment (such as the grates on Désarnod's furnaces, evoked by Olivier Jandot) – in short, a concretization of the object. This double movement, which is the keystone of invention, gives rise to a technology of the object, centred on the principle of action and on multi-functionality.[33] 'Materialization of intelligence, abstraction of things', as Daniel Roche put it:[34] these processes of intellection, which are in no way theoretical, define the operative knowledge that contributes to technical intelligence as much as engineering sciences, without ever aiming to do science.[35] They define an 'artisanal epistemology',[36] as claimed by Marco Storni, for whom the tacit knowledge of fire is shared by communities of practitioners to such an extent that it becomes stabilized, to the point of assuming a normative character.

Scholarly circles were fully aware of the sophistication, elaboration and reliability of vernacular, artisanal, working-class technical knowledge, while they failed to recognize its value.[37] Marco Beretta has shown that Lavoisier worked with craftsmen familiar with fire and flame[38] and, in the field of lighting, Benjamin Bothereau has analysed Lavoisier's borrowings from artisanal techniques, reduced to arts and appropriated.[39] For Marco Storni, the 'quasi-quantification' of combustion and heat, inscribed in the 'daily experience of practitioners', appeared to some authors of treatises and dictionaries as preferable to instruments such as the thermometer and to scholarly protocols, too rigid or speculative to apprehend fire-related phenomena. Several chapters underline the importance of the technical literature of the arts of fire, a veritable technology through the deployment of terminology, classifying even hearth fires and types of wood. While the reasons for this codification

---

33 Hilaire-Pérez L., *La pièce et le geste. Artisans, marchands et savoirs techniques à Londres au XVIIIe siècle* (Paris: 2013); Bernasconi G., *Objets portatifs au siècle des Lumières* (Paris: 2015).

34 Roche D., *La France des Lumières* (Paris: 1993) chapter 19.

35 Hilaire-Pérez, *La pièce et le geste.*

36 Smith P.H., *The Body of the Artisan. Art and Experience in the Scientific Revolution* (Chicago: 2004); Valleriani M. (ed.), *The Structures of Practical Knowledge* (Dordrecht: 2017).

37 Hilaire-Pérez – Lanoë, "Pour une relecture de l'histoire des métiers".

38 Beretta M. – Brenni P., *The Arsenal of Eighteenth-Century Chemistry: The Laboratories of Antoine Laurent Lavoisier (1743–1794)* (Leiden: 2022).

39 Bothereau B., "Rationalisation des discours et pratiques de la nouveauté: la lanterne à 'réverbères' au Concours Sartine (Paris, 1763–1766)", in Baudry J. – Blanc J. – Hilaire-Pérez L. – Ratcliff M. – Wenger S. (eds.), *Produire du nouveau? Arts, techniques, sciences en Europe (1400–1900)* (Paris: 2022) 113–126.

of practices are manifold, as are the audiences involved, these reductions to art are part of the capture of field knowledge by learned elites, which went hand-in-hand with a denial of the abstract skills of practitioners, accused of keeping their procedures secret and of not understanding or even verbalizing their techniques.[40] The essays in this volume bear the hallmarks of historiographical debates on this subject. Andrew M.A. Morris, whose text is part of the reflections on the "Industrial Enlightenment," notes that the engineer John Smeaton was inspired by the workers' know-how, which he transferred to the sciences, thus identifying principles of action that he intended to substitute for tacitly acquired dexterity. Cyril Lacheze, on the other hand, points out that mastering fire, furnaces and even the effects of oxygen with bellows is an ancient art, requiring a physical, sensitive commitment that owes nothing to the technosciences. Far from the image of "trading zones," Jean Cantelaube and Daniel Fischer have shown that in the nineteenth century, mining engineers tried to master the fire skills of Pyrenean forgers, as the only way to improve processes and control this workforce, but also to legitimize scholarly expertise.[41]

The transfer of practical knowledge of fire to the scholarly world takes yet other forms, such as the role of operative schemes at the very heart of the production of theoretical knowledge. Stefano Salvia, evoking Galileo's thermoscopes and the beginnings of heat measurement, places this evolution in a broader context of the perception of fire no longer as a substance or element, but as an operative agent, called upon in multiple fabrications that also define experimental scholarly practices, notably alchemical ones (heating, drying, rarefying, calcining, dilating, etc.). Hannah Elmer's study of purification by fire in eighteenth century chemistry treatises confirms this dimension, highlighting the Paracelsians' interest in fire as an active principle and in the 'aggregate of operations' in which it participates, a direct echo of the complexity of fire's uses in productive environments. Purification is based on a chain of operations, on the assembly and coordination of gestures, on a technical relationship that must be mastered. Purifying agents such as certain remedies and the philosopher's stone are themselves highly complex mixtures of purified ingredients. The transition from 'fire as substance' to 'fire as function', to the caloric principle, is inseparable from a world of heterogeneous, the composite and

---

40   Fischer D., *Philippe Frédéric de Dietrich. Un entrepreneur des savoirs au XVIIIe siècle* (Paris: 2022).

41   Fischer, *Philippe Frédéric de Dietrich*; Cantelaube J., "Les *Annales des mines* et la forge à la catalane", in Bret P. – Chatzis K. – Hilaire-Pérez L. (eds.), *La presse et les périodiques techniques en Europe (1750–1950)* (Paris: 2009) 71–88.

CONCLUSION: THE TECHNICITY OF FIRE IN MODERN EUROPE

the synthesis, which is that of the practitioner. Bérengère Pinaud's text hints at the Paracelsian tradition and its links with the practice of the apothecaries and their intense use of instruments, as well as links with eighteenth-century chemistry, as shown by Marco Beretta, who recalls the importance of fire instrumentation, with its many pipes and bellows activating combustion, at a time when fire was marginalized in the laboratory.[42] As Simon Werrett reminds us, open fire and its "impoliteness," its "fanaticism," and its links with the sensory universe and the "vulgar" are thus expelled from scholarly space. This is the result of a maximization of instrumentation, which makes fire an agent and an operating principle, inseparable from the closed, mastered fire, sustained by a sophisticated technical culture that for a long time provided its material and intellectual resources to modern science.[43]

This rich volume marks a milestone in historiography: it attests to the vigour and inventiveness of studies at the interface of science and technology, a veritable crucible for the renewal of history, today.[44] The study of the construction of fire as an epistemic object is nourished by a range of practices in which field knowledge plays a central role. The chemical knowledge of fire goes forward at the same time that the number of furnaces and kilns multiplies, but not without calling fire into question as an "epistemological obstacle." The increasingly technical relationship with fire opened the door to scientific knowledge, while the affects and emotions associated with the spectacle of the flame were internalized and repressed. Did this entail the end of reverie, passion, and enthusiasm in favour of knowledge, and thus the marginalization of fire? Nothing is less certain, as evidenced by the fact that, still in the early twentieth century, the comparison with the open fire fueled debates on the acceptance of electricity.[45] We should not forget that, in *Les choses banales*, Daniel Roche noted that 'progress' 'has as its horizon a complex psychology'.[46]

---

42 The reference is to the paper "Heat and Fire in Lavoisier's Experimental Techniques," presented at the workshop *Fire Management in the Early Modern Age: Knowledge, Technology, Economy* (University of Neuchâtel: 2021), but not included in this volume.

43 Meyer G., *Un savant dans son siècle: Alexandre T. Vandermonde* (1735–1796) (PhD dissertation, University of Nantes: 2021).

44 Hilaire-Pérez L. – Lanoë C. (eds.), *Les sciences et les techniques, laboratoire de l'histoire. Mélanges en l'honneur de Patrice Bret* (Paris: 2022).

45 Kay M., "'Arise, Shine, for thy Light Has Come': Electricity and Authority in British Jewish Domestic and Communal Life in the Early Years of Electrification", conference paper presented at the workshop *Technique et religion: cultures techniques, croyances, circulations de l'Antiquité à nos jours*, University of Paris, 1–3 September 2021.

46 Roche, *Histoire des choses banales* 149.

Bibliography

Benoit S., *D'eau et de feu: forges et énergie hydraulique, XVIIIe–XXe siècle. Une histoire singulière de l'industrialisation française* (Rennes: 2020).

Beretta M. – Brenni P., *The Arsenal of Eighteenth-Century Chemistry: The Laboratories of Antoine Laurent Lavoisier (1743–1794)* (Leiden: 2022).

Bernasconi G., "Cuisine et cultures du feu. De l'âtre au 'feu enveloppé' (XVIIIe–début du XIXe siècle)", *Food and History* 20.2 (2022) 109–128.

Bernasconi G. – Carnino G. – Hilaire-Perez L. – Raveux O. (eds.), *Les Réparations dans l'histoire. Cultures techniques et savoir-faire dans la longue durée* (Paris: 2021).

Bernasconi G., *Objets portatifs au siècle des Lumières* (Paris: 2015).

Bothereau B., "Illuminated Publics: Representations of Street Lamps in Revolutionary France", *Technology and Culture* 61.4 (2020) 1045–1075.

Bothereau B., "Rationalisation des discours et pratiques de la nouveauté: la lanterne à 'réverbères' au Concours Sartine (Paris, 1763–1766)", in Baudry J. – Blanc J. – Hilaire-Pérez L. – Ratcliff M. – Wenger S. (eds.), *Produire du nouveau? Arts, techniques, sciences en Europe (1400–1900)* (Paris: 2022) 113–126.

Bertucci P., *Artisanal Enlightenment: Science and the Mechanical Arts in Old Regime France* (New Haven: 2017).

Brewer J., *Volcanic: Vesuvius in the Age of Revolutions* (New Haven – London: 2023).

Cantelaube J., "Les *Annales des mines* et la forge à la catalane", in Bret P. – Chatzis K. – Hilaire-Pérez L. (eds.), *La presse et les périodiques techniques en Europe (1750–1950)* (Paris: 2009) 71–88.

Camolezi M., "Technique, *Technics, Technik*, substantifs: les mots à travers les dictionnaires du XIXe siècle", *Artefact* 15 (2021) 61–106.

Carnino G. – Hilaire-Pérez L. – Pautet S., "Avant-propos", in Carnino G. – Hilaire-Pérez L. – Pautet S. (eds.), *Technique et religion. Cultures techniques, croyances, circulations de l'Antiquité à nos jours* (forthcoming).

Fischer D., *Philippe Frédéric de Dietrich. Un entrepreneur des savoirs au XVIIIe siècle* (Paris: 2022).

Fox C., *The Arts of Industry in the Age of Enlightenment* (New Haven – London: 2010).

Franchina L., "'Bien marcher sur le feu' à l'île de La Réunion", *Techniques & Culture* 78 (2022) 50–65.

Guffroy Y., *Représenter l'invention: étude de l'évolution du dessin d'objet technique en Angleterre (ca. 1750–1850)* (PhD dissertation, École polytechnique fédérale de Lausanne: 2023).

Harris J.R., "Skills, Coal and British Industry in the Eighteenth Century", *History* 61.202 (1976) 167–182.

Hendriksen M.M. – Verwaal R., "Boerhaave's Furnace: Exploring Early Modern Chemistry through Working Models", *Berichte zur Wissenschaftsgeschichte* 43.3 (2020) 385–411.

Herr-Laporte C., *Être à l'heure: une histoire de la coordination temporelle dans les transports terrestres en France au cours d'un long XVIIIe siècle* (forthcoming).

Hilaire-Pérez L., "'Techno-esthetics' of the Smithian economy. Valeur et fonctionnalité des objets dans l'Angleterre des Lumières", *Revue de synthèse* 135.4 (2012) 495–524.

Hilaire-Pérez L., *La pièce et le geste. Artisans, marchands et savoirs techniques à Londres au XVIIIe siècle* (Paris: 2013).

Hilaire-Pérez L. – Lanoë C. (eds.), *Les sciences et les techniques, laboratoire de l'histoire. Mélanges en l'honneur de Patrice Bret* (Paris: 2022).

Hilaire-Pérez L. – Lanoë C., "Pour une relecture de l'histoire des métiers: les savoirs des artisans en France au XVIIIe siècle", in Milliot V. – Minard P. – Porret M. (eds.), *La grande chevauchée. Faire de l'histoire avec Daniel Roche* (Genève: 2011) 334–358.

Hilaire-Pérez L. – Thébaud-Sorger M., "Les techniques dans l'espace public. Publicités des inventions et littérature d'usage en France et en Angleterre au XVIIIe siècle", *Revue de synthèse* 2 (2006) 393–428.

Hilaire-Pérez L. – Thébaud-Sorger M., "Risque d'incendie en milieu urbain et *industrious revolution*: le cas de Londres dans le dernier tiers du XVIIIe siècle", *Le Mouvement Social* 249 (2014) 21–39.

Jandot O., *Les délices du feu. L'homme, le chaud et le froid à l'époque moderne* (Ceyzérieu: 2018).

Kaplan S.L., *Les ventres de Paris. Pouvoir et approvisionnement dans la France d'Ancien Régime* (Paris: 1984).

Kay M., "'Arise, Shine, for thy Light Has Come': Electricity and Authority in British Jewish Domestic and Communal Life in the Early Years of Electrification", conference paper presented at the workshop *Technique et religion: cultures techniques, croyances, circulations de l'Antiquité à nos jours*, University of Paris, 1–3 September 2021.

Lanoë C., "'Mettre les gants en couleurs'. La construction des savoirs des gantiers-parfumeurs, XVIe–XVIIIe siècles", *Zilsel* 2 (2021) 219–236.

Lanoë C., *Les ateliers de la parure, France XVIIe–XVIIIe siècle* (Ceyzérieu: 2024).

Lanoë C., "Réflexions à propos du 'secret' de fabriquer 'les perruques au métier' (XVIIe–XVIIIe siècles)", in Depretto L. – Renoux C. – Speroni C. – Vickermann-Ribémont G. (eds.), *Cultures du secret à l'époque moderne Raisons, espaces, paradoxes, fabriques* (Paris: 2023) 309–326.

Lyon-Caen N. – Morera R., *À vos poubelles citoyens! Environnement urbain, salubrité publique et investissement civique (Paris, XVIe–XVIIIe siècle)* (Ceyzérieu: 2020).

Meyer G., *Un savant dans son siècle: Alexandre T. Vandermonde (1735–1796)* (PhD dissertation, University of Nantes: 2021).

Montègre G., *L'expérience des Lumières. Voyages, savoirs, diplomaties et émotions dans l'Europe du XVIIIe siècle* (HDR-Habilitation à diriger des recherches, University of Nice: 2023).

Pyne S., *Fire in America: A Cultural History of Wildland and Rural Fire* (Seattle, WA: 1982).

Pyne S., *Fire. A Brief History* (Seattle, WA: 2001).

Pyne S., *The Pyrocene. How We Created an Age of Fire, and What Happens Next* (Berkeley: 2021).

Roche D., *La France des Lumières* (Paris: 1993).

Roche D., *Le peuple de Paris. Essai sur la culture populaire au XVIIIe siècle* (Paris: 1981).

Roche D., *Histoire des choses banales. Naissance de la consummation dans les sociétés traditionnelles (XVIIe–XIXe siècle)* (Paris: 1997).

Smith P.H., *The Body of the Artisan. Art and Experience in the Scientific Revolution* (Chicago: 2004).

Thébaud-Sorger M., "Capturing the Invisible: Heat, Steam and Gases at the Crossroads of Inventive Practices, Place and Social Milieu in France and Great Britain in the Second Half of the Eighteenth Century", in Roberts L.L. – Werrett S. (eds.) *Compound Histories: Materials, Governance and Production, 1760–1840* (Leiden: 2017) 85–105.

Thébaud-Sorger M., "Changing Scale to Master Nature: Promoting Small-scale Inventions in Eighteenth-century France and Britain", *Technology and Culture* 61.4 (2020) 1076–1107.

Thébaud-Sorger M., *Une histoire des ballons. Invention, culture matérielle et imaginaire, 1783–1909* (Rennes: 2009).

Traversier M., *L'harmonica de verre et miss Davies. Essai sur la mécanique du succès au siècle des Lumières* (Paris: 2021).

Valleriani M. (ed.), *The Structures of Practical Knowledge* (Dordrecht: 2017).

Van der Straeten J. – Weber H., "Technology and its Temporalities, a Global Perspective", in Carnino G. – Hilaire-Pérez L. – Lamy J., *Global History of Techniques (19th–21st Centuries)* (Turnhout: 2023) 261–280.

Vérin H., *La gloire des ingénieurs. L'intelligence technique du XVIe au XVIIIe siècle* (Paris: 1993).

Viguié R., *Bien au chaud. Histoire du chauffage au XXe siècle* (Paris: 2023).

Index Nominum

Aristotle   6, 70, 339
Arnaud, E.R.   95–96
Avicenna   57, 59, 62

Beethoven (van), Ludwig   1*n*3
Bélidor, Bernard Forest de   226, 235
Biancani, Giuseppe   53
Biet, Claude   134–135
Biringuccio, Vannoccio   337
Blake, William   18
Blount, Thomas   103–104
Boerhaave, Herman   85, 93, 106–109, 112, 271, 351
Borda (de), Jean-Charles   165
Boyle, Robert   80, 164
Bruno, Giordano   63, 65–66, 332
Burke, Edmund   338

Cadet de Vaux, Antoine-Alexis   163, 166–167
Cahuzac, Jean-Louis de   335
Castelli, Benedetto   46–50, 52–59, 77–82
Castiglione, Baldassarre   339
Cesarini, Ferdinando   49–50, 52, 57–58, 68, 80
Chambers, Ephraim   104
Charas, Moyse   126, 136
Charles II (King of England)   26
Cocchini, Giuseppe   82
Coga, Arthur   342
Cookworthy, William   233
Cosimo II (Grand Duke of Tuscany)   47
Costa de Beauregard, Alexis   154

Dallowe, Timothy   106
De Clave, Estienne   105
De Renou, Jean   39
De Sgobbis, Antonio   102
Delle Colombe, Ludovico   77–79, 81–82
Democritus   64–65, 75
Desaguliers, John Theophilus   226, 235, 252
Descartes, René   64
Di Grazia, Vincenzo   77, 81
Dini, Piero   73
Duhamel du Monceau, Henri-Louis   154

Elzevier, Lodewijk   86

Fahrenheit, Daniel Gabriel   81, 85, 106–107
Farey Jr., John   233
Ferdinand II (Grand Duke of Tuscany)   52–53
Formey, Jean Henri Samuel   149
François Marin   101
Franklin, Benjamin   158, 177–178, 182–183

Galen   40, 57–58, 62, 75
Galilei, Galileo   7, 49–54, 56–59, 62–64, 66–75, 77–78, 80–81, 85, 356
Gardeton, César   157
Gauger, Nicolas   12, 158–159, 172–178, 187
George II (King)   17
Giunti, Cosimo   74, 77
Glaser, Christophe   96–99
Godbid, William   80*n*56
Goethe, Johann Wolfgang von   1*n*3
Goudsblom, Johan   4
Grassi, Orazio   52, 68, 80

Handel, George Frideric   17
Harel, Charles   163, 166–168
Harris, John   104, 341
Hartley, David   14, 257–158, 259, 263–270, 273–274
Hesiod   1
Hoffmann Johann, Moritz   106

James, Robert   98–99
Janvier, Antide   161

Kepler, Johannes   75
Kessler, German Franz   10, 12, 148

Latapie, François de Paule   349
Lavoisier, Antoine-Laurent   6, 17, 50, 73*n*44, 85, 158, 179, 355
Le Cointe, Jourdan   154
Le Febvre, Nicaise   25–33, 35–42, 44–45, 97–98
Lemery, Nicolas   95–96, 105, 125–126, 132, 159

INDEX NOMINUM

Locke, John (1632–1704)   69
Louis XIV (King of France)   26
Ludolff, Christian Friedrich   228

Magalotti, Lorenzo (1637–1712)   82
Mahon, Charles   14, 257–259, 263–264, 266–270, 273
Major, Thomas   7
Mascardi, Giacomo   86
Mayow, John   339
Menon, Joseph   101
Mersenne, Marin   81–82
Mignot de Montigny, Etienne   165

Newcomen, Thomas   13, 239
Newton, Isaac   27, 67n33, 174, 176

Paracelsus   31–32, 34, 102
Parent, Antoine   237
Parmentier, Antoine   166
Pascal, Blaise   59
Pepys, Samuel   14
Piroux, Augustin   15, 271–275
Pitot, Henri   235
Pitt, Moses   267
Pivati, Gianfrancesco   99–100
Plato (Aristocles) of Athens   1, 64
Polo, Niccolò   58
Pott, Johann Heinrich   111, 149

Rey, Jean   81–82
Rouelle, Guillaume-François   108–109, 117, 122, 124, 126–129
Rousseau, François   123
Ruault, Nicolas   81
Ruland, Martin   102–103

Sagredo, Giovanni Francesco   49, 52, 58–59, 62, 76, 81
Salmon, William   111
Salviati, Filippo   76
Sanctorius, Santorio   49, 57–60, 62–71, 73, 75, 79, 81, 85
Sarcina, Giacomo   57
Sarpi, Paolo   58
Second, Joseph   122
Sendivogius, Michael   23, 25, 27, 29–30, 31–33, 35, 38–39, 42–43, 45
Sennert, Daniel   103
Simienowicz, Casimir   339
Smeaton, John   13, 225–245, 247–252, 256
Smollett, Tobias   338
Socrates   1
Somasco, Giacomo Antonio   57
Sprat, Thomas   341

Tampach, Gottfried   75
Teniers the Younger, David   7
Thompson, Benjamin   158, 178–179, 181
Torricelli, Evangelista   55–57, 59, 84–85

Van Helmont, Joan Baptista   39n45, 40, 42n52
Vanière, Thomas-Ignace   157, 160, 162
Venel, Gabriel-François   94, 109–111, 150, 154–155, 161
Villiers (de), Jacques-François   164
Viviani, Vincenzo   49, 52–53, 56, 58, 82
Von Pfeiffer, Johann Friedrich   155

Watson, William   229
Watt, James   225–226, 239
Willis, Thomas   340
Wilson, George   98, 105